Long-Term Effects of Disinfection Changes on Water Quality

Awwa Research Foundation

About the Awwa Research Foundation

The Awwa Research Foundation (AwwaRF) is a member-supported, international, nonprofit organization that sponsors research to enable water utilities, public health agencies, and other professionals to provide safe and affordable drinking water to consumers.

The Foundation's mission is to advance the science of water to improve the quality of life. To achieve this mission, the Foundation sponsors studies on all aspects of drinking water, including supply and resources, treatment, monitoring and analysis, distribution, management, and health effects. Funding for research is provided primarily by subscription payments from approximately 1,000 utilities, consulting firms, and manufacturers in North America and abroad. Additional funding comes from collaborative partnerships with other national and international organizations, allowing for resources to be leveraged, expertise to be shared, and broad-based knowledge to be developed and disseminated. Government funding serves as a third source of research dollars.

From its headquarters in Denver, Colorado, the Foundation's staff directs and supports the efforts of more than 800 volunteers who serve on the board of trustees and various committees. These volunteers represent many facets of the water industry, and contribute their expertise to select and monitor research studies that benefit the entire drinking water community.

The results of research are disseminated through a number of channels, including reports, the Web site, conferences, and periodicals.

For subscribers, the Foundation serves as a cooperative program in which water suppliers unite to pool their resources. By applying Foundation research findings, these water suppliers can save substantial costs and stay on the leading edge of drinking water science and technology. Since its inception, AwwaRF has supplied the water community with more than $300 million in applied research.

More information about the Foundation and how to become a subscriber is available on the Web at **www.awwarf.org.**

Long-Term Effects of Disinfection Changes on Water Quality

Prepared by:
John E. Dyksen, **Catherine Spencer**, **Robert Hoehn**, **Jonathan Clement**, and **Jessica Brandt-Edwards**
Black & Veatch

Melinda J. Friedman, **Amie Hanson**, and **Ronalee Spellecacy**
HDR Engineering, Inc.

Michael J. McGuire
McGuire Malcolm Pirnie

Philip C. Singer
University of North Carolina at Chapel Hill

and

Anne Camper
Center for Biofilm Engineering, Montana State University

Sponsored by:
Awwa Research Foundation
6666 West Quincy Avenue, Denver, CO 80235-3098

and

U.S. Environmental Protection Agency
Cincinnati, OH

Published by:

Awwa Research Foundation

American Water Works Association

IWA Publishing

DISCLAIMER

CONTENTS

TABLES

FIGURES

FOREWORD

The Awwa Research Foundation is a nonprofit corporation that is dedicated to the implementation of research efforts to help utilities respond to regulatory requirements and traditional high-priority concerns of the industry. The research agenda is developed through a process of consultation with subscribers and drinking water professionals. Under the umbrella of a Strategic Research Plan, the Research advisory Council prioritizes the suggested projects based upon current and future needs, applicability, and past work; the recommendations are forwarded to the Board of Trustees for final selection. The foundation also sponsors research projects through the unsolicited proposal process; the Collaborative Research, Research Applications, and Tailored Collaboration programs; and various joint research efforts with organizations such as the U.S. Environmental Protection Agency, the U.S. Bureau of Reclamation, and the Association of California Water Agencies.

This publication is a result of one of these sponsored studies, and it is hoped that its findings will be applied in communities throughout the world. The following report serves not only as a means of communicating the results of the water industry's centralized research program but also as a tool to enlist the further support of the nonmember utilities and individuals.

Projects are managed closely from their inception to the final report by the foundation's staff and large cadre of volunteers who willingly contribute their time and expertise. The foundation serves a planning and management function and awards contracts to other institutions such as water utilities, universities, and engineering firms. The funding for this research effort comes primarily from the Subscription Program, through which water utilities subscribe to the research program and make an annual payment proportionate to the volume of water they deliver, and consultants and manufacturers subscribe based on their annual billings. The program offers a cost-effective and fair method for funding research in the public interest.

A broad spectrum of water supply issues is addressed by the foundation's research agenda: resources, treatment and operations, distribution and storage, water quality and analysis, toxicology, economics, and management. The ultimate purpose of the coordinated effort is to assist water suppliers to provide the highest possible quality of water economically and reliably. The true benefits are realized when the results are implemented at the utility level. The foundation's trustees are pleased to offer this publication as a contribution toward that end.

Walter J. Bishop
Chair, Board of Trustees
Awwa Research Foundation

Robert C. Renner, P.E.
Executive Director
Awwa Research Foundation

ACKNOWLEDGMENTS

The authors wish to acknowledge the financial support of the Awwa Research Foundation and the US Environmental Protection Agency. Without their support, this project would not have been possible. This research has been guided from the outset by an experienced Project Advisory Committee (PAC) whose expertise and advice were invaluable. The authors wish to thank these individuals for their input throughout the project and for their time and efforts:

Darren Lytle, U.S. Environmental Protection Agency, Cincinnati, Ohio
Gerald Speitel, Jr., University of Texas at Austin
Kevin Wattier, Long Beach Water Department, California

A very special thanks goes to Djanette Khiari, AwwaRF Project Officer, for her guidance, expertise and patience throughout this project. Her direction and advice on this project was very much appreciated. A number of water utilities offered their time and information for the data collection, the workshop, and the case studies. The authors gratefully acknowledge these efforts:

Aqua New Jersey, Phillipsburg, New Jersey
Brick Township Municipal Utilities Authority, New Jersey
Chesterfield County Utilities, Virginia
Clayton County Water Authority, Georgia
Concord, New Hampshire
Downey, California
Flagstaff, Arizona
Greater Vancouver Regional District, Burnaby, British Columbia, Canada
Gulf Coast Water Authority, Texas City, Texas
Indiana-American Water Company, Muncie, Indiana
Lincoln Water System, Nebraska
Littleton Water Department, Massachusetts
Massachusetts Water Resources Authority, Boston, Massachusetts
Metropolitan Water District of So. California, LaVerne, California
Milwaukee Water Works, Wisconsin
Newport News Waterworks, Virginia
Passaic Valley Water Commission, Clifton, New Jersey
PA-American Water Company, Wilkes-Barre, Pennsylvania
Philadelphia Water Department, Pennsylvania
San Louis Obispo County, California
San Francisco Public Utilities Commission, Burlingame, California
Southeast Morris County Municipal Utilities Authority, Cedar Knolls, New Jersey
Southern Nevada Water Authority, Las Vegas, Nevada
Tampa Bay Water, Clearwater, Florida
United Water New York, West Nyack, New York
Washington Aqueduct, Washington, DC
A special thanks to Tampa Bay Water for hosting the project workshop.

- Changes in aesthetic quality - taste and odor

This project focuses on water quality impacts that might occur from making a disinfectant change.

USE OF THE DOCUMENT – APPLICATION POTENTIAL

The applications potential for the products of this research project is substantial, both for immediate and long-term use by water industry managers and operations staff. The results of this research project are particularly applicable to utilities for:

- Evaluating current changes
- Making operational adjustments
- Considering future changes
- Designing future process changes
- Diagnosing water quality problems

PROJECT APPROACH

To document the impacts of disinfection changes on distribution system water quality, the following overall approach was used:

- Conduct literature review
- Develop issue papers
- Collect available data from participating utilities
- Analyze and summarize the available data

The first step in this project was to conduct a comprehensive literature review of information on changing disinfectants and the resultant effects on distribution system water quality. In conjunction with the literature review, a series of issue papers were prepared on each category of impacts – microbial quality, chemical quality (DPBs), chemical quality (corrosion, metal release, color, and particle characteristics), and aesthetic quality.

A project workshop was held over a one and one half day period on May 5 & 6, 2004 at the offices of Tampa Bay Water, one of the participating utilities. The workshop participants included Project Team members, participating utilities, and other invited utilities. The overall purpose of the workshop was to ensure that the subsequent data collection and analyses were done properly and with sufficient input on the key variables of interest from the entire Research Team and the other parties involved.

In the next step, actual water quality impacts experienced by various utilities that had made disinfection changes were collected and summarized. To facilitate the collection of data on water quality impacts, a questionnaire was developed and sent to each utility. Utilities that participated in this part of the project included those that changed from chlorine to chloramines, chlorine to ozone, chlorine to chlorine dioxide and chlorine to UV. The utilities represented locations from around the United States and from Canada. Also, various types of water supplies and treatment processes are represented. Twenty six (26) utilities participated in the data collection.

The data analyses serve as the basis for much of the research project. The objective was to assimilate the numerical data and lessons learned from the participating utilities. After the data were collected, they were analyzed and graphed using time-series plots. Also, statistical analyses were performed on selected data. The observations gleaned from these analyses are summarized below for each potential disinfectant change.

CHLORINE TO CHLORAMINES

A significant number of utilities have switched from using chlorine to chloramines for secondary disinfection. The use of chloramines can provide a number of benefits relative to maintaining water quality in the distribution system, particularly maintaining a residual throughout the distribution system. Of the utilities that participated in the project, eleven (11) reported switching from chlorine to chloramines for secondary disinfection. Based on the findings of the issue papers and the information provided by the participating utilities, the following general observations can be made regarding switching from chlorine to chloramines for secondary disinfection:

- Impacts on Microbial Quality
  - Overall, there is reported an improvement in microbial quality based on lower heterotrophic plate count (HPC) levels, better control of biofilms (based on lower HPC levels) and coliform bacteria, and a more persistent and higher disinfectant residual (higher chloramine residual compared to chlorine residual) throughout the distribution system.
  - Initial HPC levels could increase because of dislodging of biofilms on the pipe walls. Ultimately, possibly immediately or after 1 to 2 years, HPC levels appear to decrease.
  - HPC analyses should be based on the use of R2A agar, as the use of standard agar will yield levels that are too low to detect a difference between before and after the conversion.
  - Adverse impacts on coliform levels were not observed, and coliform positives may be reduced.
  - Maintenance of a higher and more persistent disinfectant residual throughout the distribution system is more likely with chloramines compared to free chlorine.
  - One study found that chloramines provided better control of *Legionella* in buildings.
  - Nitrification is a potential problem in systems converting to chloramines.
- Impacts on Chemical Quality - DBPs
  - Overall, disinfection by-product levels should be reduced with the use of chloramines compared to the use of free chlorine.
  - Lower total trihalomethane (TTHM) formation potential should be expected – with both average and peak TTHM values being reduced.
  - Lower haloacetic acid (HAA5) levels might be expected system wide, however, increases may be observed at specific locations that were previously susceptible to biodegradation of HAA species.
  - One utility that switched to chloramines did not observe high cyanogen chloride levels after the switch.

- Nitrosodimethyl amine (NDMA) levels did not significantly increase for the two utilities that tested for this parameter.
- Impacts on Chemical Quality - Corrosion, Metal Release, Color, and Particle Characteristics
  - A reduction in corrosion and metal release may be realized as exhibited by lower discolored water and red water complaints.
  - It appears from this project that lead and copper levels are not necessarily adversely affected with the conversion from free chlorine to chloramines; however, in two instances lead levels were higher after the switch from chlorine to chloramines. In one case, lead levels exceeded the standard, while in the other case lead levels remained below the standard.
- Impacts on Aesthetic Quality - Taste and Odor
  - A reduction in taste and odor complaints may be realized.
  - The taste and odor descriptors probably will change from chlorinous taste and odors to other types of descriptors.

CHLORINE TO OZONE

A number of utilities have switched from using chlorine to ozone for primary disinfection. Of the utilities that participated in the project, seven (7) reported switching from chlorine to ozone for either pre-oxidation or primary disinfection. Based on the findings of the issue papers and the information provided by the participating utilities, the following observations can be made regarding a utility switching from chlorine to ozone for secondary disinfection:

- Impacts on Microbial Quality
  - Overall, microbial quality could be reduced based on higher HPC levels associated with more degradable organic matter in the bulk water in the distribution system. This could be countered by treatment techniques such as biological filtration designed for removing this organic matter.
  - HPC analyses should be based on the use of R2A agar, since standard agar may be insensitive to detect differences before and after a conversion.
  - No adverse impacts on coliforms were observed.
- Impacts on Chemical Quality - DBPs
  - Overall, a reduction in disinfection by-product levels was observed with the use of ozone compared to the use of chlorine.
  - Lower THM formation potential was observed – with both average and peak THM values being reduced.
  - Lower HAA levels were observed.
- Impacts on Chemical Quality - Corrosion, Metal Release, Color, and Particle Characteristics
  - There is insufficient information to make any conclusions regarding corrosion and metal release.
- Impacts on Aesthetic Quality - Taste and Odor
  - Reductions in taste and odor complaints were observed.

CHLORINE TO CHLORINE DIOXIDE

Only two of the participating utilities reported a disinfectant change from chlorine to chlorine dioxide. Based on the findings of the issue papers and the information provided by the participating utilities, the following observations can be made regarding a utility switching from chlorine to chlorine dioxide for either primary or secondary disinfection:

- Impacts on Microbial Quality
  - Overall, with use of chlorine dioxide as a secondary disinfectant, there could be a degradation in microbial quality based on higher HPC levels resulting from more degradable matter in the bulk water in the distribution system.
  - The use of chlorine dioxide as a primary disinfectant should provide better microbial quality in the distribution system based on lower HPC levels.
  - HPC analyses should be based on the use of R2A agar, as the use of standard agar will yield levels that are too low to detect a difference between before and after the conversion.
  - There should be no adverse impact on coliforms, and coliform positives may be reduced.
- Impacts on Chemical Quality - DBPs
  - Overall, disinfection by-product levels should be reduced with the use of chlorine dioxide compared to the use of chlorine.
  - Lower THM formation potential should be expected – with both average and peak THM values being reduced.
  - Lower HAA levels should be expected.
- Impacts on Chemical Quality - Corrosion, Metal Release, Color, and Particle Characteristics
  - There is insufficient information to make any conclusions regarding corrosion and metal release.
- Impacts on Aesthetic Quality - Taste and Odor
  - There should be a reduction in taste and odor complaints.

CHLORINE TO UV

The use of ultraviolet (UV) radiation for primary disinfection has received much attention over the past 5 to 10 years. The initial interest in UV resulted from its ability to inactivate *Cryptosporidium parvum* oocysts. More recently, more stringent disinfection by-product regulations have resulted in the need for water utilities to critically evaluate their disinfection practices to reduce DBPs in the distribution system. UV has been investigated because it does not produce DBPs of concern. However, it does not produce a residual, and so there is still a need for a secondary disinfectant such as free chlorine or chloramines. Of the utilities that participated in the project, five (5) reported switching from chlorine to UV for primary disinfection. Based on the findings of the issue papers and the information provided by the participating utilities, the following observations can be made regarding a utility switching from chlorine to UV for primary disinfection:

- Impacts on Microbial Quality
  - UV treatment followed by chemical disinfection may reduce the HPC levels entering the distribution system.

- The use of UV in combination with a chemical disinfectant may not result in the use of a lower disinfectant dose to maintain a target residual in the distribution system.

- Impacts on Chemical Quality - DBPs
  - The use of UV in combination with chlorine should not significantly change THM and HAA levels in the distribution system.
- Impacts on Chemical Quality - Corrosion, Metal Release, Color, and Particle Characteristics
  - No data were available to draw conclusions in this area. However, none of the participating utilities reported any negative impacts.
- Impacts on Aesthetic Quality – Taste and Odor
  - No data were available to draw conclusions in this area. However, none of the participating utilities reported any negative impacts.

FUTURE RESEARCH NEEDS

The results of this project have led to an identification of future research that is required to better understand the long-term impacts of changing disinfectants on distribution system water quality. Data gaps and research needs were identified for each of the major changes in disinfection strategy. There are many data gaps regarding the use of booster chlorination, and research needs in this area also were identified.

Overall, more long-term data are needed to adequately assess the impacts of changing disinfectants. Very little data were available regarding the impacts on particle characteristics and corrosion. Research is needed to determine under what conditions switching to chloramines adversely impacts lead levels. Also, because of the relative infancy of the technology, very little data are available on the impacts from switching to UV. A detailed list of research needs is included in Chapter 7.

CHAPTER 1
INTRODUCTION

HISTORICAL PERSPECTIVE

Chlorine has been used in the treatment of drinking waters in the United States since 1908 (Leal 1909). Its use has included disinfection, taste and odor control, oxidation of iron and manganese and of some organic chemicals. Chlorine also has been used to control microbiological growth in the distribution system.

Since the Safe Drinking Water Act of 1974 and its Amendments, utilities around the country have made changes to their disinfection strategies to meet more stringent regulations for either microbial contaminants or disinfection by-products (DBPs). Specifically, with the promulgation of the Surface Water Treatment Rule and the Stage 1 Disinfectant/Disinfection By-Products Rule, utilities investigated the use of alternatives to chlorine. Consequently, ozone and chlorine dioxide have been tested and applied at treatment plants for primary disinfection, and the use of chloramines has increased for secondary disinfection.

Even more utilities are changing or considering changing, disinfectants either for enhanced disinfection of microbial agents or to reduce the formation of DBPs in accordance with the Long Term Stage 2 Enhanced Surface Water Treatment Rule and the Stage 2 Disinfectant/Disinfection By-Product Rule. The future Ground Water Rule also may result in changes in disinfection strategies for ground water systems. While it is well accepted by many in the drinking water community that many of these changes will have beneficial impacts, it is not well known how these changes in disinfection will affect distribution system water quality in the long term.

PURPOSE AND SCOPE

The objective of this project is "to document the long-term effects of disinfectant changes on distribution system water quality." The project involved collecting and analyzing available utility and other data on changing disinfectants, and summarizing the data so that utilities may use the information when selecting effective disinfection strategies and when diagnosing water quality problems that may be caused by disinfectant changes.

The primary objective of this project is the documentation of changes in various water quality parameters in the distribution system resulting from disinfectant changes. This project focuses primarily on trends in water quality changes. This project does not specifically address certain problems that can arise from switching disinfectants. For example, it is well known that switching to chloramines can create problems when the water is used for special purposes. Water containing chloramines is not suitable for kidney dialysis treatment and is toxic to aquatic organisms. This report does not address these types of issues.

Disinfectant Changes

This research project focuses on the effects of altering a typical disinfection scheme using chlorine disinfection to one of the following disinfectants:

- Chloramines

- Ozone
- Chlorine Dioxide
- Ultraviolet light (UV)

The use of booster chlorination also was reviewed.

A matrix of the type of disinfection changes that a utility might make and that were reviewed as part of this project is shown below in Table 1.1. The old disinfection scheme would include using chlorine as both a primary and secondary disinfectant. The changes that were reported typically included a change in either the primary or secondary disinfectant and not both at the same time.

Table 1.1
Potential disinfection changes

Old Disinfection Scheme	New Primary Disinfectant	New Secondary Disinfectant
Chlorine	Chlorine	Chloramines
Chlorine	Ozone	Chlorine
Chlorine	Ozone	Chloramines
Chlorine	Chlorine dioxide	Chlorine
Chlorine	Chlorine dioxide	Chloramines
Chlorine	UV	Chlorine
Chlorine	UV	Chloramines

Water Quality Impacts

It is well understood that changes in disinfection as described above can have impacts on the treatment process and finished water quality. What is less understood is how these disinfection changes can affect distributed water quality. There are several major classes of impacts that a change in disinfection can have on a distribution system:

- Changes in microbial quality
- Changes in chemical quality
  - DBPs
  - Corrosion, metal release, color-producing compounds and particle characteristics
- Corrosion, metal release, color-producing compounds and particle characteristics
- Changes in aesthetic quality - taste and odor

This project focuses on water quality impacts that might occur from making a disinfectant change.

USE OF THE DOCUMENT – APPLICATION POTENTIAL

The applications potential for the products of this research project is substantial, both for immediate and long-term use by water industry managers and operations staff. In response to recently promulgated and upcoming federal drinking water regulations, many water utilities have changed, or are considering changing their disinfection practices, especially relative to the control of disinfection by-products. One of the challenges of making a disinfection change is that associated with balancing competitive needs for water quality maintenance, such as maintaining an adequate chlorine residual while minimizing DBP formation. Consequently, from an immediate perspective, utilities that have already made changes may use the results of this project for assessing current issues, and in the long-term, utilities that are considering changes will find the results from this project beneficial in their decision-making process. The results of this research project are particularly applicable to utilities:

Evaluating current changes – Utilities that have already made disinfection changes may utilize these results to monitor water quality changes, so they can be prepared to deal with potential impacts before they become critical customer relations issues, or worse, regulatory violations.

Making operational adjustments - Utilities that have already made disinfection changes may use the results to make modifications to their operations to mitigate adverse water quality impacts that have been observed by utilities that participated in this study.

Considering future changes - Utilities that are considering disinfection changes may utilize these results in their decision-making process, particularly to determine the pros and cons regarding water quality issues associated with each disinfectant alternative they might consider.

Designing future process changes - Utilities that make future changes may use these results to assist in designing the changes in their disinfection process to avoid or minimize any adverse water quality impacts.

Diagnosing water quality problems - If a utility has made a disinfection change, the results of this project may be used to diagnose whether specific water quality problems could be associated with the disinfection change.

CHAPTER 2
PROJECT APPROACH

To document the impacts of disinfection changes on distribution system water quality, the following overall approach was used:

- Conduct literature review
- Develop issues papers
- Collect available data from participating utilities
- Analyze and summarize the available data

Each of these steps for documenting water quality changes is described below.

LITERATURE REVIEW

The first step in conducting this research project was to conduct a comprehensive literature review of information on changing disinfectants and the resultant effects on distribution system water quality. The objective of the literature review was to document and summarize the state of knowledge of the effects of the various disinfectants on the four major water quality impacts. The Project Team utilized AwwaRF project reports, AWWA journal articles, conference proceedings, and articles from other publications.

A comprehensive bibliography was developed that summarized available resources. The bibliography was then used by the project team to develop the issue papers. The following databases were used to develop the bibliography and prepare the literature review: AWWA's June 2003 Waternet, EPA's website and Online Library Search Engine, and the American Society of Civil Engineers Civil Engineering Database. Using these databases, it was possible to search materials from 76 publishers and 61 journals. Over 200 articles and presentations were found covering these topics. A Microsoft Excel spreadsheet was developed to summarize each of the materials, including a general summary of the article, presentation, or document. This spreadsheet is included in Appendix A.

Some of the AwwaRF project reports that provided pertinent information included:

- Assessment of Seasonal Chlorination Practices and Impacts to Chloraminating Utilities
- Case Studies of the Impacts of Treatment Changes on Biostability in Full-Scale Distribution Systems
- Chloramine Decomposition in Distribution System and Model Waters
- Guidance Manual for Maintaining Distribution System Water Quality
- Optimizing Chloramine Treatment
- Impact of UV Disinfection on Biological Stability
- Use of Chlorine Dioxide and Ozone for Control of Disinfection By-Products

As a result of the literature review, the following gaps in available information were identified:

- Booster Chlorination – Little information was available on water quality impacts, aside from residual management, associated with implementing booster chlorination.
- Ozone – There was little information on ozone's impacts on particulate matter or metals uptake/release in the distribution system.
- UV – There was little information available on UV's impacts on corrosion or particulate matter in the distribution system.

Where possible, the Project Team obtained the materials that were found during the literature search. These documents were kept in a series of project notebooks. The results of the literature review were used to prepare a series of issue papers on potential water quality impacts.

DEVELOPMENT OF ISSUE PAPERS

In conjunction with the literature review, a series of issue papers were prepared on each category of impacts – microbial quality, chemical quality (DPBs), chemical quality (corrosion, metal release, color, and particle characteristics), and aesthetic quality. The scope of each issue paper is summarized below:

- **Changes in Microbial Quality**
  This paper focused on key aspects of degradation with regard to microbiological quality. Issues associated with coliform bacteria, nitrification, heterotrophic bacteria, and biofilms were addressed. This paper also reviewed potential revisions to the Total Coliform Rule (TCR).
- **Changes in Chemical Quality - DBPs**
  This paper focused on bringing together critical information on how changes in disinfection practices can lead to long term changes in the formation and decay of disinfection by-products within distribution systems. This paper synthesized information with respect to the research that has been done and the critical gaps that must be filled to address needs particularly covered in the Stage 2 Disinfectant/Disinfection By-Product Rule (D/DBP Rule).
- **Changes in Chemical Quality - Corrosion/Metal Release/Color/Particle Characterization**
  This paper assembled knowledge regarding the effects of the various disinfectants on release of metals from piping systems. Specifically, the interactions between biofilms, metallic scales, and bulk water quality were examined. Emphasis was paid to recent research indicating the possible release of arsenic particles and scales following disruptions in the equilibrium between pipe surfaces and the bulk water flow. This paper also examined the changes that can occur in particles as a result of disinfectant changes and the potential impacts on distribution system water quality. Recent studies have indicated that distribution system materials may concentrate metals from the bulk water and become a reversible reservoir. Examples are arsenic, aluminum, radium, lead and copper.
- **Changes in Water Aesthetics - Taste and Odor**
  This paper incorporated existing knowledge regarding the effects of changes in disinfection practices on aesthetic issues. Specific attention was paid to impacts on delivered water tastes and odor.

The issue papers address key factors affecting water quality changes, positive impacts on water quality, potential adverse impacts on water quality, and information gaps and data needed to fill those gaps. The four issue papers are included in Appendices C, D, E, and F, respectively.

PROJECT WORKSHOP

A project workshop was held over a one and one half day period on May 5 and 6, 2004 at the offices of Tampa Bay Water, one of the participating utilities. The workshop participants included Project Team members, participating utilities, and other invited utilities. The overall purpose of the workshop was to ensure that the subsequent data collection and analyses were done properly and with sufficient input on the key variables of interest from the entire Research Team and the other parties involved. Specifically, the objectives of the workshop were to:

- Present the existing state of the knowledge on water quality changes resulting from disinfectant changes, i.e, the findings of the issue papers.
- Review the experiences of selected utilities that have made disinfectant changes.
- Identify knowledge and data gaps that will need to be filled in during subsequent tasks.
- Review overall research approach in light of the existing knowledge and data, and make any necessary modifications to assure that the project goals are attained.

The workshop agenda included four major blocks of time:

1. Review of project scope and objectives
2. Presentations of the four issue papers including a summary presentation
3. Presentations by the utilities on their experiences with disinfection changes
4. Breakout groups to review the presented material and to identify knowledge/data gaps and what is needed to fill those gaps

The results of the workshop are presented in a report that is included in Appendix F.

DOCUMENTATION OF ACTUAL IMPACTS

Participating Utilities

Following the preparation of the issue papers and the convening of the workshop, actual water quality impacts experienced by various utilities that had made disinfection changes were collected and summarized. Utilities that participated in this part of the project included those that changed from chlorine to chloramines, chlorine to ozone, chlorine to chlorine dioxide and chlorine to UV. The utilities represented locations from around the United States and from Canada. Also, various types of water supplies and treatment processes are represented. Twenty six (26) utilities participated in the data collection including:

Utility

Aqua New Jersey
Brick Township Municipal Utilities Authority, NJ
Chesterfield County Utilities, VA
Clayton County Water Authority, GA
Concord, NH
Downey, CA
Flagstaff, AZ
Greater Vancouver Regional District, BC
Gulf Coast Water Authority, TX
Indiana-American Water Company, IN
Lincoln Water System, NE
 East Water Treatment Plant (WTP)
 West WTP
Littleton Water Department, MA
Massachusetts Water Resources Authority, MA
Metropolitan Water District of So. California, CA
Milwaukee Water Works, WI
Newport News Waterworks, VA
 Harwood's Mill WTP
 Lee Hall WTP
Passaic Valley Water Commission, NJ
PA-American Water Company, PA
Philadelphia Water Department, PA
San Francisco Public Utilities Commission, CA
San Louis Obispo County, CA
Southeast Morris County Municipal Utilities Authority, NJ
So. Nevada Water Authority, NV
Tampa Bay Water, FL
United Water New York, NY
Washington Aqueduct, Washington DC

Data Questionnaire

As part of the project, information was collected from the participating utilities and several other utilities identified during the workshop to document available data regarding the impacts of changing disinfectants on water quality in the distribution system. To facilitate the collection of data on water quality impacts, a questionnaire was developed and sent to each utility. The questionnaire is included as Appendix G. Data that were requested included treatment plant size, source of supply, treatment processes, raw and finished water quality, and distribution system water quality. The utilities were asked if they had observed any changes in water quality after they had changed their disinfection strategy. If they had observed a change, they were requested to provide data (if available) for any parameters that changed – one year of data before the change and two years of data after the change (or any data they had).

DATA ANALYSES

The data analyses serve as the basis for much of the research project. The objective was to assimilate the numerical data and lessons learned from the participating utilities. After the data were collected, they were analyzed and graphed using time-series plots. Also, statistical analyses were performed on selected data. One goal of the analyses was to determine the time over which the water quality changes occurred - short or long-term. It should be noted that because of site specific conditions including water quality, piping materials and operational practices, it was not possible to draw direct utility-to-utility conclusions. Instead, the analyses were used to see if any general trends could be observed in either the various disinfectants or the impact areas. The results of the data analyses are presented in the following chapters according to the disinfectant change that was made:

Chapter 3 - Chlorine to Chloramines
Chapter 4 - Chlorine to Ozone
Chapter 5 - Chlorine to Chlorine Dioxide
Chapter 6 - Chlorine to UV

Although booster chlorination is addressed in the issue papers, it is not included in a separate chapter of the report because of the lack of utility experience with this disinfectant change. Chapter 7 includes a list of future research needs.

CHAPTER 3
CHLORINE TO CHLORAMINES

INTRODUCTION

A significant number of utilities have switched from using chlorine to chloramines for secondary disinfection. The use of chloramines can provide a number of benefits relative to maintaining water quality in the distribution system, particularly maintaining a residual throughout the distribution system. The impacts (microbial, chemical, and aesthetic) of switching from chlorine to chloramines are presented in this chapter. The potential impacts are based on the findings of the issue papers and the reported impacts are based on the findings of the utility survey. Three case studies are presented in more detail at the end of the chapter.

Of the utilities that participated in the project, eleven (11) reported switching from chlorine to chloramines for secondary disinfection. These utilities are listed in Table 3.1 along with the prior and current disinfection strategies.

Table 3.1
Participating utilities that converted from chlorine to chloramines

Utility	Capacity mgd	Source	Prior Disinfection			Current Disinfection			Date Changed
			Preox	Primary	Secondary	Preox	Primary	Secondary	
Brick Twp MUA, NJ	15	River/ GW	ClO_2	Cl_2	Cl_2	ClO_2	Cl_2	NH_2Cl	May 2004
Chesterfield Co, VA	12	Reservoir	Cl_2	Cl_2	Cl_2	Cl_2	Cl_2	NH_2Cl	Mar 1998
Concord, NH	10	Lake	Cl_2	Cl_2	Cl_2	Cl_2	Cl_2	NH_2Cl	May 1996
IN-American Water	21	River/GW	Cl_2	Cl_2	Cl_2	Cl_2	Cl_2	NH_2Cl	Dec 1992
MWDSC	1,670	Reservoir	none	Cl_2	Cl_2	none	Ozone	NH_2Cl	1985
Newport News, VA	54	Reservoir	none	Cl_2	Cl_2	none	O_2/Cl_2	NH_2Cl	1998
Philadelphia, PA	270	River	none	Cl_2	Cl_2	$KMnO_4$	Cl_2	NH_2Cl	1975
San Francisco, CA	310	Reservoir	none	Cl_2	Cl_2	none	Cl_2	NH_2Cl	Feb 2004
Tampa Bay Water, FL	15	GW	none	Cl_2	Cl_2	none	Cl_2	NH_2Cl	2004
PA-American Water	4	Reservoir	$KMnO_4$	Cl_2	Cl_2	$KMnO_4$	Cl_2	NH_2Cl	2002 and 2004
Washington DC	240	River	Cl_2	Cl_2	Cl_2	Cl_2	Cl_2	NH_2Cl	Nov 2000

Notes: "preox" denotes preoxidation "GW" denotes groundwater supply
"Cl_2" denotes use of free chlorine
"NH_2Cl" denotes use of chloramines
"ClO_2"denotes use of chlorine dioxide
"O_2" denotes use of ozone
"$KMnO_4$" denotes potassium permanganate

IMPACTS ON MICROBIAL QUALITY

This section provides a description of microbiological changes associated with changing disinfectant practices. First, potential impacts are discussed. Potential impacts are those that have been reported previously in the literature, or are those impacts that are generally accepted as valid by the water industry. Second, reported impacts are discussed. These are actual microbiological impacts that were reported by the utilities that participated in this research study.

Potential Impacts

The potential impacts of switching from a free chlorine residual in the distribution system to a monochloramine residual were identified based on a review of numerous investigations. It is important that the reader understand that the exact conditions under which the data were collected may vary significantly from utility to utility. Water quality conditions, including temperature, organic matter, pH, etc. interact with chloramines and the microbial populations. Therefore, it may be difficult to extrapolate from one system to another. The method of determining bacterial numbers is also very important.

Positive Impacts

- Overall, monochloramine is thought to provide better control of distribution system biofilms compared to free chlorine, particularly with respect to heterotrophic plate count (HPC) and coliform.
- In one study on the occurrence of *Legionella* in hospitals, the authors claim that there may be up to a 90% reduction in infections caused by this organism if the disinfectant residual is monochloramine instead of free chlorine (Kool et. al, 1999).

Adverse Impacts

- The distributed water has a lower capacity for inactivating suspended organisms. Consequently, utilities generally tend to have a higher monochloramine residual in their distribution systems than a free residual.
- The presence of excess ammonia associated with the over addition of ammonia during monochloramine formation or decay of monochloramine in the distribution system can lead to nitrification in the distribution system and subsequent loss of monochloramine residual, increased HPCs, and formation of biofilms.

Reported Impacts

The impacts on microbial water quality reported by the participating utilities that converted from using chlorine to chloramines are described in detail as case studies at the end of this Chapter and in Appendix H. The impacts that were reported included changes in HPC levels, coliform counts, and disinfectant residuals. Also, a special study was conducted at one utility to determine the impacts on *Legionella* in the distribution system. A summary of the data analyses is presented below and in Table 3.2.

HPC Levels

- Three of the utilities reported that the switch to chloramines did not affect HPC levels in the distribution system even during warm water conditions.
- Four of the utilities reported a decrease in HPC levels after the conversion. HPC measurements were made using R2A agar both before and after the conversion.
- One utility reported that, before the conversion, HPC levels averaged 1,200 colony forming units per milliliter (cfu/ml), with some sites in excess of 6,000 cfu/ml. Within three months after the conversion, no sites had levels exceeding 3,000 cfu/ml, and within two years the average HPC level had dropped to 65 cfu/ml.
- One utility reported that they did not see an impact on HPC levels based on the use of SPC agar, but did see a significant drop in HPC levels based on the use of R2A agar.
- Two utilities reported some sampling locations where HPC levels increased initially, but within one to two years after the conversion HPC levels dropped below the before conversion levels.
- One utility reported an increase in HPC levels by an order of magnitude after converting from chlorine to chloramines.

Coliform Levels

- Several of the utilities reported no change in coliform levels after switching to chloramines – levels were acceptable before and after the conversion.
- One utility reported a drop in coliform positives after switching to chloramines.
- Another utility reported that coliform occurrence continued to spike during the summer months, but the peak levels were lower after conversion to chloramines.

Disinfectant Residual

- Two utilities reported that free chlorine residuals were difficult, and at times, impossible to maintain throughout the distribution system. After the switch to chloramines, they were able to maintain residuals throughout the system.

Legionella

- One utility reported the results of a special study that investigated *Legionella* colonization in 53 buildings before and after the switch to chloramines. The findings of this study indicated a drop in percent of buildings colonized from 70 % using chlorine to 9 % using chloramines.
- Also, average *Legionella* concentrations in water heaters dropped from 271 cfu/ml when using chlorine to 10 cfu/ml when using chloramines.

Table 3.2
Summary of microbiological and residual data from utilities

Utility	Residuals, mg/L		Total coliform %		HPC levels, cfu/ml	
	Cl_2	NH_2Cl	Cl_2	NH_2Cl	Cl_2	NH_2Cl
Brick, NJ	NA	1.68	0	0	5	7
Concord, NH	0.56	1.70	0	0	1200	65
IN-American	23%<1.0	5%<1.0	0.60	0.36	420	102
PA- American	1.2	1.5	0	0	77	13
Philadelphia	NA	>0.5	NA	0.2	NA	0.3% > 100
San Francisco	0.3	2.1	0.2	0.1	200	50
Washington, DC	2.19	3.30	0.99	1.21	8	46

Notes: 1. "NA" denotes data not available.
2. Data indicate disinfectant residuals, total coliform levels, and HPC levels under chlorine disinfection and under chloramine disinfection.
3. Data represent averages unless indicated otherwise.
4. Total coliform are percent positive.

IMPACTS ON CHEMICAL QUALITY - DBPs

Potential Impacts

Most utilities that have switched from a free chlorine residual to a monochloramine residual have done so primarily to reduce DBP formation and to comply with current and future DBP regulations.

Positive Impacts

- Monochloramine will not produce appreciable additional TTHMs or trihaloacetic acids, but will react with residual organic carbon to some degree to produce dihaloacetic acids, such as dichloroacetic acid. TTHMs and HAA5 will not change appreciably after primary disinfection when employing combined chlorine as a secondary disinfectant. It should be noted that the impact of monochloramine on DBP formation will depend on the extent of chlorination that occurs prior to ammonia addition.
- Some continuing formation of other organic halides will occur, measurable as total organic halides (TOX), which are as yet unidentified. However, the extent of TOX formation from combined chlorine will be appreciably less than TOX production from free chlorine.

Adverse Impacts

- In a limited survey intended to assess the occurrence and stability of other halogenated DBPs of potential toxicological concern that have recently been identified, it has been found that some of these newly identified DBPs were present to a significant degree in chloraminated distribution systems. These included iodo-trihalomethanes and dichloroacetaldehyde.
- One recently identified DBP of concern associated with conversion to combined chlorine is nitrosodimethyl-amine (NDMA), and the possible formation of other nitrosamines can be anticipated. NDMA is associated to a greater degree with the use of combined chlorine than with the use of free chlorine.
- Haloacetic acid species are known to be biodegradable. It is possible that use of chloramines could reduce the amount of biological activity within the distribution system, which in turn, could result in increased HAA5 levels at specific locations.
- Cyanogen chloride is a disinfectant by-product that is expected to occur at higher concentrations in systems that use chloramines.

Reported Impacts

The reported impacts on chemical (DBP) water quality from all eleven of the participating utilities that converted from using chlorine to chloramines are described in detail in case studies at the end of this chapter or in Appendix H. The impacts that were reported included changes in TTHM and HAA5 levels. Also, one utility reported levels of cyanogen chloride in the distribution system. Table 3.3 presents a summary of DBP by-product data from several of the participating utilities. A summary of the data analyses is presented below.

TTHM Levels

- All of the participating utilities reported a drop in TTHM levels as a result of switching from chlorine to chloramines.
- A few of the utilities already had low TTHM levels, and they reported even lower levels after the conversion.
- In some cases, the reduction in TTHMs was significant - greater than 50 percent.
- In all cases, the switch to chloramines enabled the utility to reduce TTHM levels to below the current standard of 80 μg/L.

HAA5 Levels

- The impact on HAA5 levels was not as significant as the impact on TTHM levels. In most cases, HAA5 levels were already below the 60 μg/L standard.
- Most utilities that reported a decrease in HAA5 levels indicated about a 10 percent reduction.
- One utility reported a 30 percent reduction in HAA5 levels.

- Although system wide decreases in HAA5 levels were typically observed, HAA5 levels remained the same or even increased slightly at specific locations within some distribution systems using chloramines. This could be a result of decreased biodegradation of HAA5 under chloramination conditions.

Cyanogen Chloride

- Only one utility conducted monitoring of cyanogen chloride for 18 months under the Information Collection Rule (1997-1998). Overall, for three plant finished waters (entry points) and three distribution points representative of maximum travel times, the cyanogen chloride ranged from 2-8 nanograms per liter (ng/L), with the highest concentrations occurring at the maximum travel time locations.

NDMA

- Two utilities tested for NDMA. One did not detect any in the distribution system after the conversion to chloramines. They reported no significant impact from the conversion to chloramines on NDMA levels; most were non-detect and the highest was 4 ng/L.

Table 3.3
Summary of disinfection by-product data from utilities

Utility	TTHM, μg/L		HAA5, μg/L		Other data
	Cl_2	NH_2Cl	Cl_2	NH_2Cl	
Brick, NJ	54	34	33	30	ClO_2 0.6 →0.46 mg/L
Chesterfield Co, VA	70	40	NA	NA	
Concord, NH	70	30	NA	NA	
IN-American	100	63	88	51	Org DBPs 52→19 μg/L
Newport News, VA	90	60	60	50	
Philadelphia, PA	NA	45	NA	30	Cyanogen chloride 2-8 ng/L
PA-American	24	22	19	13	
Tampa Bay, FL	83	31	NA	NA	
San Francisco, CA	80	37	30	25	See Table 3.8
Washington DC	80	45	56	36	

In 2001-2002, a survey (Barrett et. al, 2003) was conducted of NDMA occurrence in treated waters of 21 North American drinking water treatment plants. Quarterly samples of treatment plant influent, finished, and distribution system waters were collected and analyzed for NDMA as part of an American Water Works Association Research Foundation and Water Environment Research Foundation jointly funded study. Sampled systems were located in seven states of the United States and four provinces of Canada and comprised systems with pristine sources as well as systems downstream of wastewater treatment plants. Various treatment processes and disinfectants were represented as were various detention times and pipe materials in the distribution system. Four quarters of samples were analyzed for most of the systems. The results for distribution system samples are shown in Table 3.4.

Table 3.4
NDMA results for distribution systems with free chlorine and chloraminated residuals

NDMA Level (ng/L)	Free Chlorine (# of results)	Chloramines (# of results)
0 – 2.5	26	30
2.5 – 5	0	12
5 – 7.5	0	4
7.5 – 10	0	1
10 – 12.5	2	0
12.5 – 15	0	0
15 – 17.5	0	2
17.5 – 20	0	0
20 – 22.5	0	0
22.5 – 25	1	0
Total number of samples	29	50

The summary and conclusions reported for this study and these particular samples and conditions were as follows:

- Distribution system samples generally had higher levels of NDMA than plant effluent samples, except where effluent water was blended with other water prior to distribution.
- NDMA levels above the method reporting limit (0.6 to 1.0 ng/L) were found in disinfected (with chlorine and chloramines) samples originating from rivers, lakes, and groundwater.
- The median for drinking water treatment plant effluent samples, whether chlorinated or chloraminated, was less than 1 ng/L, and the median for treated drinking water distribution system samples was less than 2 ng/L for chloraminated water and less than 1 ng/L for chlorinated water.
- Fifty percent of the data was equal to or below 3 ng/L for chloraminated drinking water distribution system samples and below 2 ng/L for chlorinated drinking water distribution system samples.

IMPACTS ON CHEMICAL QUALITY - CORROSION, METAL RELEASE, COLOR, AND PARTICLE CHARACTERISTICS

Potential Impacts

Overall, there are two primary water quality effects from chloramines; the impact of altered oxidizing conditions in the distribution system (either increased or decreased depending on historical free chlorine residual levels) and the impact that nitrification may have in the system with respect to changes in pH, alkalinity, dissolved oxygen and loss of free chlorine residual.

Positive Impacts

- Iron - Free chlorine has been found to be significantly more reactive with iron scales than monochloramine (Valentine, et. al, 2000). Maintaining an oxidized insoluble pipe surface using monochloramines could reduce red water problems.
- Copper - Chloramines have demonstrated the ability to control biofilms more successfully than free chlorine. Thus increased mitigation of microbially mediated copper corrosion would be expected with the change to chloramines from free chlorine.
- Lead - A possible benefit associated with the conversion to chloramine from free chlorine would be if biofilm were reduced on the lead surface. As biofilms are controlled, lead corrosion control that relies on the formation of insoluble films may also demonstrate improved solid stability.
- Particle Changes - The use of chloramine may slow the manganese oxidation rate sufficiently for manganese to pass through the distribution system un-oxidized and thus reducing the formation of discolored water.
- Overall, with a more persistent residual, one might observe a higher redox potential in portions of the distribution system that previously had no chlorine residual.

Adverse Impacts

- Iron – With the potential reduction in biofilms, there may be a consequential instability of the iron pipe surface, resulting in iron release and discolored water.
- Copper - In a situation where chloramines were introduced into a free residual depleted area, the increase in redox potential may cause a shift from Cu^{+1} to Cu^{+2} solids and increase solubility. Also, decay of chloramines results in the release of ammonia and could increase the solubility of copper in the system by promoting the formation of copper-ammonia complexes.
- Lead - The potential exists for a water system that converted from free chlorine to chloramine to have increased lead levels because of changes in Pb oxidation state of Pb solubility – controlling solids.
- Particle Changes - Theory has not yet been formulated to elucidate what may be expected with regard to particle transformation when a utility changes from free chlorine to chloramines for disinfection.

Reported Impacts

The reported impacts on chemical water quality (corrosion/metal release/color/particle characteristics) of the participating utilities that converted from using chlorine to chloramines are described in detail in Appendix H. The impacts on corrosion and metal release generally were reported in terms of discolored water complaints. Three utilities submitted information on lead levels before and after the switch to chloramines. A summary of the data analyses is presented in Table 3.5. Five utilities reported data with regard to iron corrosion (red or discolored water complaint records), lead 90th percentile levels, and copper 90th percentile copper levels.

Discolored Water

- The utilities that submitted discolored water complaint data generally indicated a reduction in complaints after switching to chloramines.
- Two utilities reported that discolored water complaints initially increased for several months after the switch, but then dropped to levels that were lower than before the switch.
- One utility reported discolored water complaints during chloramine conversion due to main flushing.

Lead Levels

- Most of the utilities indicated no impact on lead levels. One utility has been using chloramines for over 10 years and has not seen an increase in lead levels over this period.
- One utility reported a decrease in lead levels, resulting in compliance with the Lead and Copper Rule (LCR).
- Another utility reported an increase in lead levels, placing them out of compliance with the LCR.

Table 3.5
Summary of corrosion data from utilities

Utility	Red Water, complaint numbers		Lead, µg/L		Copper, mg/L	
	Cl_2	NH_2Cl	Cl_2	NH_2Cl	Cl_2	NH_2Cl
Brick, NJ	134	29	NA	NA	NA	NA
Concord, NH	11	18	35	20	0.125	0.085
Philadelphia	Lower with NH_2Cl		NA	9-15	NA	0.3
Washington DC	NA	NA	8	75	NA	NA
San Francisco	NA	NA	7	11.5	0.12	0.09

"NA" denotes data not available

In addition to the information from the participating utilities, a telephone survey of chloraminating utilities was conducted to obtain additional information regarding any observed changes in lead levels following chloramines conversion. The findings are summarized below:

- Regional Engineers were contacted at each of the five regional offices of the State of California Department of Health Services Drinking Water Program. In Region III, the City of Coalinga converted to chloramines in November 2005, and conducted an initial round of lead testing in January 2006. No change in average lead levels was observed; however, the percent of samples with non-detected lead levels increased, from 50 percent before to 70 percent after conversion.

- Other California utilities have recently converted to chloramines, including Irvine Ranch Water District (1998), City of Newport Beach (2003), and Castaic Lake Water Agency (2005). These utilities have not yet collected sufficient information to assess any potential changes in lead levels.
- In Hawaii, the Maui Department of Water Supply (MDWS) uses chloramines in the Upper Kula System. In 2001, a spike in lead levels was observed, and MDWS began adding zinc orthophosphate for corrosion control for the Upper Kula water supply. Numerous customer complaints about skin irritations prompted the Department of Water Supply to replace the phosphates with soda ash in 2004. Recent water quality testing has shown lead levels to be consistently below action levels.

IMPACTS ON AESTHETIC QUALITY - TASTE AND ODOR

Potential Impacts

Overall, a utility should experience fewer taste and odor (T&O) complaints with the switch from a free chlorine residual to a monchloramine residual.

Positive Impacts

- There should be a reduction in T&O complaints, particularly chlorinous odors.
- There should be fewer tastes associated with formation of chlorinated organic or inorganic by-products, such as chlorophenolic odors.
- More persistent residuals should result in more consistent tastes and odors in the distributed water.

Adverse Impacts

- Chloramines cannot oxidize algal or microbial odors.
- Possible formation of odors and tastes from dichloramine and trichloramine formation.
- Tastes and odors can be generated from anaerobic conditions associated with nitrification.

Reported Impacts

The reported impacts on aesthetic water quality of the participating utilities that converted from using chlorine to chloramines are described in detail in Appendix H. Five utilities provided numerical data while others provided anecdotal information. The impacts that were reported included taste and odor complaints. A summary of the data analyses is presented below and in Table 3.6.

Taste and Odor

- Overall, the utilities reported a reduction in the number of taste and odor complaints as a result of the switch to chloramines.
- Several of the participating utilities documented not only the number of complaints but also the type of complaint. The general descriptor of taste and odor complaints changed from chlorinous to musty, earthy or metallic. It is possible that these other types of tastes and odors were present when using chlorine but were masked by the chlorinous tastes and odors. With the elimination of the chlorinous tastes and odors, the other descriptors were detected by customers.

Table 3.6
Taste and odor data from utilities

Utility	Complaint Numbers, Cl_2	Complaint numbers, NH_2Cl
Brick, NJ	100 in 16 months	13 in 4 months
Concord, NH	20 per year	9 in year 1, 24 in year 2
Newport News	41 per year	20 in year 1, 23 in year 2
PA – American	1 per year	1 per year
Tampa, FL	17 in 5 months	6 in 7 months

CASE STUDIES

More detailed information is presented in the following paragraphs for three of the participating utilities - Brick Township Municipal Utilities Authority (MUA) in New Jersey, the San Francisco Public Utilities Commission in California and the Washington Aqueduct, District of Columbia, United States.

Brick Township (NJ) MUA

Brick Township MUA is located in southern New Jersey along the coastal plain. The MUA serves 80,000 residents and sells water to Point Pleasant and Point Pleasant Beach. The 16-MGD facility obtains the majority of its source water from the Metedeconk River. Water is also obtained on a supplemental basis from a groundwater source in the Potomac-Raritan-Magothy Aquifer.

A conventional rapid mix, flocculation, sedimentation and filtration treatment train provides the basic treatment sequence. Alum is used as the coagulant and chlorine is added for disinfection at pre- and post- filter locations. Lime and carbon dioxide are added as pretreatment chemicals at the rapid mix; this provides the basis for corrosion control in the system. Raw water from the Metedeconk River and groundwater wells are also treated with chlorine dioxide, which is added as a preoxidant and aeration is applied to help remove hydrogen sulfide.

Microbial Quality

Brick Township MUA collects samples at about 80 locations for HPC analysis. Before the conversion to chloramines, HPC levels typically were very low – below 15 cfu/ml as shown on Figure 3.1. A few sites had HPC levels between 15 and 150 cfu/ml. After the conversion to chloramines, HPC levels generally remained low but varied with sample location. As shown on Figure 3.2, HPC levels at this location dropped from as high as 60 cfu/ml to less than 5 cfu/ml. At another site as shown on Figure 3.3, HPC levels were zero when using chlorine, and increased to between 3 and 10 cfu/ml after the conversion. At a third site shown on Figure 3.4, HPC levels went from less than 10 cfu/ml to as high as 300 cfu/ml .

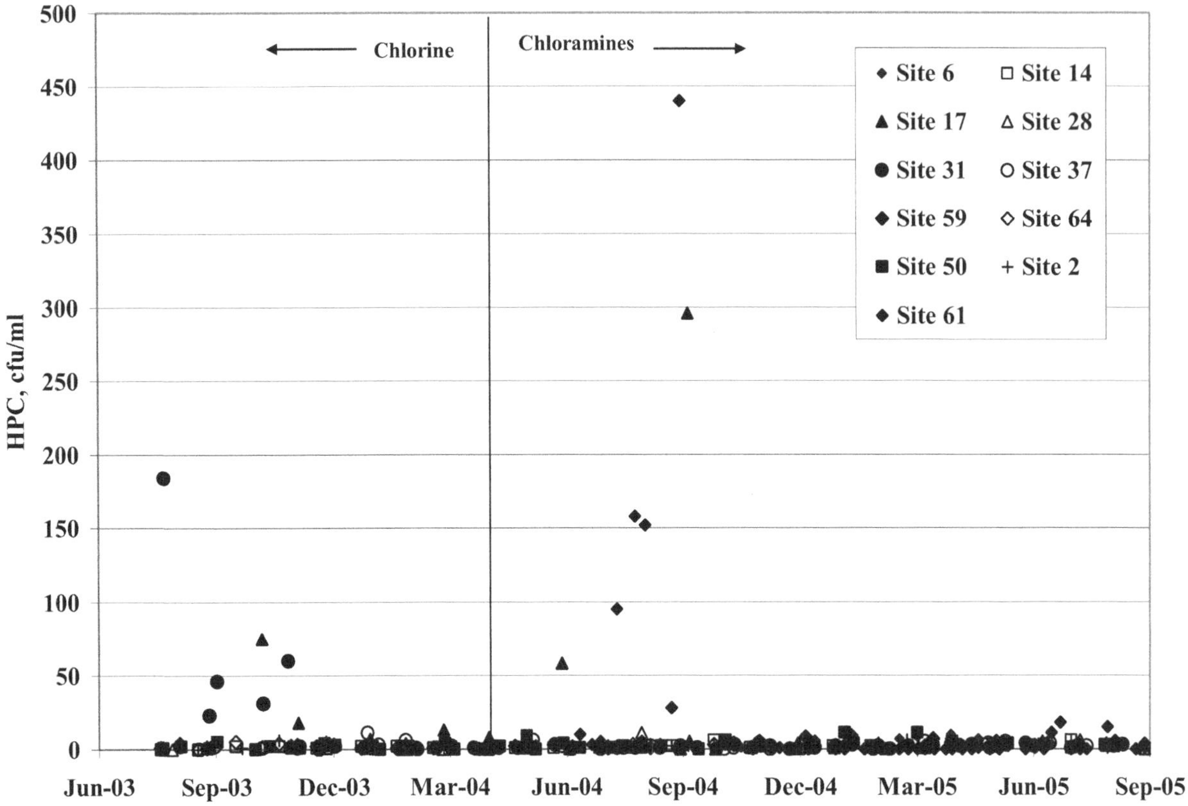

Figure 3.1 Brick MUA HPC data

Confounding the HPC data was a change in heterotrophic plate count methodology from plate count agar to R2A concurrent with the conversion to chloramines. The Brick laboratory director noted that the R2A medium improved recovery of bacteria, with typical numbers doubling compared to the standard plate counts but overall numbers were still well below the regulatory limit. High HPC numbers at Site 61 were traced to sample collection at an outside spigot rather than inside the building.

Analysis of the HPC data was performed using a conservative variation of Student's t test at the 95% confidence level. The variances could not be pooled so the larger standard deviation and corresponding degrees of freedom were used to look at differences in HPC levels before and after the conversion to chloramines. No differences could be found in the overall data, though specific sites demonstrated short term differences after the conversion. The results of the analysis are plotted on Figure 3.5.

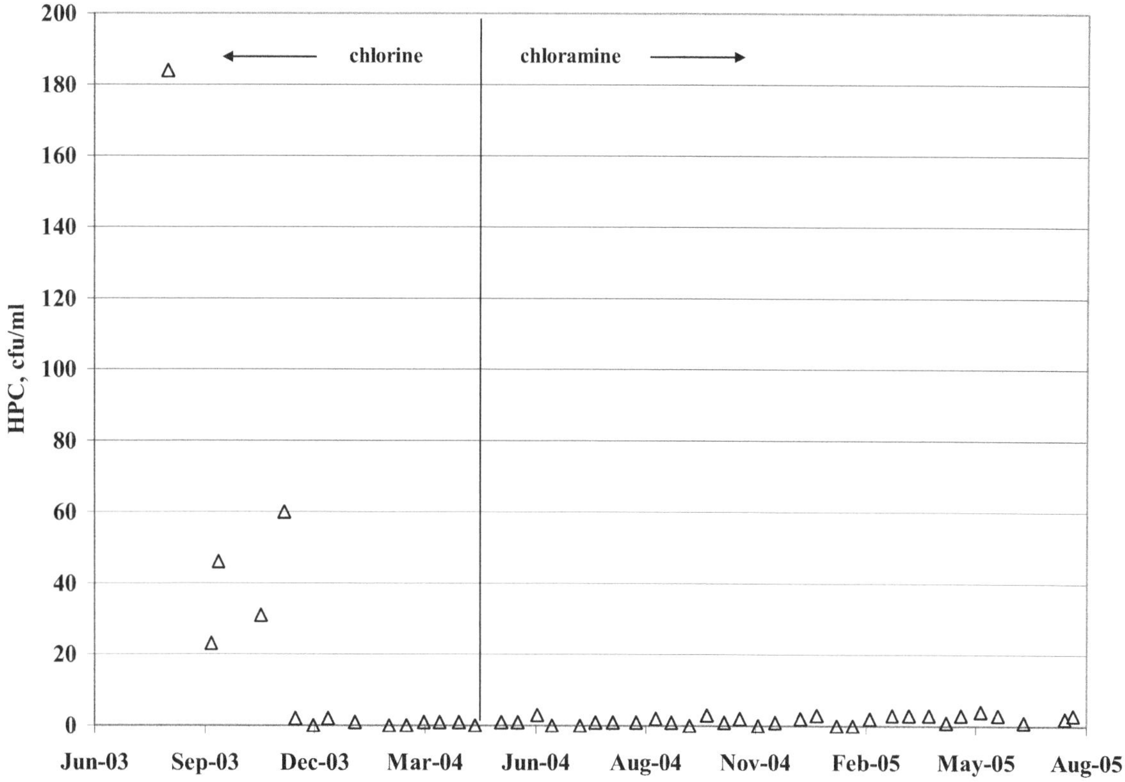

Figure 3.2 Brick MUA site 28 HPC data

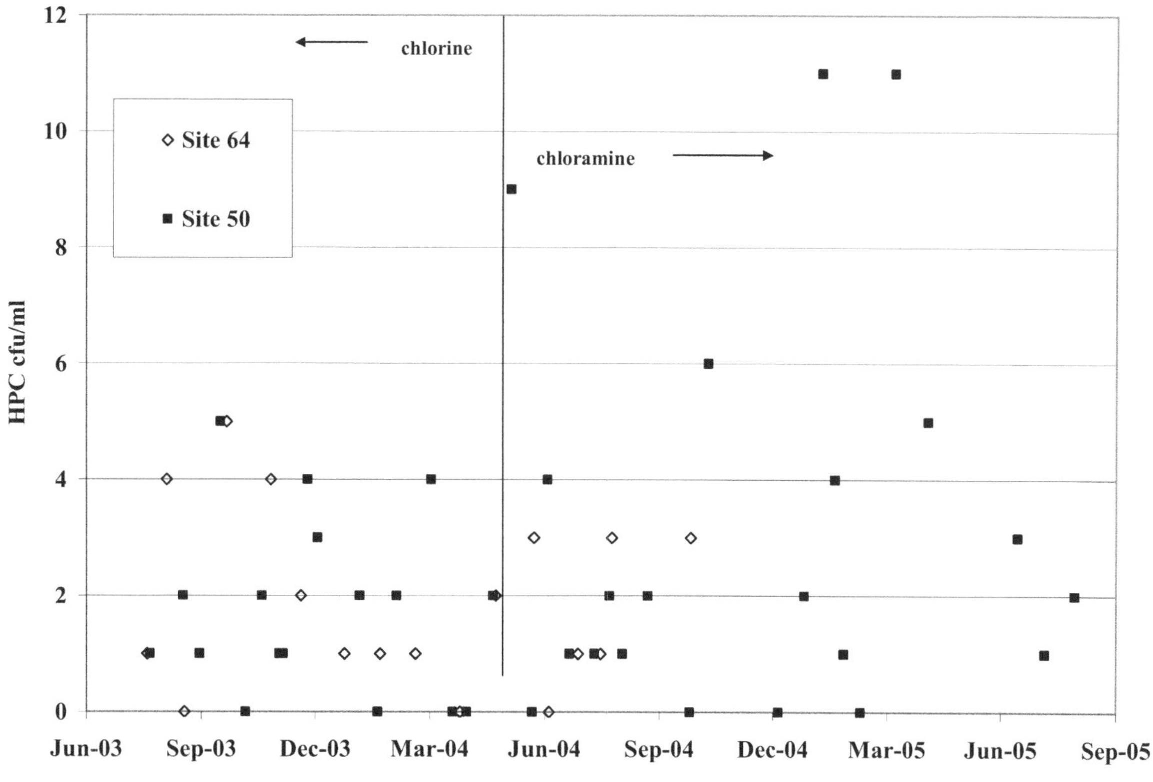

Figure 3.3 Brick MUA sites 50 and 64 HPC data

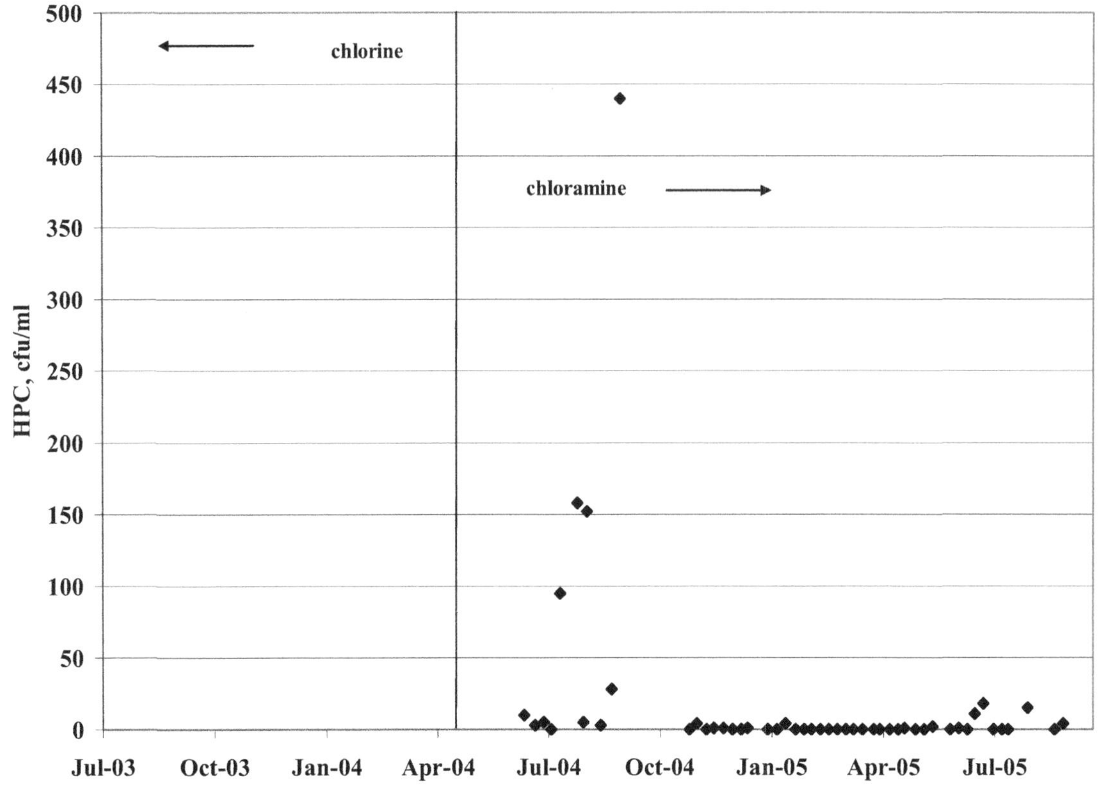

Figure 3.4 Brick MUA site 61 HPC data

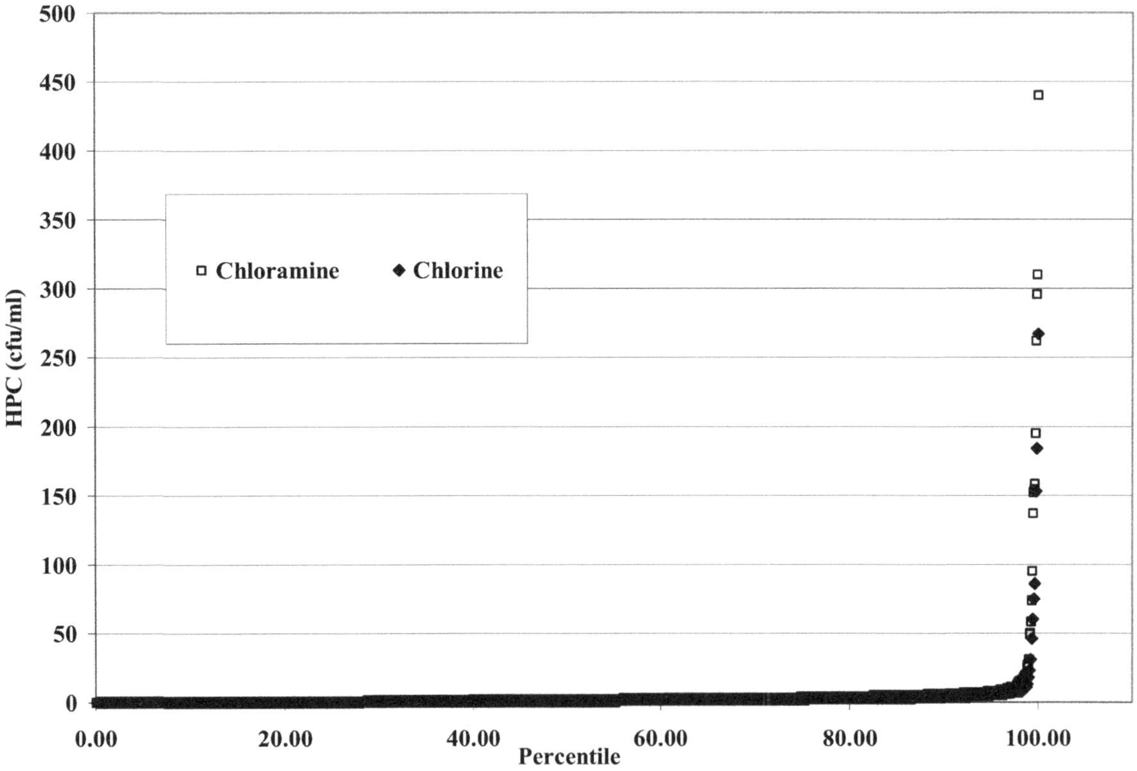

Figure 3.5 Cumulative frequency analysis of Brick HPC data

Chemical Quality – DBPs

Brick MUA provided TTHM and HAA5 data from 13 sites throughout their distribution system. TTHM data for seven of the thirteen sites are shown on Figure 3.6. The data show that the switch to chloramines resulted in an overall reduction in TTHMs from about 80-100 μg/L to 40-60 μg/L. Figure 3.7 shows TTHM levels at two locations, Sites 54 and 80, illustrating the drop in TTHMs at both sites.

Analysis of the TTHM data before and after conversion to chloramines revealed a significant difference at the 95% confidence level between the two sets of data. Once again, the variances could not be pooled so the larger standard deviation and corresponding degrees of freedom were used to look at the data. The conversion to chloramines made significant reductions in TTHM levels, as shown on Figure 3.8.

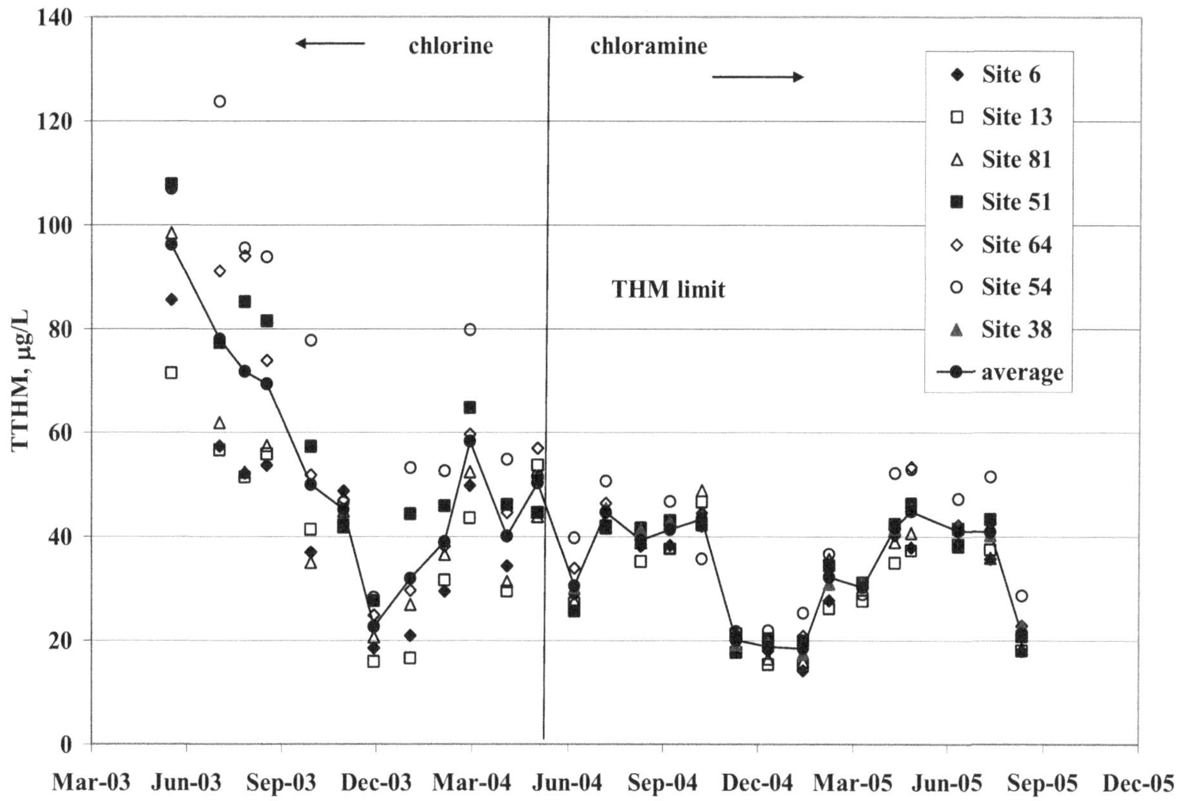

Figure 3.6 Brick MUA TTHM data

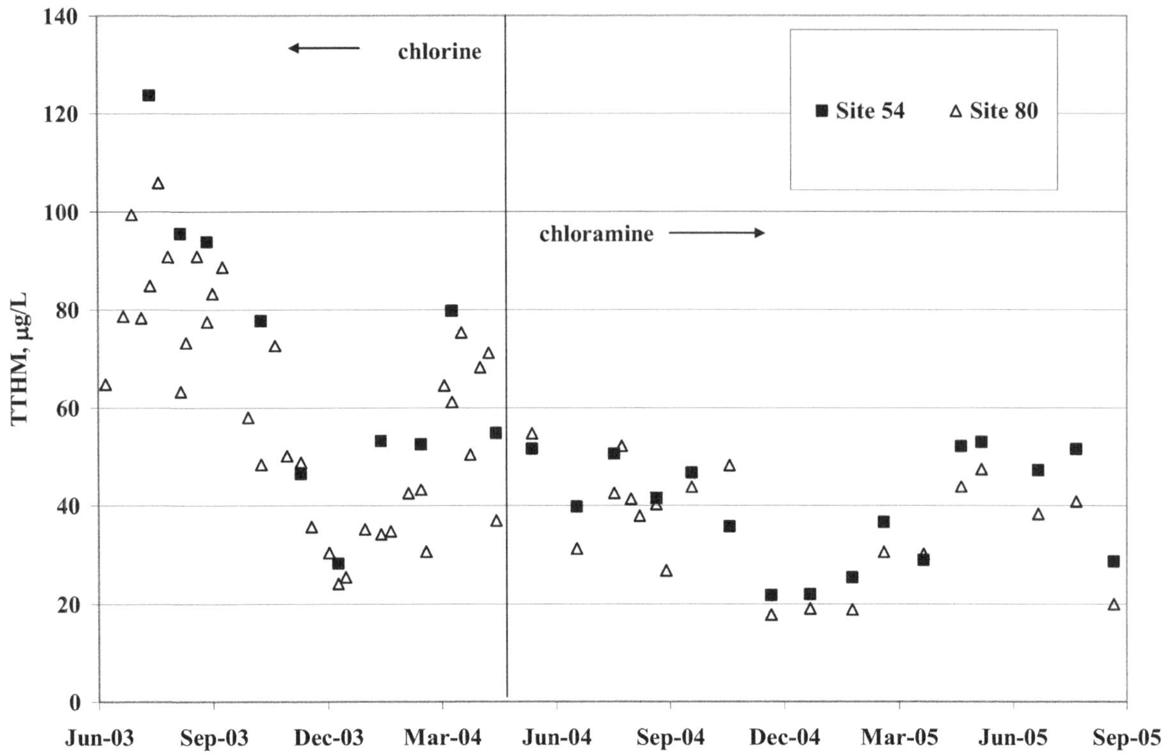

Figure 3.7 TTHM levels at site 54 and site 80, Brick MUA

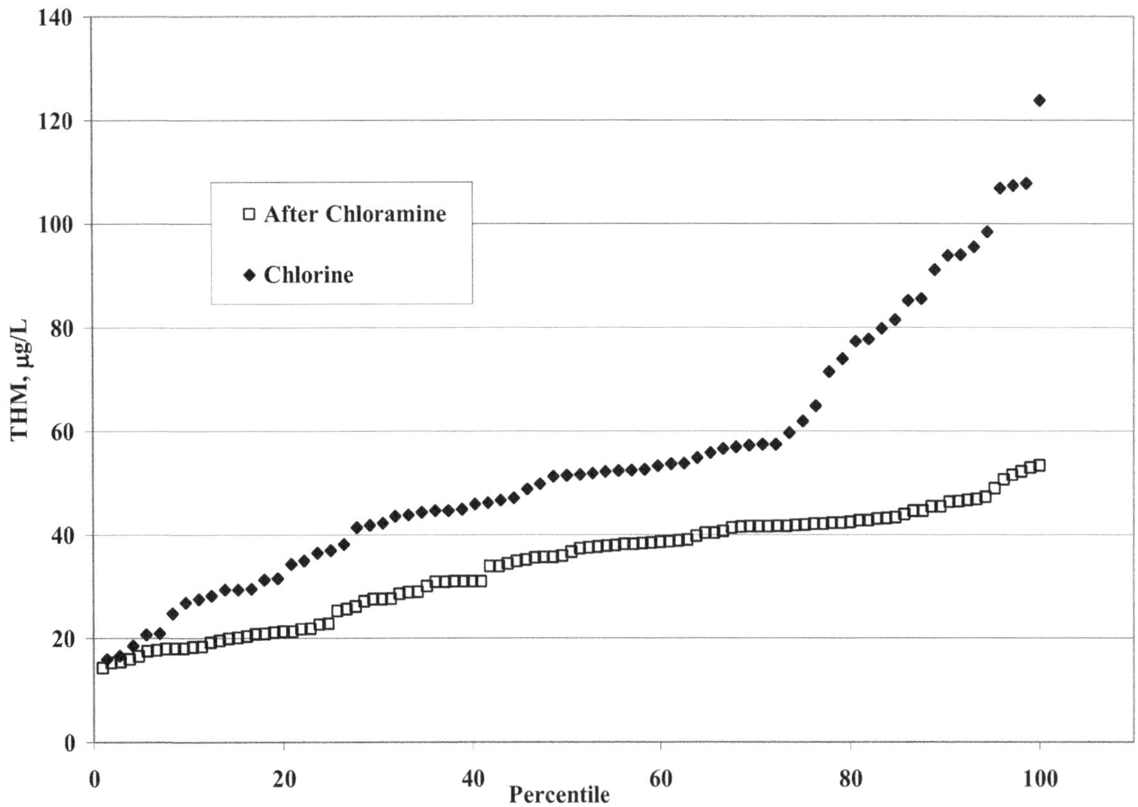

Figure 3.8 Cumulative frequency analysis of Brick TTHM data

HAA5 data for seven of the thirteen sites are shown on Figure 3.9. The data show that HAA5 levels appear to have dropped at many locations; however, at some locations there was no change in HAA5 levels. Figure 3.10 shows one location where HAA5 levels dropped (summer and fall) as a result of the switch to chloramines. Figure 3.11 shows two other locations where HAA5 levels did not appear to change after the switch to chloramines. It is possible that the presence of chloramines reduced biological activity at these sites and therefore reduced biodegradation of the HAA5 species.

Analysis of the HAA5 data before and after conversion to chloramines revealed no significant differences at the 95% confidence level between the two sets of data. The statistical methods used were the same as for the TTHM data, but the results were quite different. For Brick, the conversion to chloramines did not reduce HAA5 levels significantly. The results are presented on Figure 3.12.

Brick performs a monthly analysis of chlorite concentrations at selected sites in the distribution system. Figure 3.13 presents the distribution system chlorite data. Brick reports that the chlorite concentrations are stable in the system, changing little from the levels at the point of entry into the distribution system.

Brick also sent out six samples for N-nitrosylmethylamine (NDMA) analysis. All of the Brick samples were non-detectable for this disinfection by-product of chloramination.

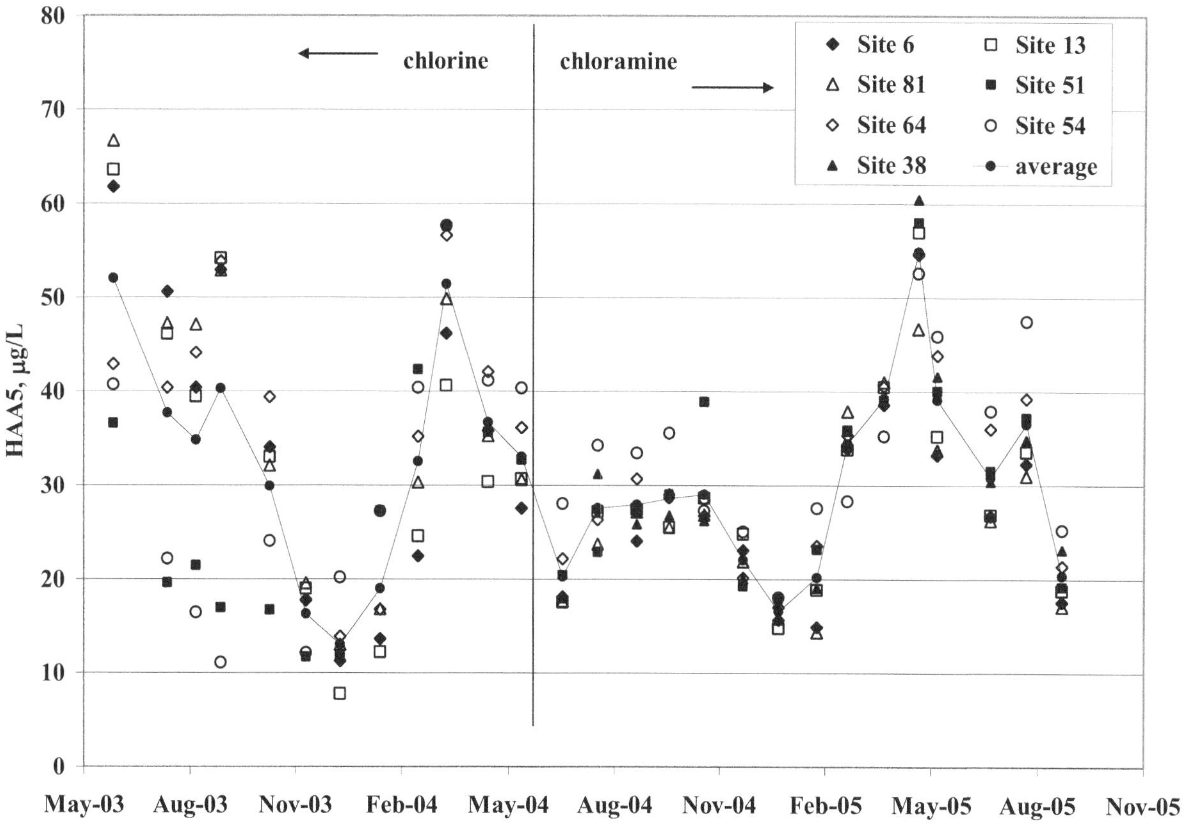

Figure 3.9 Brick MUA HAA5 data

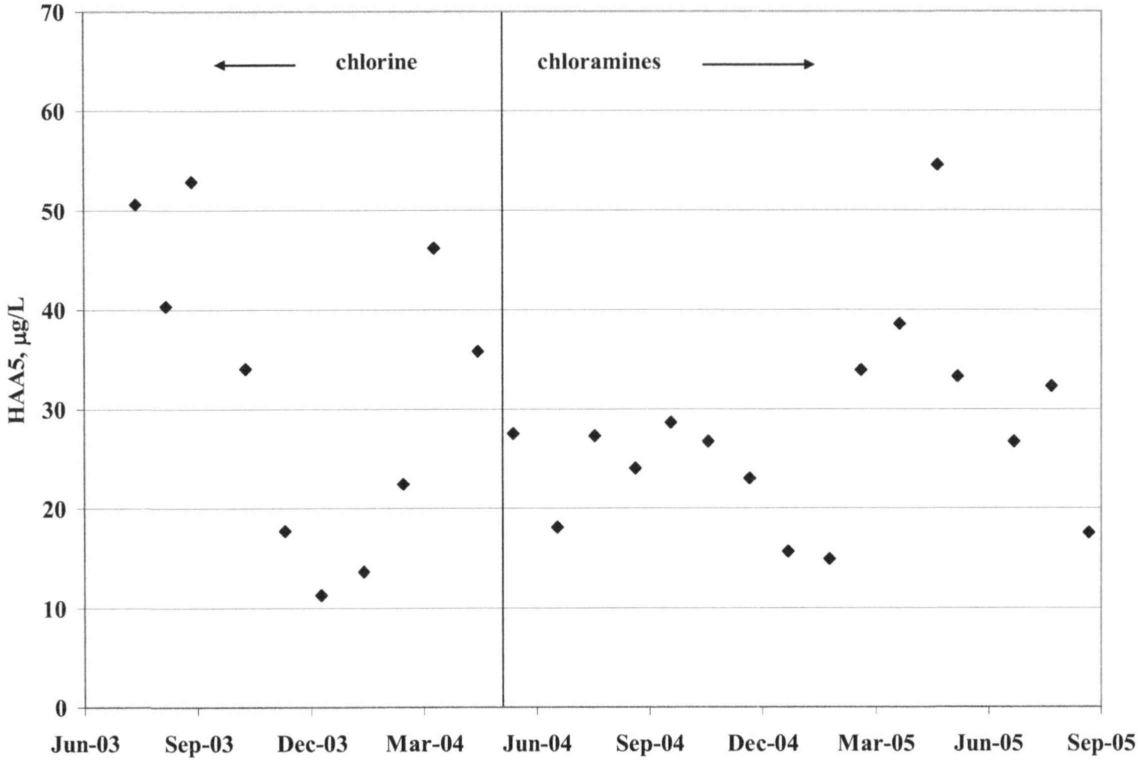

Figure 3.10 HAA5 data for site 6, Brick MUA

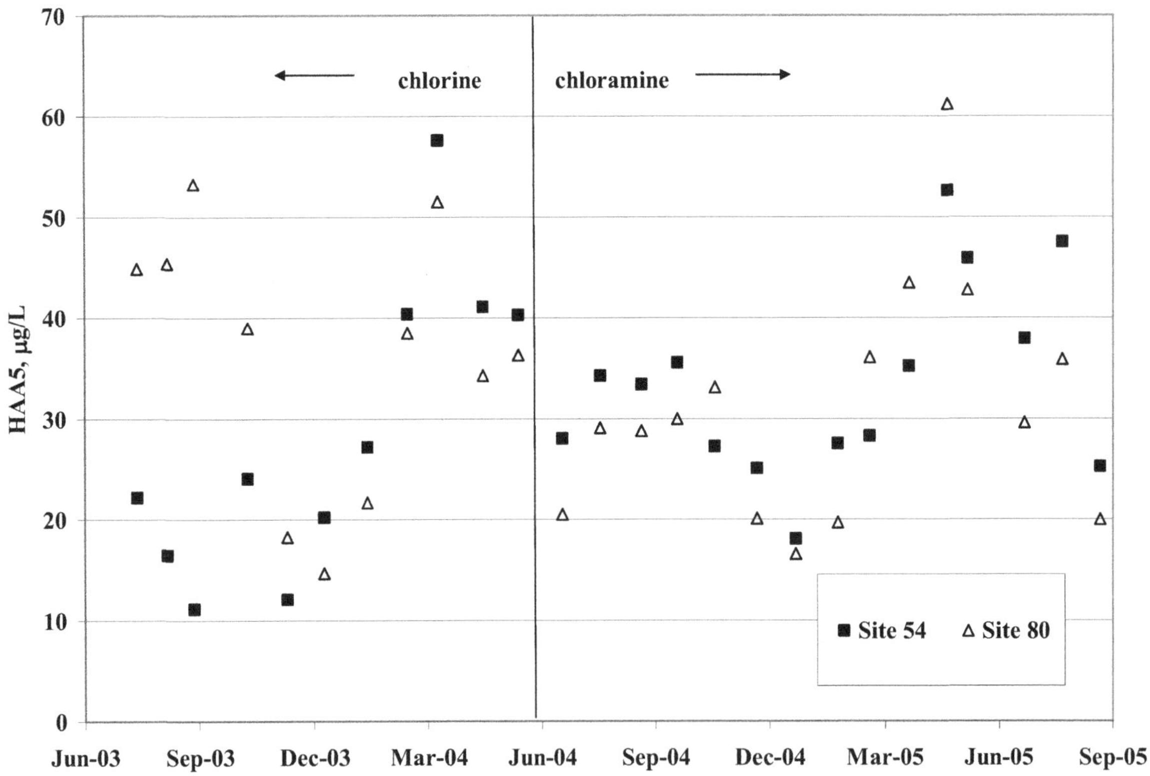

Figure 3.11 HAA5 data for sites 54 and 80, Brick MUA

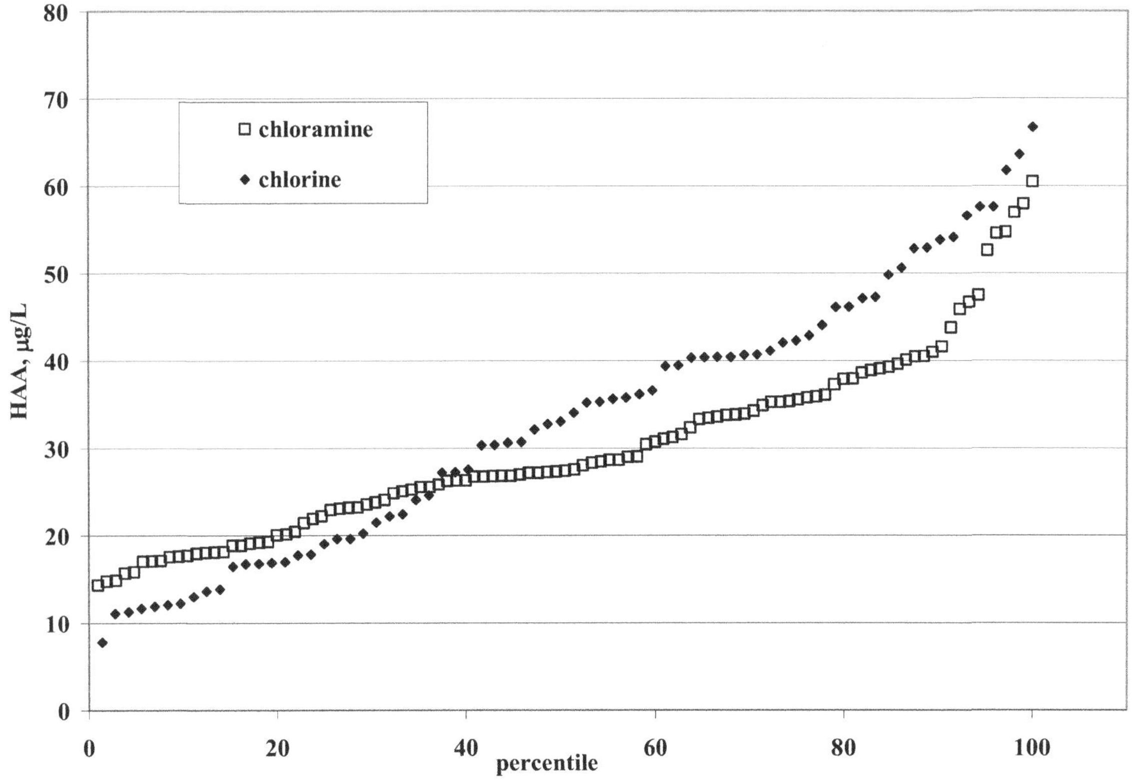

Figure 3.12 Cumulative frequency analysis of Brick HAA5 data

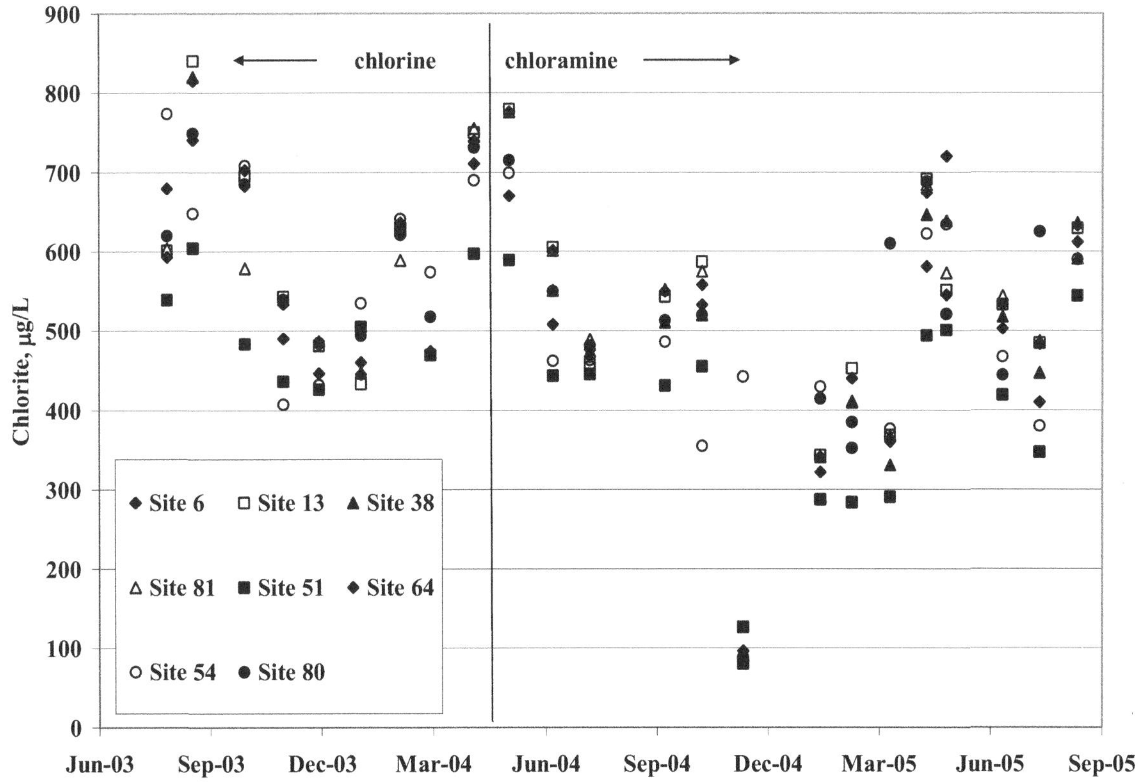

Figure 3.13 Brick MUA chlorite data

Chemical Quality – Corrosion, etc.

Brick MUA records customer complaint data and differentiates between discolored water and taste and odor complaints. A record of discolored water complaints for the year prior to the disinfection conversion is shown on Figure 3.14. In general, the number of complaints appear to have dropped after the switch from chlorine to chloramines, possibly indicating less corrosive water with respect to iron release. Discolored water complaint numbers have varied from a low of 39 to a high of 332 per year in the past eight years.

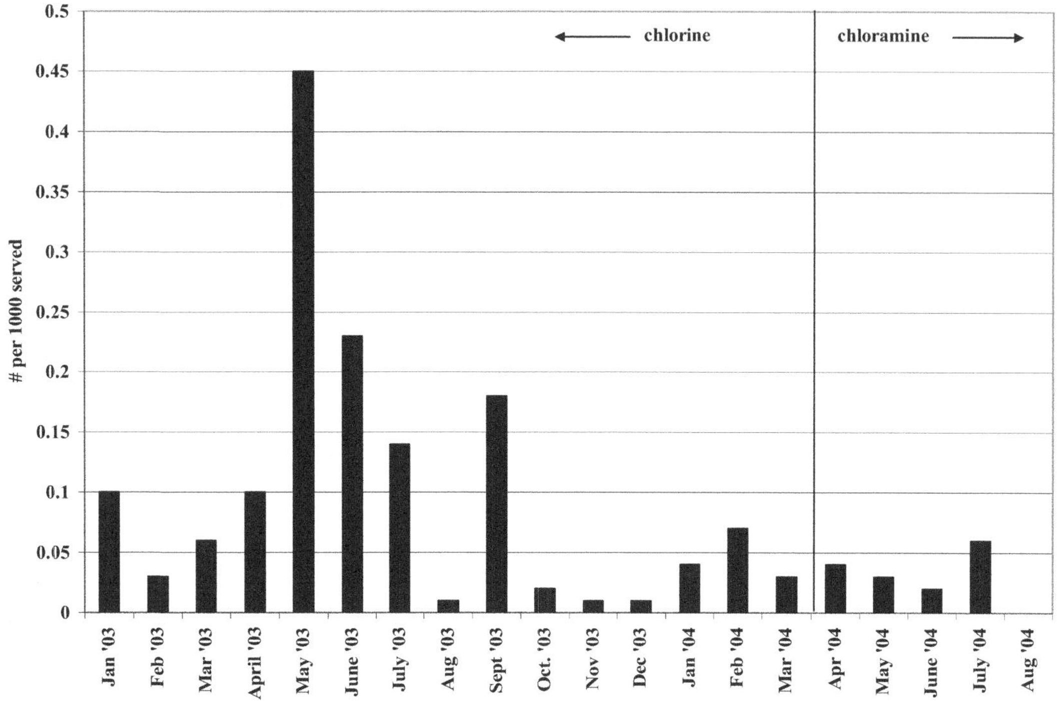

Figure 3.14 Brick MUA discolored water complaints

Aesthetic Quality

Brick MUA not only records customer water quality complaints, but differentiates between taste and odor complaints, and also differentiates between types of taste and odor complaints (i.e., chlorinous, earthy, musty, fishy, etc.). Figure 3.15 shows the number of customer complaints regarding the taste of the water before and after the switch to chloramines. No taste complaints were recorded for 2003. Only three taste complaints (one fishy, one chemical, and one metallic) were recorded in 2004 for the 80,000 customers before the conversion to chloramines. After the conversion to chloramines, one complaint of metallic taste and five "off taste" or "bad taste" complaints were received in the first few months after the conversion.

Figure 3.16 shows customer complaints regarding the odor of the water. There was an apparent reduction in odor complaints after the switch to chloramines, and also a change in the type of odors described. With the use of chlorine, the description of the odor was mostly chlorinous or musty. After switching to chloramines, the predominant descriptor has been earthy though other customers cited swampy or fishy odors. It could be that these other types of odors were present with the use of chlorine, but were masked by the chlorine odor.

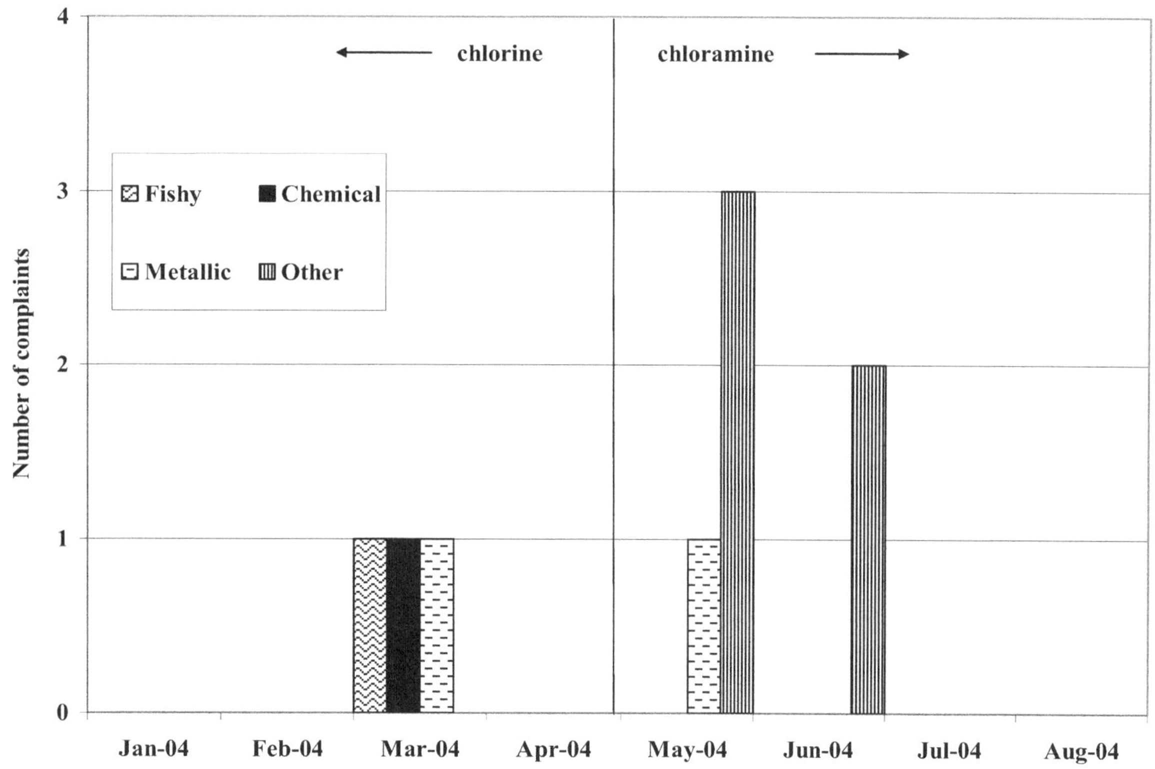

Figure 3.15 Brick MUA taste complaint data

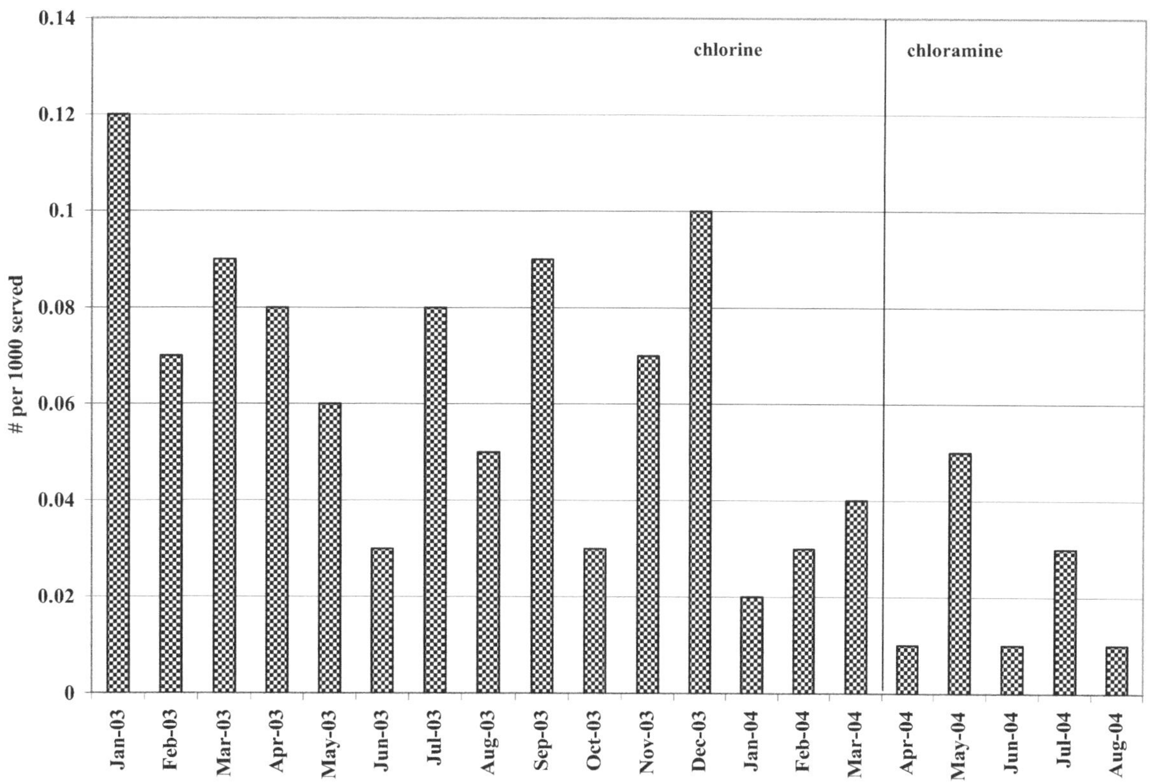

Figure 3.16 Brick MUA odor complaints

San Francisco Public Utilities Commission (SFPUC)

The SFPUC regional water system provides water to 2.4 million people in San Francisco, Santa Clara, Alameda and San Mateo counties. Eighty-five percent of the water delivered to SFPUC customers comes from Sierra Nevada snowmelt stored in the Hetch Hetchy reservoir situated on the Tuolumne River in Yosemite National Park. This water supply meets all federal and state requirements for remaining exempt from filtration. Primary disinfection of the Hetch Hetchy supply is achieved using chlorine, and chloramines are used for secondary disinfection. SFPUC converted from free chlorine to chloramines for secondary disinfection during February, 2004. Lime is used for corrosion control in Hetch Hetchy Aqueduct.

The remaining 15 percent of water comes from runoff in the Alameda and Peninsula watersheds. Water from the Alameda watershed is treated at the Sunol Valley Water Treatment Plant, which is comprised of conventional treatment and sodium hydroxide for corrosion control. Chlorine is used for predisinfection and primary disinfection. Hetch Hetchy and SVWTP waters are blended, chloraminated, fluoridated, and pH is adjusted with sodium hydroxide for corrosion control. Water from the Peninsula watershed is treated at the Harry Tracy Water Treatment Plant, which consists of direct filtration (coagulation, flocculation, filtration) and sodium hydroxide for corrosion control. Ozone is used for pre-disinfection, chlorine and ozone for primary disinfection, and chloramines for secondary disinfection.

It should be noted that SFPUC city distribution system was designed with an emphasis on fire protection after the 1906 earthquake had engulfed much of the city in fire. Consequently, prior to chloramination, the city system featured oversized reservoirs with detention times as long as 180 days, and 10 chlorine booster pump stations maintained residuals. Significant operational changes were prompted by chloramination, with several supporting programs, including:

- Water age evaluations and bench scale studies to predict chloramine decay for residuals maintenance
- Conveyance evaluation and improvements
- Reservoir Cleaning
- Unidirectional Flushing
- Installation of large mechanical mixers in large oversized reservoirs
- Reservoir pumped sample systems and roof sample ports
- Chloramine Monitoring and Nitrification Response Planning
- Disinfectant conversion plan
- Reservoir drawdown studies

Prior to the unidirectional program, SFPUC's flushing program was conducted at dead ends, at routine frequencies, to improve chlorine residual levels. The primary goal of the unidirectional flushing program was to reduce chlorine demand caused by biofilm, corrosion products on pipe walls, and accumulated sediment in the distribution system.

Microbial Quality

SFPUC very recently converted from free chlorine to chloramines for secondary disinfection. Total coliform and heterotrophic plate count sampling takes place at 101 sites within the service area. The majority of these sample sites showed no coliforms over the 16 month sampling period surrounding the change. Prior to chloramination, the SFPUC enumerated total coliforms at terminal reservoirs using the Q-tray method. Figure 3.17 presents a summary of the coliform data from these sites. The majority of positives occurred during the warmer water months of 2003 though four Coliform positive samples were recorded in early 2004. Two positive samples were measured in 2005 from sites not included in Figure 3.17. (Note: no samples were collected at site MMR in 2005).

SFPUC collects speciation data on coliform positives. Since chloramines conversion, there have been very few coliform positives (4 in 2004; 2 in 2005). Since there have been so few coliform positives following the conversion, insufficient data are available to assess changes in bacteria speciation. The testing laboratory reported no significant changes in species. Table 3.7 provides a summary of total coliform positives before and after the conversion to chloramines.

Table 3.7
Summary of coliform positives before and after chloramines

Year	Total Coliform Positives	Total Samples Collected
1999	26	3861
2000	13	3794
2001	4	3689
2002	6	3711
2003	9	4331
2004 (conversion Feb 2004)	4	3876
2005	2	3897

Source: personal communication with Alan Wong, SFPUC, April 27, 2006

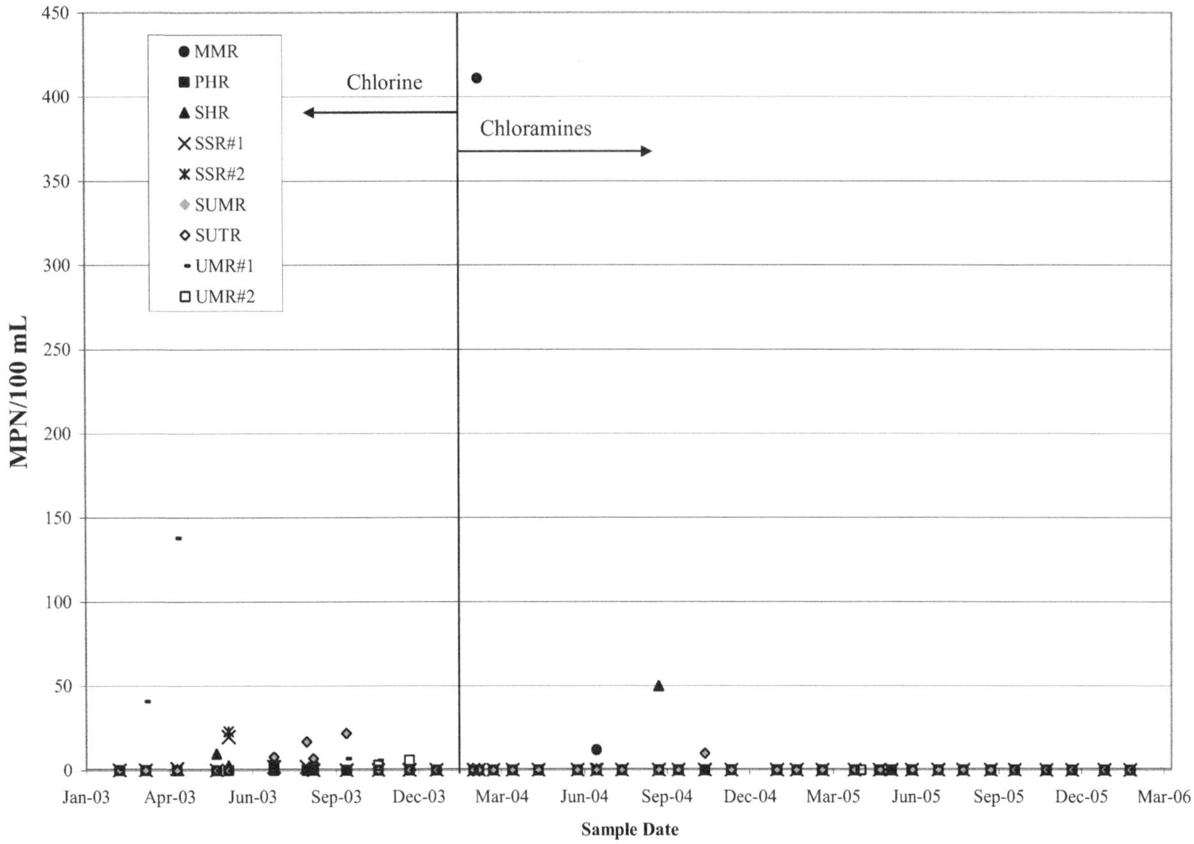

Figure 3.17 San Francisco PUC total coliform data

Figure 3.18 presents a summary of heterotrophic plate count data (7 day incubation) from six representative sample sites out of forty within the SFPUC service area that were sampled before and after the conversion to chloramines in February 2004. All samples were measured using R2A agar with detection limits of 10 to 57,000 cfu/mL. HPC levels spiked at several sample sites during the warmer water months of 2003. Mean HPC levels at these six locations decreased by more than 40% since chloramination, decreasing from a mean value of 111 cfu/mL prior to chloramination to a mean value of 63 cfu/mL after chloramination. Analysis of the heterotrophic plate count data was performed using a variation of Student's t-test at the 95% confidence level. The t-test assumed that the variances could not be pooled because the two data sets (HPCs before and after achloramination) had significantly different standard deviations. A significant difference in the means of the data sets was observed at the 90% confidence level, but not at the 95% confidence level.

Figure 3.19 shows the frequency distribution of HPC data before and after the conversion to chloramines. The 90th percentile HPC levels dropped from about 700 cfu/ml when using free chlorine to about 50 cfu/ml when using chloramines.

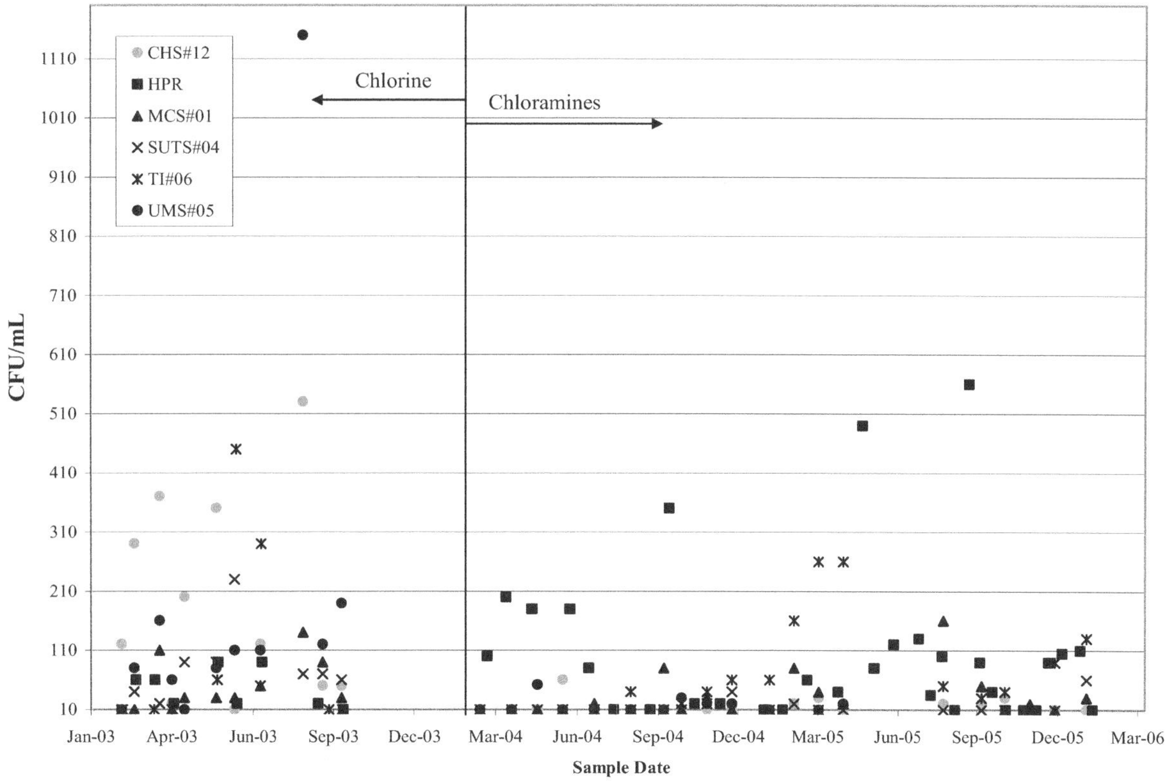

Figure 3.18 San Francisco PUC selected HPC data

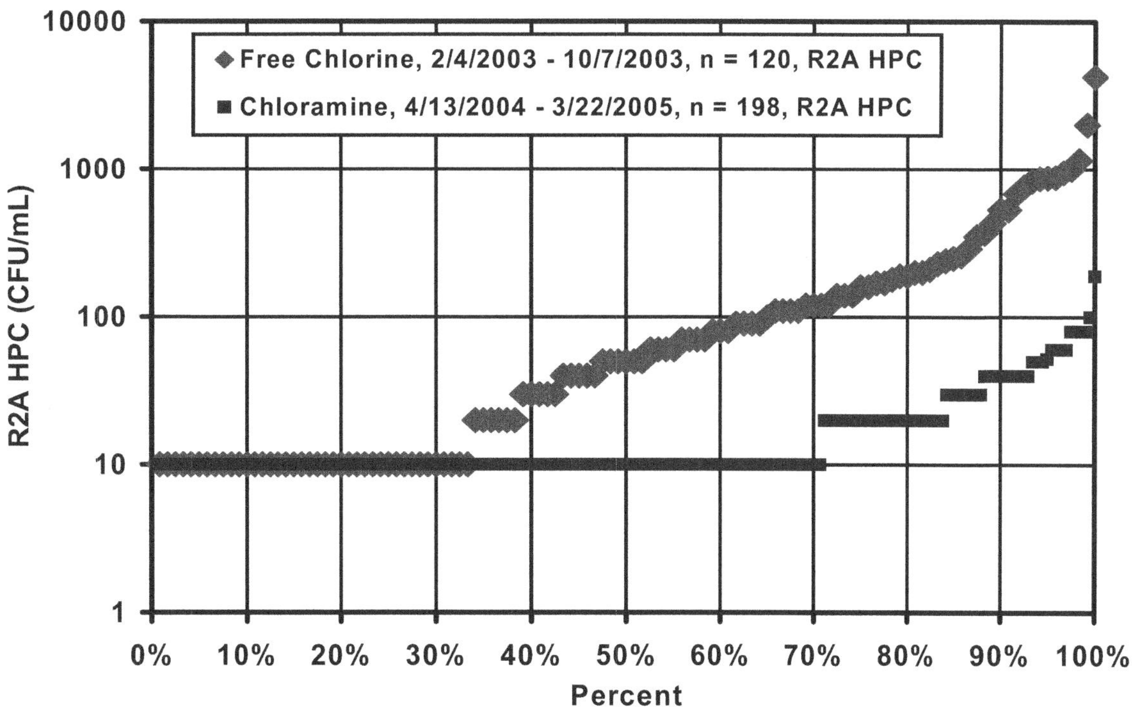

Source: Adapted from Boozarpour, et.al, 2005

Figure 3.19 San Francisco PUC cumulative frequency distribution of HPC data

Historically, samples for heterotrophic bacteria in the San Francisco Water System (SFWS) were collected only when a non-detectable chlorine residual was measured in the field. These samples were analyzed using standard Plate Count Agar (PCA) and are described under the Surface Water Treatment Rule as a substitution for a detectable disinfectant residual if results are less than 500 cfu/mL. In 2005, there were no measured non-detectable chlorine residuals in the SFWS distribution system; therefore, no HPC-PCA samples were collected.

After chloramines conversion, SFPUC implemented a special monitoring program to assess any potential impacts to water quality. As part of this monitoring program, SFPUC collected 462 HPC-R2A samples in the SFWS in 2005. R2A media differs from PCA media in that it is a low nutrient agar that is incubated at a lower tempurature for a longer period of time (25 °C vs. 35 °C, 7-days vs. 2-days). The purpose of using R2A and the alternative incubation parameters was to facilitate the growth of stressed heterotrophic bacteria that may be in the sample but not normally counted using the PCA agar. R2A agar tends to yield a higher growth of bacterial colonies and thus higher values than PCA analyses. No significant variations in sampling techniques occurred before or after chloramines conversion.

San Francisco PUC also noted a more stable distribution system residual when using chloramines as indicated below in Table 3.8. Disinfectant boosting was not needed with the use of chloramines. Disinfectant residuals at City TCR taps are shown on Figure 3.20.

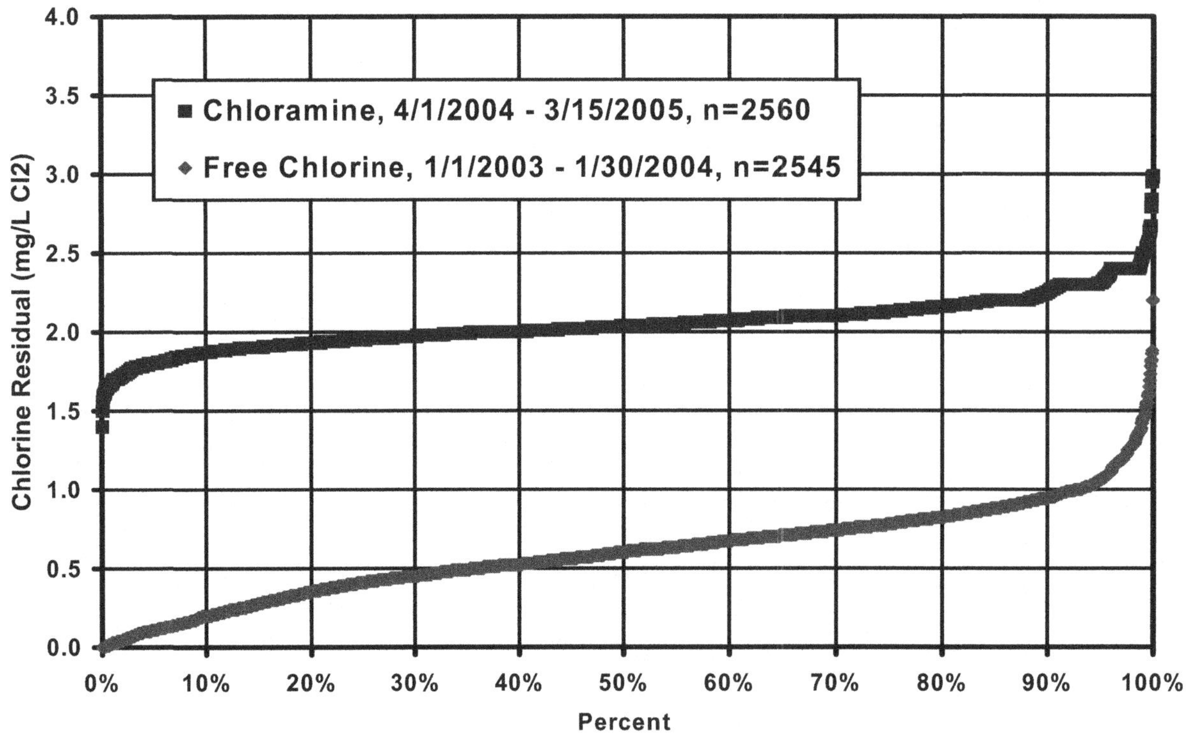

Source: Boozarpour, et. al, 2005

Figure 3.20 San Francisco PUC cumulative frequency distribution of residual data

Table 3.8
San Francisco PUC distribution system residuals

Table 3.8 San Francisco PUC distribution system residuals	Chlorine	Chloramine
	Chlorine	Chloramine
Disinfectant Residuals at TCR Sampling Points	40% below target of 0.5 mg/L Cl_2	Virtually all above target of 1.5 mg/L Cl_2
Disinfectant Residual Targets	> 0.5 mg/L at Reservoir outlets >0.2 mg/L at TCR sites	> 1.8 mg/L at Reservoir outlets >1.5 mg/L at TCR sites
Booster Stations	17 Booster locations	0 Booster stations
Median Disinfecant Residuals within Reservoirs	0.3 mg/L	2.1 mg/L

Source: Boozarpour, et. al, 2005

Prior to their conversion to chloramines, the San Francisco PUC was contacted by the Center for Disease Control (CDC) to conduct a study of *Legionella* in large building water systems before and after the conversion. Other team members in this project included the San Francisco Department of Health, the California Department of Health Services, and the California Emerging Infections Program. Samples were collected from water heaters at 53 building sites throughout the City. Three sampling rounds were conducted before the conversion and three rounds after the conversion. The results of the sampling and analyses are shown in Figure 3.21 and summarized as follows:

- Pre-chloramination: 70% of the building water systems were colonized on one or more occasions; 50% were colonized in all three rounds; 24% of all samples actually grew *Legionella*. Average hot water heater concentrations were 271 cfu/mL.
- Post-chloramination: only 5 of 53 buildings were ever colonized in all three rounds; <1% of all samles grew *Legionella*. Average hot water heater concentrations were reduced to 10 cfu/mL.

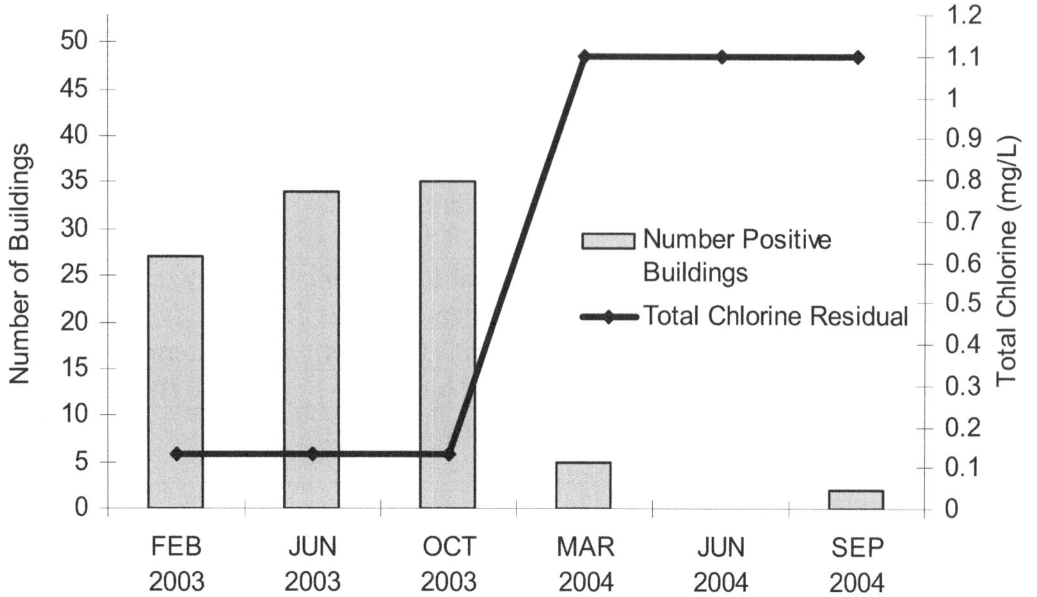

Source: Salerno, personal communication, 2006

Figure 3.21 San Francisco PUC *Legionella* data

Chemical Quality – DBPs

Levels of disinfection by-products were also measured before and after conversion to chloramines. Figure 3.22 presents a summary of total trihalomethane levels measured at ten sample sites within the SFPUC service area while Figure 3.23 presents running average and locational running average data from the system before and after the conversion. The data indicate TTHM levels decreased at several sampling locations soon after conversion to chloramines, and levels at all locations stabilized, measuring between 25 and 50 μg/L after February 2004, compared to 20 – 140 μg/L prior to conversion to chloramines. Analysis of the TTHM data at these ten sites before and after conversion to chloramines revealed a significant difference at the 95% confidence level between the two sets of data. The conversion to chloramines resulted in significant reductions in total trihalomethane levels. The mean TTHM level at these sites was reduced by 50% after chloramine implementation.

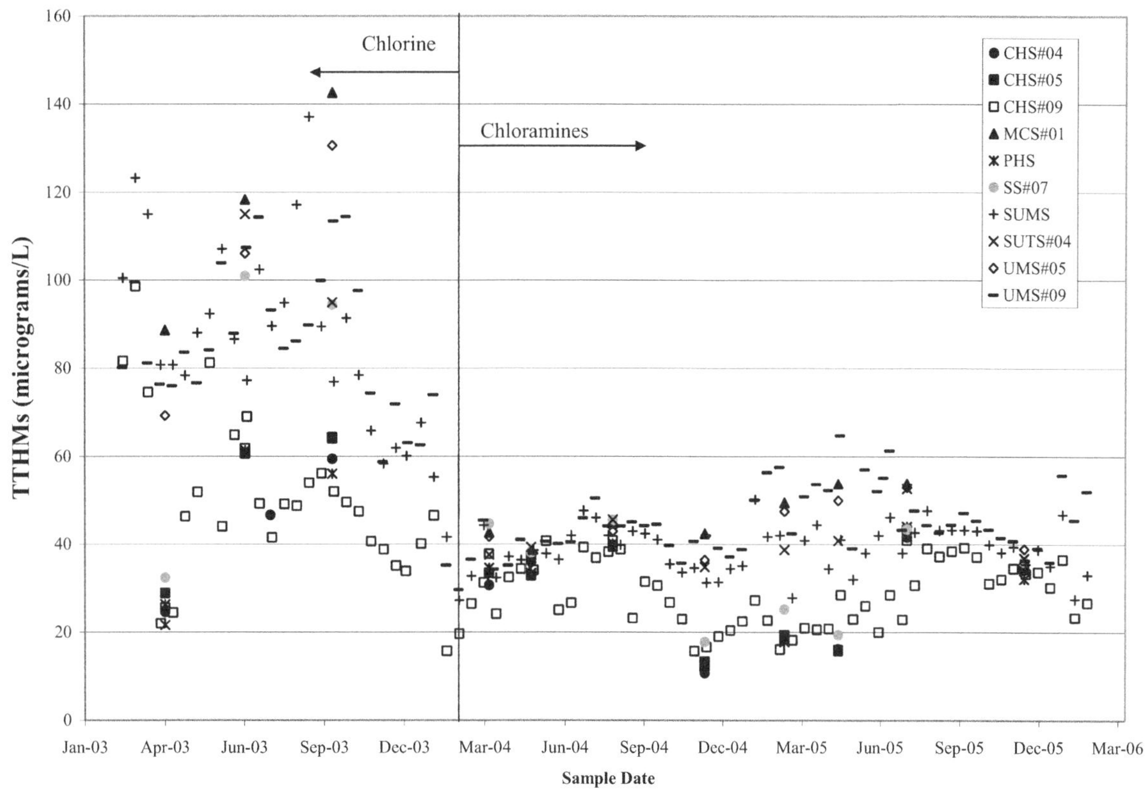

Figure 3.22 San Francisco PUC TTHM data

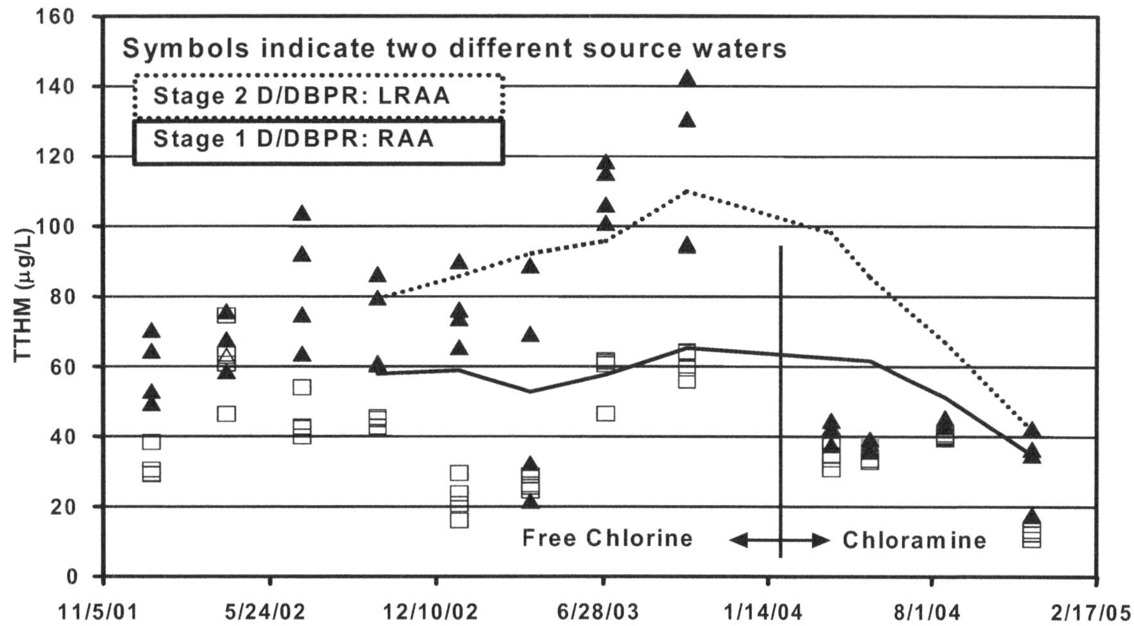

Source: Boozarpour, et. al, 2005

Figure 3.23 San Francisco TTHM average data

Figure 3.24 presents sampling results for haloacetic acids at eight sampling locations within the SFPUC service area. Statistical analysis of the HAA5 data before and after conversion to chloramines revealed no significant differences between mean levels at the 95% or 90% confidence level. It is interesting to note that HAA5 levels may have increased at certain locations (such as CHS#04). It is possible that increased HAA5 levels are associated with decreased biodegradation of HAA5, which may have been occurring prior to the conversion to chloramines. Continued sampling and correlation with historical heterotrophic bacterial levels will likely clarify this trend.

The impacts of the change to chloramination on San Francisco PUC's ability to meet the Stage 1 and Stage 2 D/DBP Rules are shown on Figure 3.23 for TTHMs and on Figure 3.25 for HAA5. The use of chloramines will enable the utility to meet the new Stage 2 requirements for both TTHMs and HAA5s.

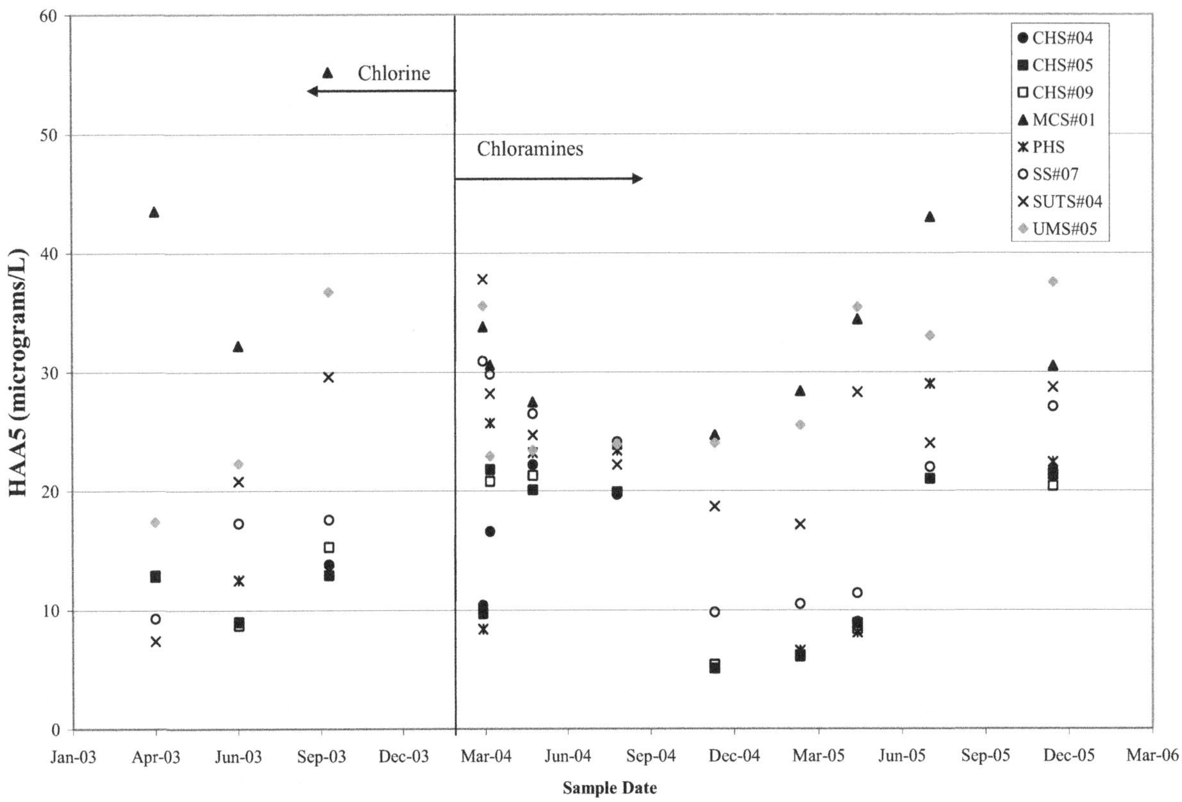

Figure 3.24 San Francisco PUC HAA5 data

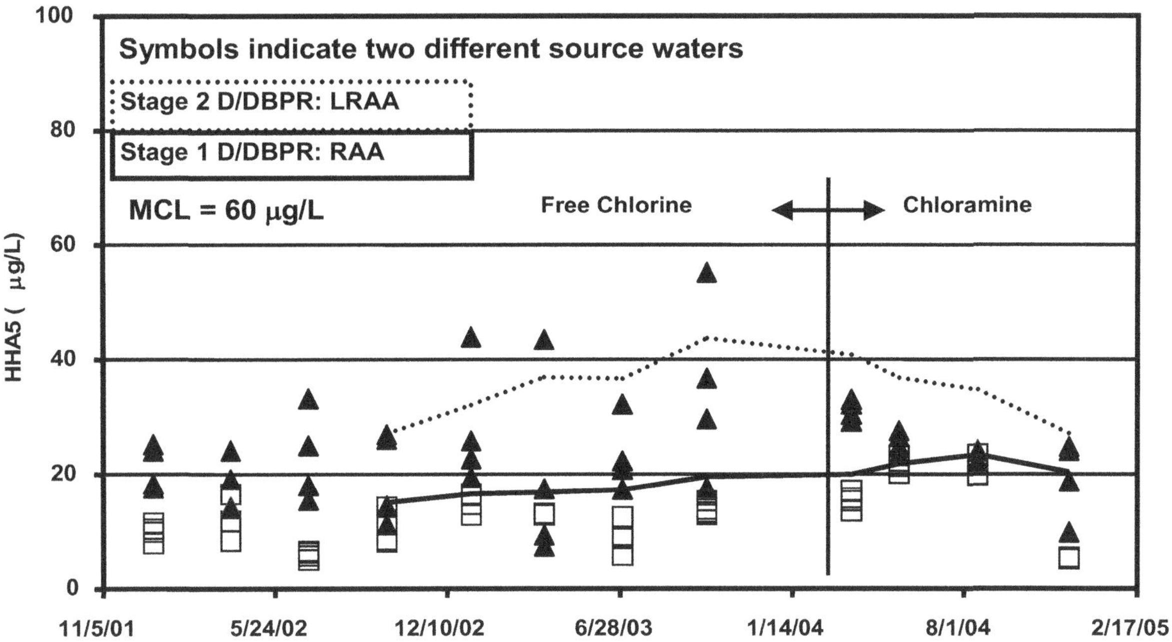

Source: Boozarpour, et. al, 2005

Figure 3.25 San Francisco HAA5 average data

The San Francisco PUC also reported NDMA levels before and after the conversion to chloramines. The results are shown in Table 3.9 and indicate no significant impact on NDMA.

Table 3.9
NDMA data for San Francisco

Sample Date	1999	Oct-00	Apr-04	Aug-04	Nov-04	Mar-05
Sample Site Name	Free Chlorine		Chloramine			
	Hetch Hetchy Source					
Free Chlorinated Alameda East	<1,<2	NA	<2	<2	<2	<2
	Sunol Source					
Distribution System	<1,<2	<1.0	<2	<2	<2	<2
	Harry Tracy Source					
Distribution System	1.3, <2	<1.0	<2	<2	2.1	2.9

Note: All values in ng/L. California NDMA Notification Level (NL) = 10 ng/L
Source: Adapted from Boozarpour, et. al, 2005

Chemical Quality – Corrosion, etc.

The San Francisco PUC reported that lead levels after the conversion to chloramination were the same as when using chlorine. The results are shown on Figure 3.26.

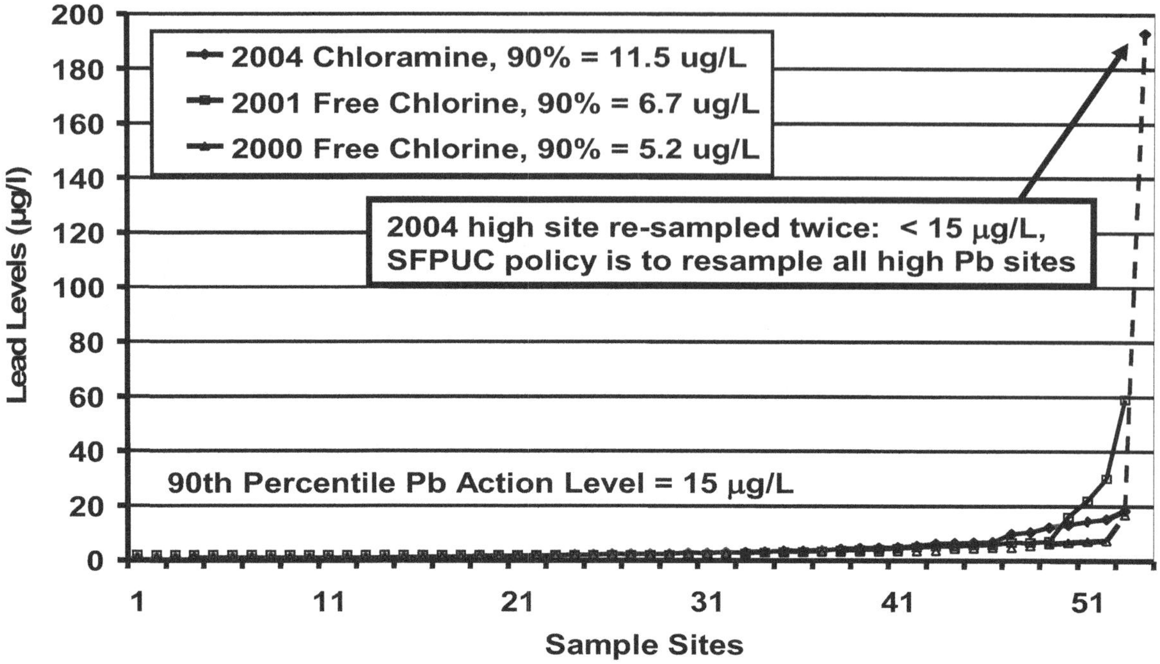

Source: Adapted from Boozarpour, et. al, 2005

Figure 3.26 San Francisco PUC frequency distribution for lead data

Aesthetic Quality

The SFPUC maintained excellent records of consumer complaints during the first month of conversion. A total of 269 calls were received during this time period. Inquiries were most frequent during the first few days of start-up, coinciding with the SFPUC's multiple press releases, and then tapered down over the following weeks. Calls were received, responded to, and logged by Communications, Customer Services, Operations dispatch and Water Quality staff.

Of these calls, only about 11% were aesthetic complaints (some not related to chloramines); all customers with aesthetic complaints received a personal follow-up call and/or visit by Water Quality staff to investigate the customer's water quality situation. Other calls received included inquiries about the following: fish (21%); other pets and plants (11%); human health concerns (18%); treatment/removal options (14%); requests for additional information (6%); plumbing (4%); general, scientific and other inquiries (15%).

Figures 3.27 and 3.28 present information on SFPUC's customer complaints before and after implementing chloramination. SFPUC tracks the type of complaint, differentiating between dirty water, taste and odor, color, illness-related, and "other" complaints. Figure 3.27 shows all of the complaints SFPUC received between January 2003 and January 2006. Figure 3.28

presents the number of complaints associated with taste and odor. According to the data, the average number of complaints per month was 43 before and 48 after chloramination. The average number of taste and odor complaints was 4 per month before chloramination and 7 per month after chloramination. It is unknown if these differences are significant and what is the cause of them.

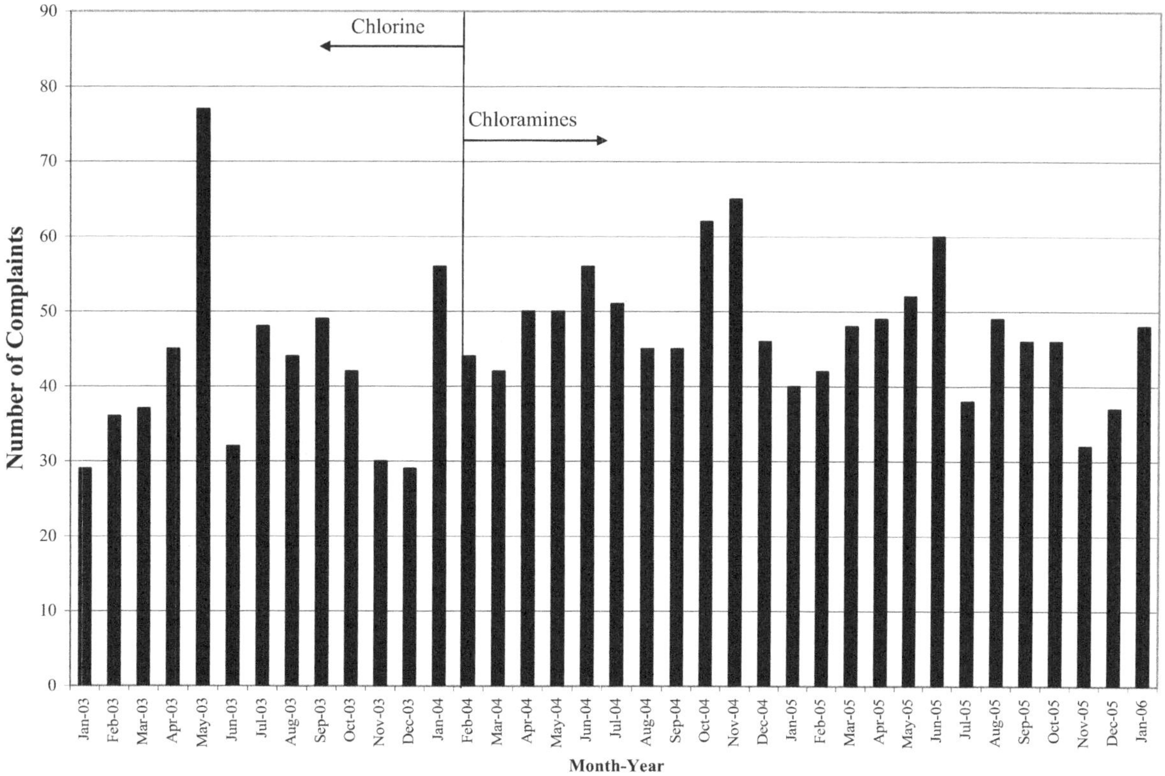

Source: Boozarpour, et. al, 2005

Figure 3.27 San Francisco PUC total number of water quality complaints per month

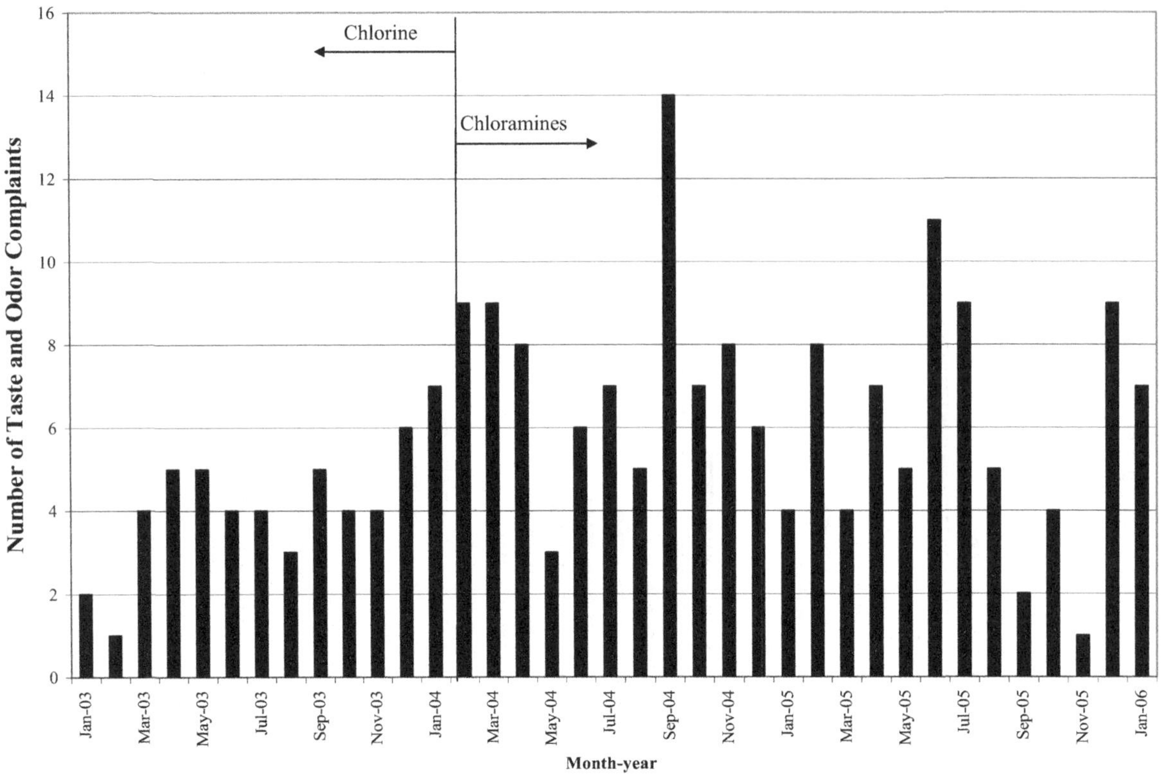

Source: Boozarpour, et. al, 2005

Figure 3.28 San Francisco PUC taste and odor complaints

Washington Aqueduct, Washington D.C.

Washington Aqueduct, a branch of the Army Corps of Engineers, operates the two water treatment plants that serve the 900,000 people in the District of Columbia while the Washington Aqueduct and Sewer Authority operates the distribution system facilities. The Dalecarlia and MacMillan water treatment plants provide approximately 180 million gallons a day of potable water wholesale to the Washington Water and Sewer Authority (WASA) for the District of Columbia as well as Arlington County, and the City of Falls Church, VA. The Washington Aqueduct laboratory also serves as the contract laboratory for WASA, i.e., WASA collects monitoring samples within the distribution system and Washington Aqueduct analyzes them. Free chorine residuals had been targeted at 5 mg/L leaving the plants and at greater than 3 mg/L in the distribution system since 1996 in response to violations of the Total Coliform Rule. In November 2000, Washington Aqueduct converted from chlorine to chloramines for secondary disinfection at both water treatment plants.

Microbial Quality

WASA routinely collects heterotrophic plate counts at forty six sites in the distribution system. Within a year after the conversion to chloramines, significant (greater than 1 log) increases in average heterotrophic plate counts were observed at thirty five of the forty six sites, particularly during the summer and early fall. Counts at ten sites remained about the same and only at one site did the HPC counts lower. Figure 3.29 presents some typical heterotrophic plate count data.

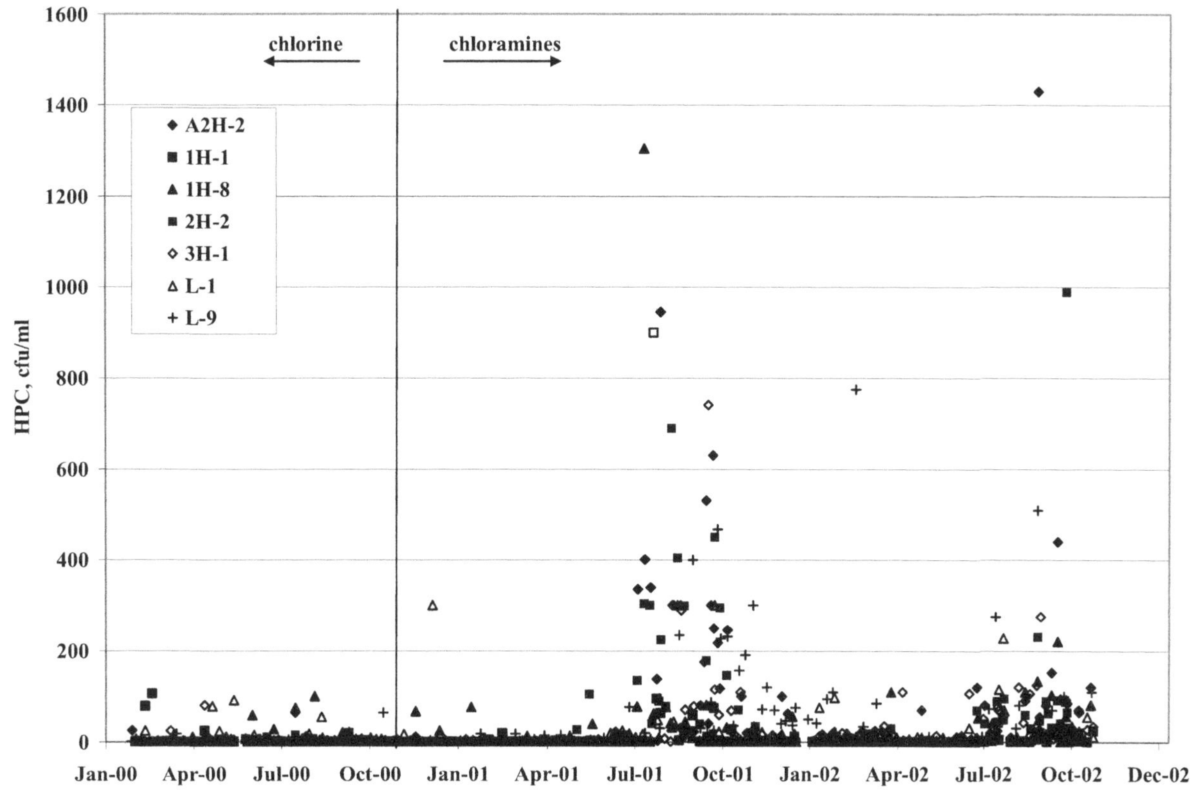

Figure 3.29 Washington DC heterotrophic plate count data

Chemical Quality – DBPs

Washington Aqueduct reported a significant decrease in the levels of trihalomethanes and haloacectic acids since the conversion to chloramines for secondary disinfection in 2000. While there is still a cyclic rise in halogenated disinfection by-product formation during warm weather, the maximum concentration of trihalomethanes is now consistently below 100 μg/l as shown in Figure 3.30.

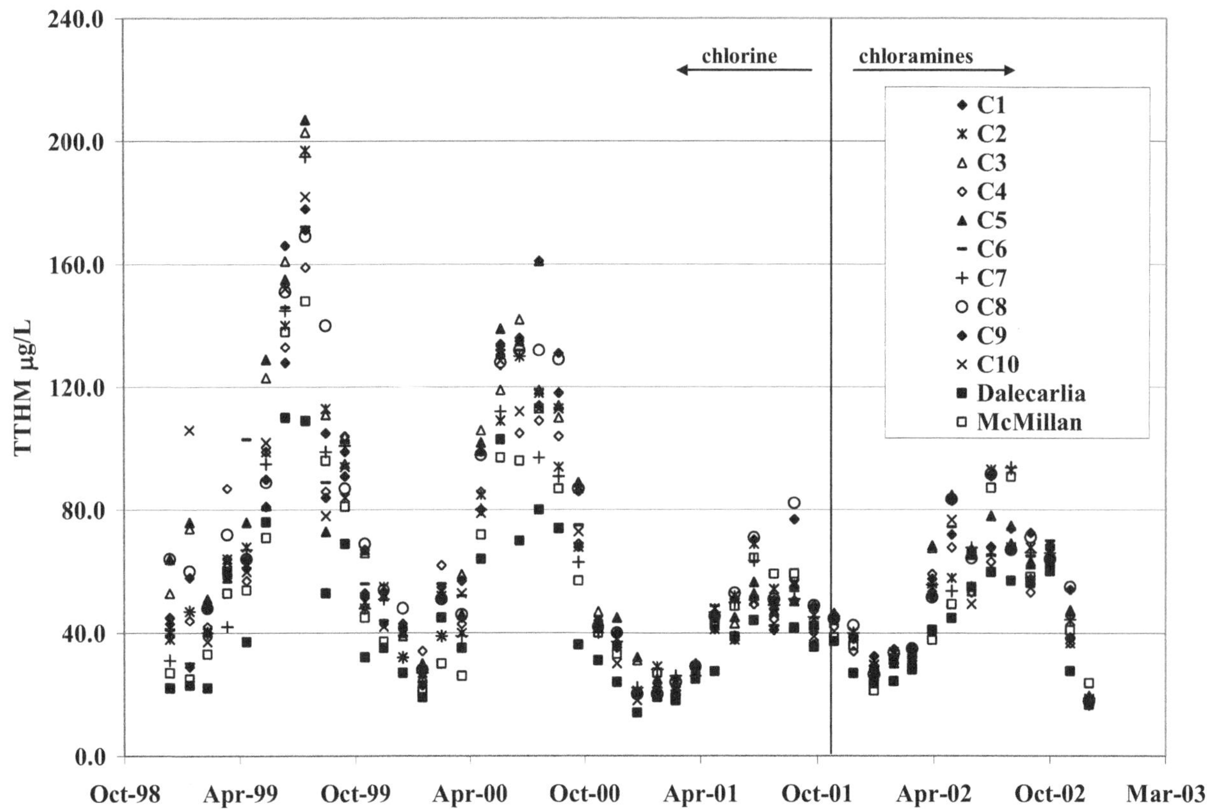

Figure 3.30 Washington DC total trihalomethane data

The maximum concentration of haloacetic acids is now consistently below 70 µg/l as shown in Figure 3.31.

Chemical Quality – Corrosion, etc.

After their conversion from chlorine to chloramines, Washington Aqueduct experienced a very significant increase in lead levels as shown on Figure 3.32. Prior to the conversion, lead levels were below the EPA action level of 15 µg/L. After the conversion, 90th percentile lead levels in the distribution system rose as high as 75 µg/L. Extensive study was conducted to determine the cause or causes of the increased lead solubility. A USEPA rsearch team (Schock et al., 2001) hypothesized that insoluble lead IV solids destabilized after the oxidation potential of the high free chlorine levels lowered with the introduction of chloramines. This hypothesis was confirmed by Switzer et al. 2006, who demonstrated that the change in lead solubility could be induced by a change in redox conditions.

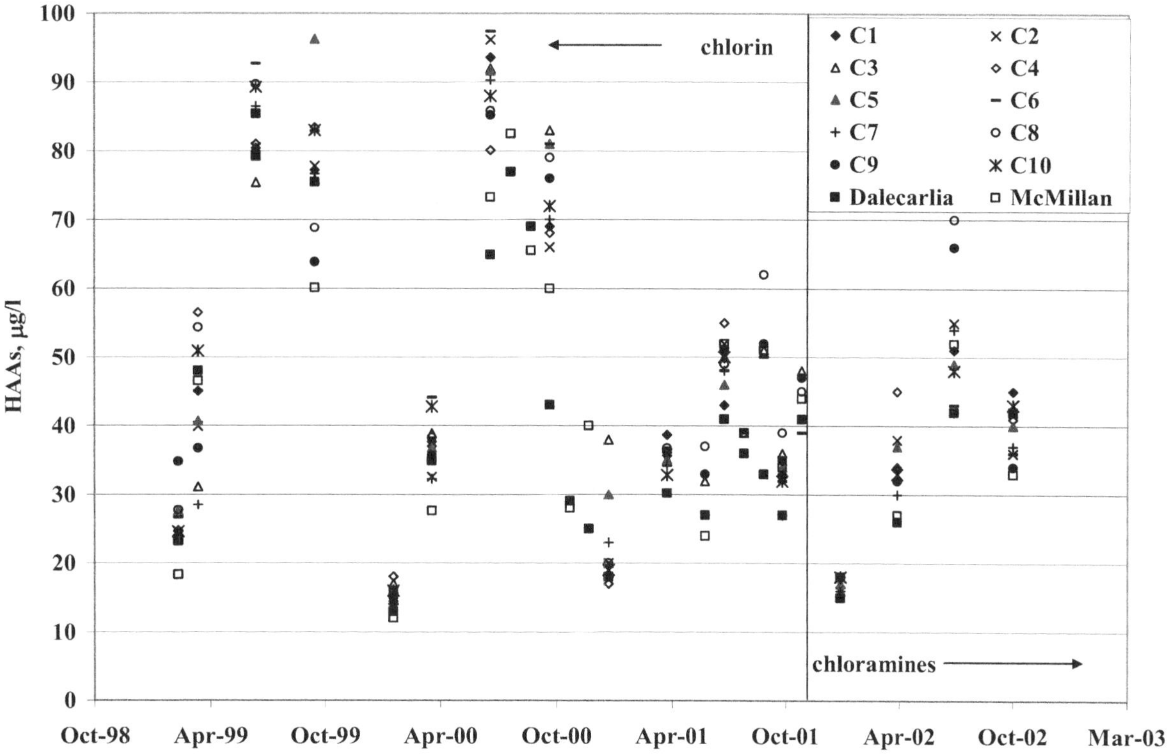

Figure 3.31 Washington DC haloacetic acid data

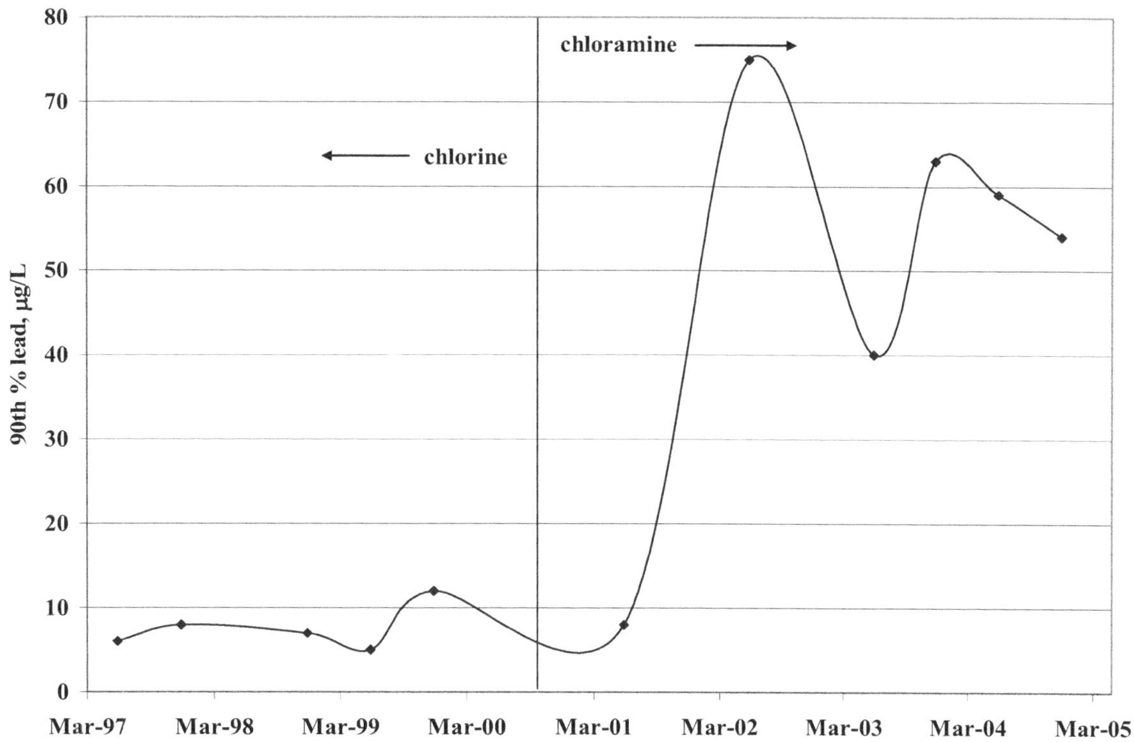

Figure 3.32 Washington DC 90th percentile lead levels

OBSERVATIONS

Based on the information provided by the participating utilities that switched from free chlorine to chloramines, the following general observations can be made regarding switching from chlorine to chloramines for secondary disinfection

Impacts on Microbial Quality

- Overall, there is a reported improvement in microbial quality based on lower HPC levels, better control of biofilms (based on lower HPC levels) and coliform bacteria, and a more persistent and higher disinfectant residual throughout the distribution system.
- HPC levels increased immediately after the conversion at several locations because of dislodging of biofilms from the pipe walls. For some systems, heterotrophic plate count levels take a number of months to decrease to pre-conversion levels or lower.
- HPC analyses should be based on use of R2A agar, as the use of standard pour plate agar will yield levels that are too low to detect a difference before and after the conversion.
- Adverse impacts on coliform levels were not observed for the majority of systems, and coliform levels may be reduced as a result of converting to chloramines. Conversion to chloramines in the absence of good distribution system maintenance, however, does not appear to be sufficient to control coliforms, as shown by the one system that had increased HPC and coliform levels after the conversion.
- Maintenance of a higher and more persistent disinfectant residual throughout the distribution system is more likely with chloramines compared to free chlorine.
- One study found that chloramines provided better control of *Legionella* in buildings.

Impacts on Chemical Quality – DBPs

- Overall, regulated disinfection by-product levels were reduced with the use of chloramines compared to the use of free chlorine.
- Lower TTHM formation potential was observed for all utilities – with both average and peak TTHM values being reduced.
- Lower HAA5 levels might be expected system wide, however, increases may be observed at specific locations that were previously susceptible to biodegradation of HAA species.
- One utility that switched to chloramines and measured cyanogen chloride did not observe unusually high cyanogen chloride levels after the switch.
- NDMA levels did not significantly increase for the two utilities that tested for this parameter.

Impacts on Chemical Quality - Corrosion, Metal Release, Color, and Particle Characteristics

- A reduction in corrosion and metal release was realized for eight of the ten utilities in this survey as exhibited by lower discolored water and red water complaints.

- It appears from this project that for most utilities, lead and copper levels are not affected with the conversion from free chlorine to chloramines. One utility that showed increased lead levels had maintained very high free chlorine residuals before the conversion. One utility had a measurable increase in 90th percentile lead after the conversion but remained in compliance with the LCR.

Impacts on Aesthetic Quality - Taste and Odor

- A reduction in taste and odor complaints may be realized.
- The taste and odor descriptors probably will change from chlorinous taste and odors to other types of descriptors.

CHAPTER 4
CHLORINE TO OZONE

INTRODUCTION

A number of utilities have switched from using chlorine to ozone. The impacts (microbial, chemical, and aesthetic) of switching from chlorine to ozone are presented in this chapter. The potential impacts are based on the findings of the Issue Papers, and the reported impacts are based on the findings of the utility survey. Two case studies are presented in more detail at the end of this chapter.

Of the utilities that participated in the project, seven (7) reported switching from chlorine to ozone for either pre-oxidation or primary disinfection. These utilities are listed in Table 4.1 along with the prior and current disinfection strategies. Detailed information on all the utilities is included in Appendix I.

Table 4.1 Participating utilities that converted from chlorine to ozone

Utility	Capacity mgd	Source	Prior Disinfection			Current Disinfection			Date Changed
			Preox	Primary	Secondary	Preox	Primary	Secondary	
Lincoln, NE East Plant	50	Groundwater	None	None	None	Cl_2	Ozone	NH_2Cl	1994
MWRA, MA	400	Reservoirs (unfiltered)	none	Cl_2	NH_2Cl	none	Ozone	NH_2Cl	2005
Milwaukee, WI	126	Lake	Cl_2	Cl_2	NH_2Cl	Ozone	Cl_2	NH_2Cl	1998
Newport News, VA	31	Reservoir	none	Cl_2	NH_2Cl	none	Ozone	NH_2Cl	2002
Passaic Valley WC, NJ	110	River	Cl_2	Cl_2	Cl_2	$KMnO^4$	Ozone	Cl_2	2003
Southern Nevada WA, NV	900	Reservoir	none	Cl_2	Cl_2	none	Ozone	Cl_2	2003
Vancover, BC, (GVRD)	317	Reservoir (unfiltered)	none	Cl_2	Cl_2	none	Ozone	Cl_2	2000

It should be noted that Lincoln Water System has two plants that feed into the same distribution system. Their West Plant (60 mgd capacity) uses aeration for pre-oxidation, chlorine for primary disinfection, and chloramines for secondary disinfection. The East Plant was added in 1994.

IMPACTS ON MICROBIAL QUALITY

Potential Impacts

Ozone is a powerful oxidizing agent that may be used as a primary disinfectant in water treatment due to its ability to inactivate cysts and oocysts. It is also used for taste, odor, and color control.

Positive Impacts

- The key positive impact of ozone has been higher microbial quality of the bulk water prior to entering the distribution system.
- Ozonation can increase microbial growth potential. However, if the water is further treated after ozonation, or if an adequate secondary residual is maintained even in the absence of subsequent treatment, it is possible that microbial growth in the distribution system can be maintained or lowered. For example, Sharp et. al, (2001) showed that the biological stability of ozonated water could be greatly enhanced if a chlorine residual was retained.
- In a survey of full scale utilities that reduced biodegradable organic content of their water after ozone application, bulk water counts were more sensitive to chlorine residual than the content of degradable organic matter (Najm et. al, 2000).

Adverse Impacts

- Ozonation may increase the biodegradabilty of the natural organic matter present in water, and this may lead to biofilm proliferation in the distribution system. Water that has been ozonated often has elevated levels of lower molecular weight organic compounds; these compounds have been associated with increased biofilm development downstream (van der Kooij et. al, 1989; Price 1994; LeChevallier et. al, 1996b).
- If biodegradable organic matter is not reduced, or if secondary disinfection is not adequate, regrowth of opportunistic pathogens may occur in biofilms. In a field study of eight distribution systems, it was shown that mycobacteria in distribution system water samples and biofilms were positively correlated with elevated assimilable organic carbon and biodegradable organic carbon levels (Falkinham et. al, 2001). Also, these organisms are resistant to disinfection by chlorine and monochloramine (Taylor et. al, 2000).

Reported Impacts

Most of the utilities that switched to ozone reported no detectable change in microbial quality as a result of the switch based on coliform and HPC data. Two utilities, Massachusetts Water Resources Authority and Southern Nevada Water Authority provided microbiological monitoring data. Two utilities reported an increase in assimilable organic carbon (AOC) as a result of the switch to ozone. AOC is a measure of the amount of biodegradable material in the water and can enhance the growth of microorganisms in the distribution system. For one utility, AOC levels increased from 150 μg/L in the raw water to 300 μg/L in the treated water after ozonation. This utility did note an increase in average HPC levels in the distribution system from 10 cfu/ml to 50 cfu/ml. For the other utility, AOC levels after ozonation were higher than typically acceptable levels, but the utility did not notice any increase in coliform positives or HPC levels or any loss of chloramine residual.

IMPACTS ON CHEMICAL QUALITY - DBPs

Potential Impacts

The use of ozone as a primary disinfectant should help to reduce chlorine-containing DBPs, but it can create brominated species when bromide is present, as well as other DBPs due to oxidation of natural organic matter.

Positive Impacts

- Ozonation does not produce chlorine-containing DBPs.
- Ozonation results in a minor reduction (on the order of 10-15%) in the TTHM and HAA5 formation potential of most waters when applied at a dose of 0.5-1.0 mg ozone per mg dissolved organic carbon (DOC) (Chang and Singer 1991, Schechter and Singer 1995, Singer et. al, 1999, Chaiket et. al, 2002). Greater reductions can be achieved when ozonation is coupled with biofiltration.
- There is a minor reduction in the subsequent dosage of chlorine that must be applied to meet a target chlorine residual following ozonation (Singer et. al, 1999). Subsequent chlorine doses are reduced noticeably when ozonation is followed by biofiltration prior to the application of chlorine.

Adverse Impacts

- Principal DBPs of concern are bromate, brominated organics (bromoform, bromoacetic acids, BrOX), and biodegradable organic material (aldehydes, acids, and aldo- and ketoacids).
- Minor increases in the concentration of chloral hydrate and chloropicrin have been observed following chlorination of ozonated waters compared to parallel experiments in chlorinated waters that were not preozonated (Logsdon et. al, 1992, McKnight and Reckhow 1992). Similarly, minor increases in the concentration of cyanogen chloride have been observed following chloramination of ozonated waters compared to parallel experiments in chloraminated waters that were not preozonated (Krasner et. al, 1991, Singer et. al, 1999).
- Bromate may be produced when bromide-containing waters are ozonated, but bromate formation can be controlled by lowering the pH of the water, or by adding ammonia, or chlorine and ammonia to tie up the bromide as bromamines. (Buffle et. al, 2003)
- Minor amounts of bromine-containing organic DBPs have been found in ozonated waters (Song et. al, 1997). More recent work by Weinberg et. al, (2002) found the presence of a number of brominated acids, such as brominated propanoic, propenoic, butanoic, butenoic, oxopentanoic, heptanoic, nonanoic, butenedioic acids, in finished water from a treatment plant using ozone and treating water with a high level of bromide.
- Ozonation results in the formation of a variety of aldehydes (e.g. formaldehyde, acetaldehyde, glyoxal and methyl glyoxal), acids (oxalic acid and formic acid), aldo- and ketoacids (pyruvic acid, ketomalonic acid). On a mass balance basis, less than

5% of the DOC measured in ozonated waters is accounted for by the known oxidation by-products identified to date. The majority of these identified by-products are biodegradable.

- Although epoxides and organic peroxides can be produced by ozonation, there is no evidence that these compounds are stable in ozonated water.

Reported Impacts

Most of the participating utilities reported a reduction in both TTHM and HAA5 levels as a result of the conversion to ozone. Information from the three utilities that provided disinfection by-product data is presented in Table 4.2. A few utilities reported no significant change in these levels. Reductions in TTHM levels ranged from as low as 15 percent to as high as 80 percent at one utility. Typical TTHM reductions were about 25 percent.

Information regarding impacts on HAA5 levels was somewhat similar to that for TTHMs. Reductions in HAA5s ranged from as low as 10 percent to as high as 75 percent. Overall, HAA5 reductions were reported to be around 30 percent, slightly higher than TTHM reductions. One utility reported a 50 percent reduction in total organic halides (TOX), indicating a significant reduction in general halogenated disinfection by-products.

Another utility investigated the formation of aldehydes and keto-acids, both by-products of ozonation. Their studies found that aldehyde levels in the finished water after ozonation were higher than those in the raw water. Keto-acid levels were about the same in both the raw water and the finished water.

Table 4.2
Summary of disinfection by-product data from utilities

Utility	TTHMs, μg/L		HAA5, μg/L	
	Cl_2	NH_2Cl	Cl_2	NH_2Cl
MWRA	70-118	8-15	20-65	7-12
Newport News, VA	55	45	61	44
Milwaukee, WI	22	10	8	4

IMPACTS ON CHEMICAL QUALITY - CORROSION, METAL RELEASE, COLOR, AND PARTICLE CHARACTERISTICS

Potential Impacts

The primary impacts from ozonation that are expected on corrosion, metal release, color and particle characteristics are: changes in the nature of particles, increased redox potential due to the introduction of oxygen through the injection of ozone, and the alteration of organic molecules and their potential effects on metal release.

Positive Impacts

- The corrosion rate of iron has been shown to increase with increasing dissolved oxygen, although some studies found that iron release decreased with higher DO during stagnation (McNeill and Edwards, 2001). The impact of ozone on red water has been studied on a model distribution system. The ozone dose was 1.5 mg/L and the oxygen level of the water increased from approximately 8 mg/L to 22 mg/L. The lack of observed iron increases may have been a result of the water being saturated with dissolved oxygen prior to ozone addition (Clement et. al, 2002).

- No expected positive impacts on copper or lead corrosion.
- While the impacts of ozone as a process to aid filtration have been studied, there has been little research on how particles have been altered and how this may affect distribution system water quality.

Adverse Impacts

- During pilot work, the combination of ozone with elevated pH resulted in a greater amount of iron corrosion compared to elevated pH treatment alone (Clement et al 2002).
- Ozone oxidizes reduced iron(II) and manganese(II) to particulate ferric hydroxide and particulate manganese dioxide, respectively. The ferric hydroxide tends to be removed easily, but the manganese dioxide tends to be colloidal in nature and has been found to be difficult to remove by filtration (Wilczak et. al, 1993). Overdosing with ozone can oxidize the manganese to permanganate (+7), and, as a consequence, the treated water will develop a pink coloration.
- Ozone could induce changes in organic molecules as stated in the above section. Organics have been noted to effect metal uptake in some cases, but no data are available on the impacts related to ozone.

Reported Impacts

Very little information was reported regarding the impacts of converting from chlorine to ozone on corrosion control and metal release. One utility reported that studies conducted on severely tuberculated pipes from the distribution system indicated a greater release of iron to the water when using ozone compared to chlorine. This could result in a greater number of red water complaints. Another utility reported that manganese levels in the finished water had increased after switching to ozone, resulting in an increase in customer complaints regarding black water. However, the utility had moved its secondary chlorination point to after the filters when it switched to ozone. When they subsequently moved the point of chlorination to just ahead of the filters, the manganese problem was eliminated.

IMPACTS ON AESTHETIC QUALITY - TASTE AND ODOR

Potential Impacts

Ozonation should effectively eliminate many odorous compounds from water, but it also can generate odorous compounds because of its ability to oxidize many organic compounds.

Positive Impacts

- Ozone oxidizes the humic substances responsible for the yellowish hue in naturally colored waters (Singer and Johnson 2003). Klimkina et. al, (1987), cited by Langlais et. al, (1991), reported color reductions ranging from 20 percent to 60 percent using high ozone dosages (8-13 mg/L).

- Lalezary et. al, (1986) and Glaze et. al, (1990) demonstrated that ozone, both alone and in combination with hydrogen peroxide (H_2O_2) and ultraviolet (UV) irradiation, effectively eliminated earthy and musty tastes and odors caused, respectively, by geosmin and 2-methlylisoborneol (MIB). Owen and Johnson (2002) demonstrated during an investigation that geosmin and 2-MIB could be effectively eliminated from raw water by ozone at dosages required for *Cryptosporidium* inactivation when 1.0 mg/L H_2O_2 was added. Lalezary et. al, (1986) also found that ozone was effective for removing another earthy-smelling compound, trichloroanisole.
- Singer and Reckhow (1999) cited the work of Hoigne and his coworkers, who demonstrated the relative effectiveness of molecular ozone and AOP for eliminating synthetic organic chemicals (SOCs) from drinking water. Some of these compounds, such as certain pesticides and industrial chemicals, are readily oxidized by ozone while others are refractory. Certain pesticides, such as aldrin and heptachlor, cause musty and chlorinous odors but are readily oxidized by ozonation (AwwaRF and Lyonnaise des Eau 1995).

Adverse Impacts

- Aldehydes causing fruity odors have been found in drinking water following ozonation at moderate dosages and at less than optimal dosages (Anselme et. al, 1988; Glaze et. al, 1990). Khiari et. al, (Khiari et. al 2002) reported that lower molecular-weight aldehydes often produced by ozonation have rancid or otherwise unpleasant odors, while the higher-molecular-weight aldehydes have pleasant odors.
- Bruchet et. al, (1992) demonstrated the potential for taste-and-odor problems caused by amino acids that are either created by ozonation or escape oxidation during the ozonation process. If chlorine is used as the final disinfectant, the amino acids may produce odorous aldehydes, some of which have very low odor threshold concentrations.
- Crozes et. al, (1997) evaluated ozone for disinfection and taste-and-odor control at a lime softening plant in Wisconsin. Earthy, musty, and wet-hay odors were reduced in intensity, but increasing the ozone dose unmasked fishy and swampy odors. At high ozone dosages, buttery and rancid odors were produced.
- Extremely odorous iodonated THMs have been found at above-threshold levels in nitrified groundwater containing as little as 0.10 mg/L iodide ion at locations in France where ozonation was practiced, and the concentrations increased with increasing ozone dosages (AwwaRF and Lyonnaise des Eaux. 1995).

Reported Impacts

Only two of the participating utilities provided data regarding the impacts of converting to ozone on the aesthetic quality of the water. None of the utilities reported an increase in taste and odors problems. One utility simply reported a reduction in taste and odor complaints. Another utility reported that taste and odor complaints dropped from several hundred before ozonation to less than 100 after the conversion.

CASE STUDIES

More detailed information is presented in the following paragraphs for two of the participating utilities - Milwaukee Water Works, Wisconsin, and Greater Vancouver Regional District, British Columbia, Canada.

Milwaukee Water Works

Milwaukee Water Works serves more than 800,000 people in the greater Milwaukee area. The Water Works obtains water from Lake Michigan and treats it at two conventional water treatment plants that were modified in 1998 to include ozone addition for primary disinfection. The Linwood Plant treats an average of 86 million gallons a day and the Howard Avenue Plant treats an average of 36 million gallons a day.

In addition, Milwaukee Water Works made the conversion to Biologically Active Filtration (BAF) between 1998 and 2000 at both water plants to complement the ozonation process.

Microbial Quality

Milwaukee Water Works reported no detectable changes in distribution system total coliform numbers or heterotrophic plate counts as a result of the change to ozone as the primary disinfectant.

Chemical Quality – DBPs

Milwaukee has done extensive disinfection by-product research at each of their plants as well as within the distribution system. The conversion from chlorine to ozone as the pre-oxidant and primary disinfectant reduced the already low levels of TTHMs and HAA5s in the Milwaukee system. Since the conversion to ozone, Milwaukee has observed 30 to 50 % reduction in TTHM and HAA5 levels and 50% reduction of total organic halides (TOX) which is a general measure of halogenated disinfection by-products. The ozone has also reduced ultraviolet absorbing organics in the raw water by 45 to 60% as measured by ultraviolet absorbance at 254 nm. Figures 4.1 and 4.2 present TTHM and HAA5 data, respectively, before and after ozone conversion. Figures 4.3 and 4.4 present the cumulative frequency distribution of the trihalomethane concentrations and haloacetic acid concentrations before and after the change to ozone from chlorine for primary disinfection.

Levels of four haloacetonitriles also decreased after the conversion to ozone as the primary disinfectant. Overall concentrations of the four studied (but not regulated) compounds were lowered to the minimum detectable level of 2 μg/L (total), or less than 0.5 μg/L each after ozonation. Figure 4.5 presents these data. Statistical analysis revealed that the decrease was significant at the 95% confidence level.

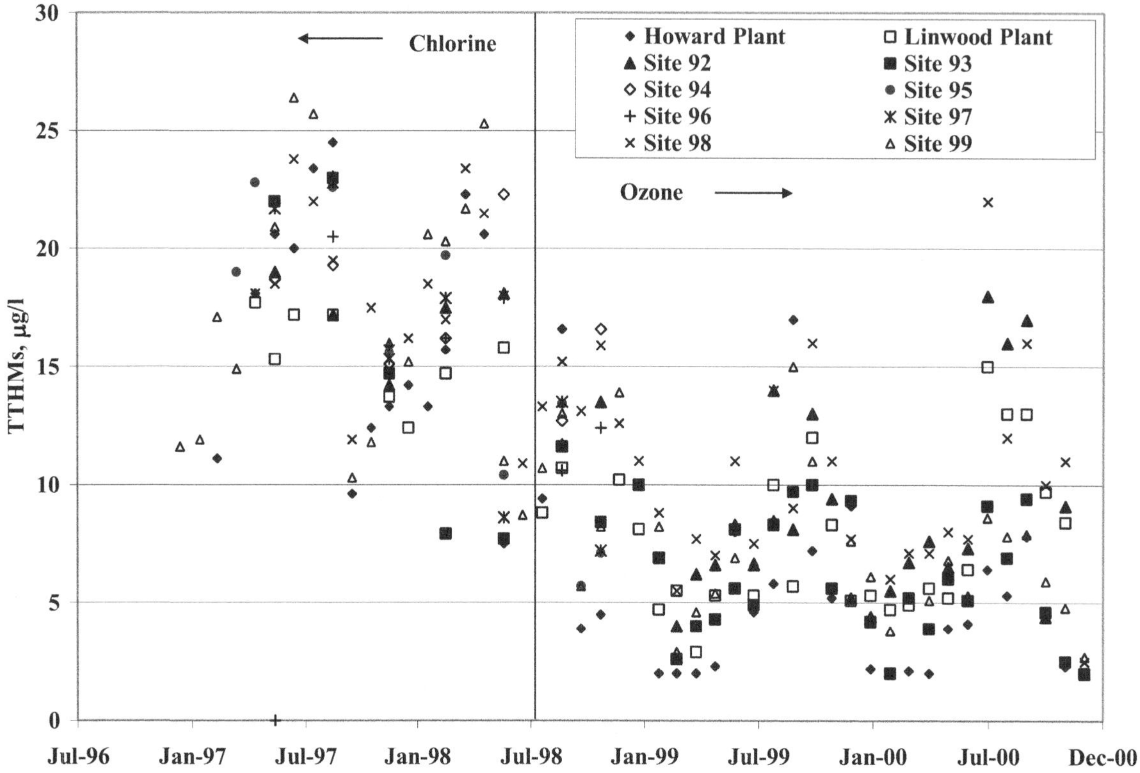

Figure 4.1 Milwaukee, WI TTHM data

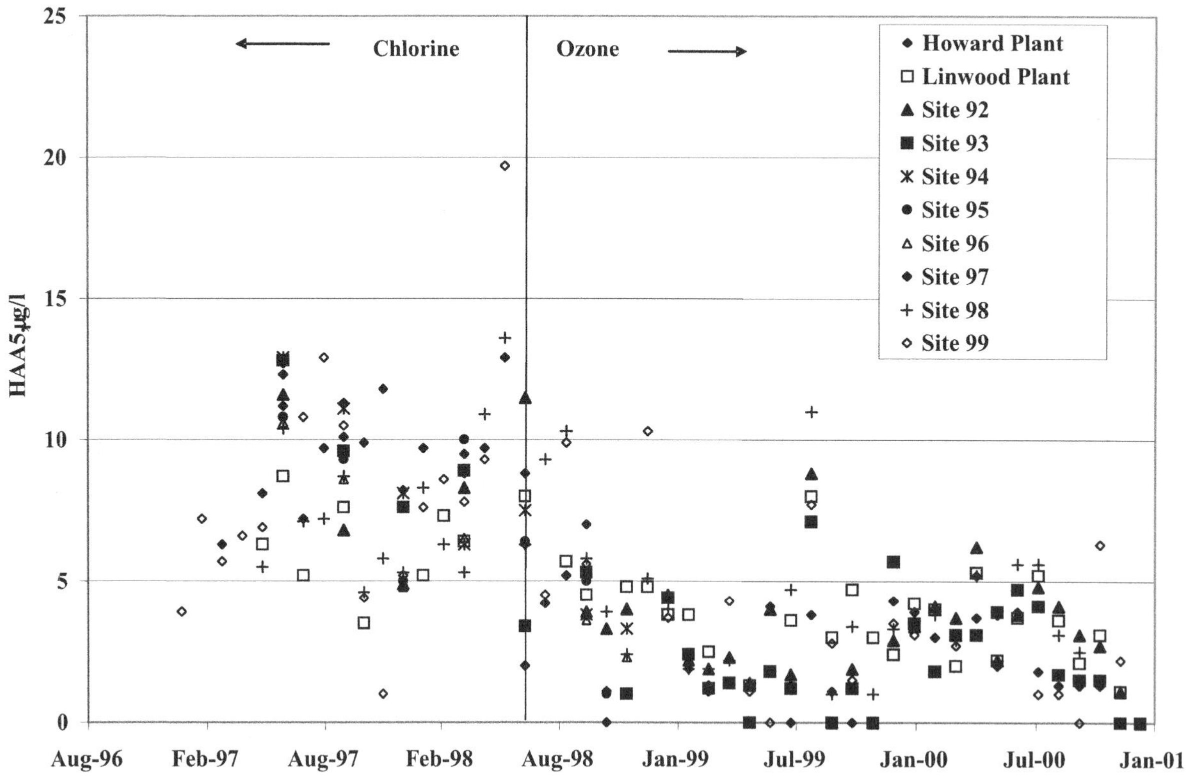

Figure 4.2 Milwaukee, WI HAA5 data

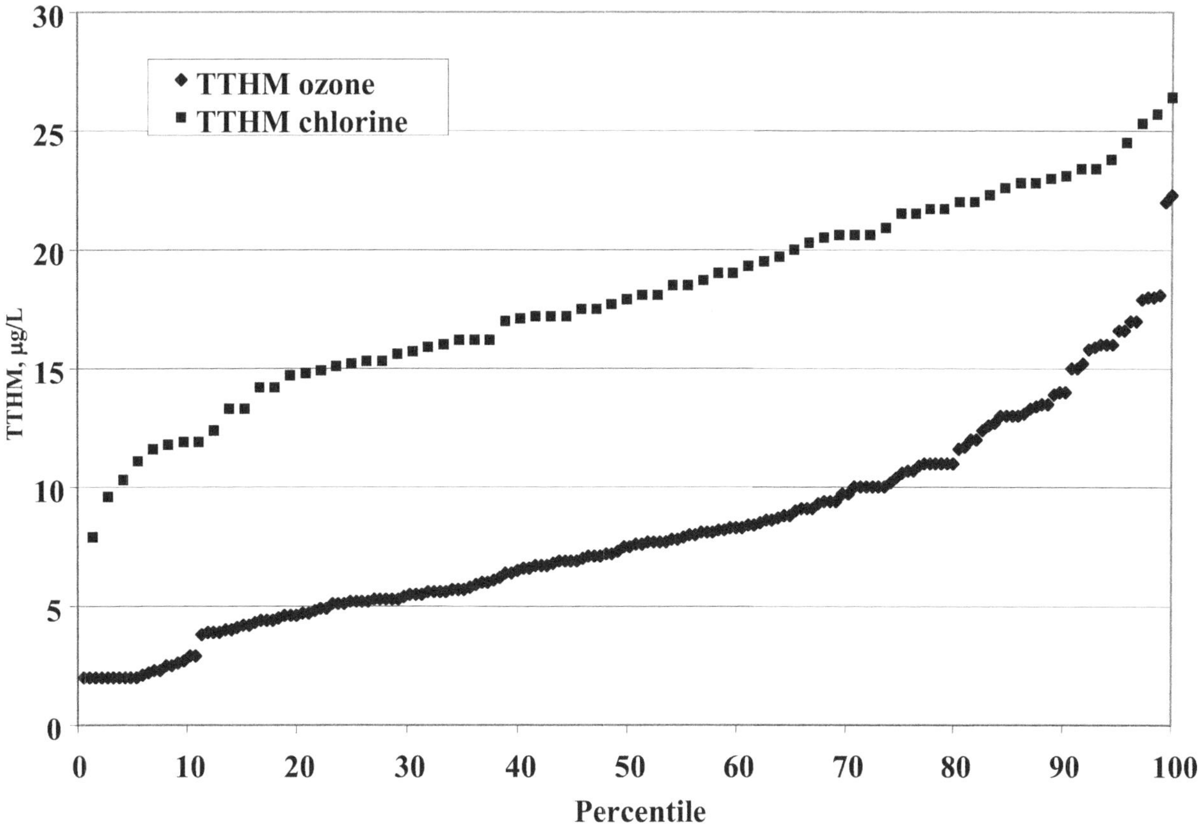

Figure 4.3 Cumulative frequency distribution of Milwaukee TTHM data

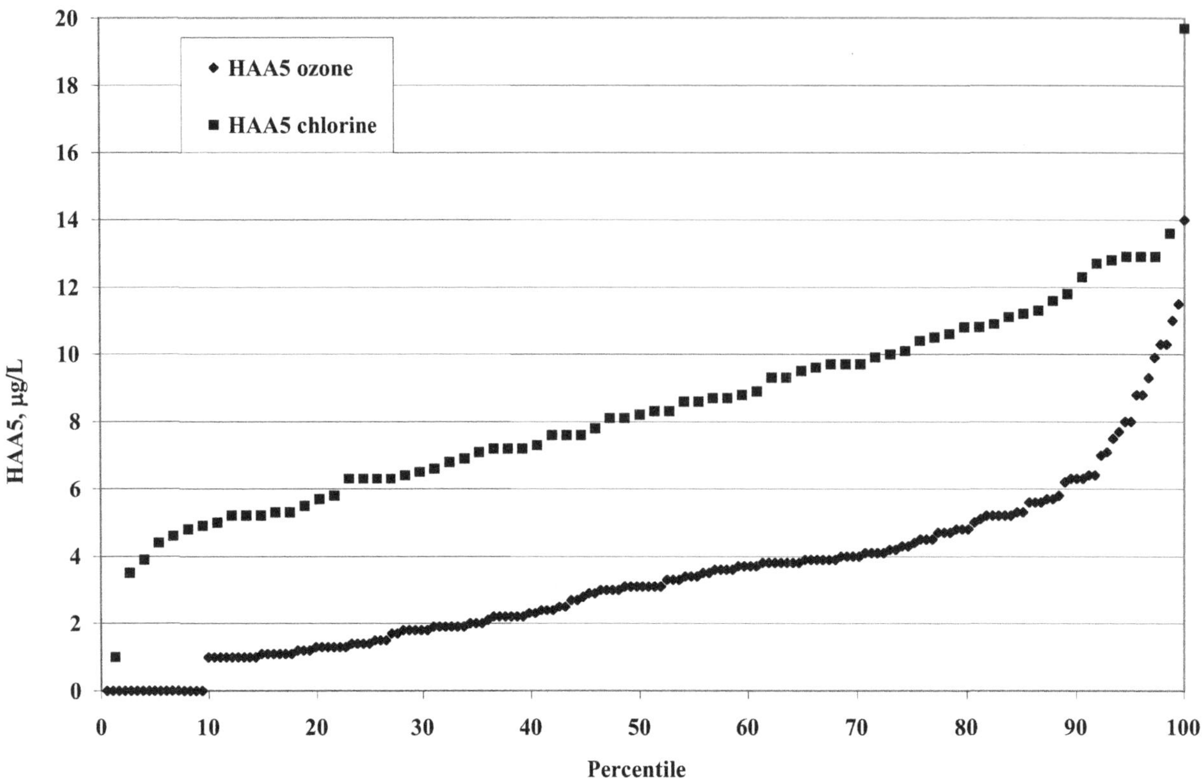

Figure 4.4 Cumulative frequency distribution of Milwaukee HAA5 data

Levels of four haloacetonitriles also decreased after the conversion to ozone as the primary disinfectant. Overall concentrations of the four studied (but not regulated) compounds were lowered to the minimum detectable level of 2 µg/L (total), or less than 0.5 µg/L each after ozonation. Figure 4.5 presents these data. Statistical analysis revealed that the decrease was significant at the 95% confidence level.

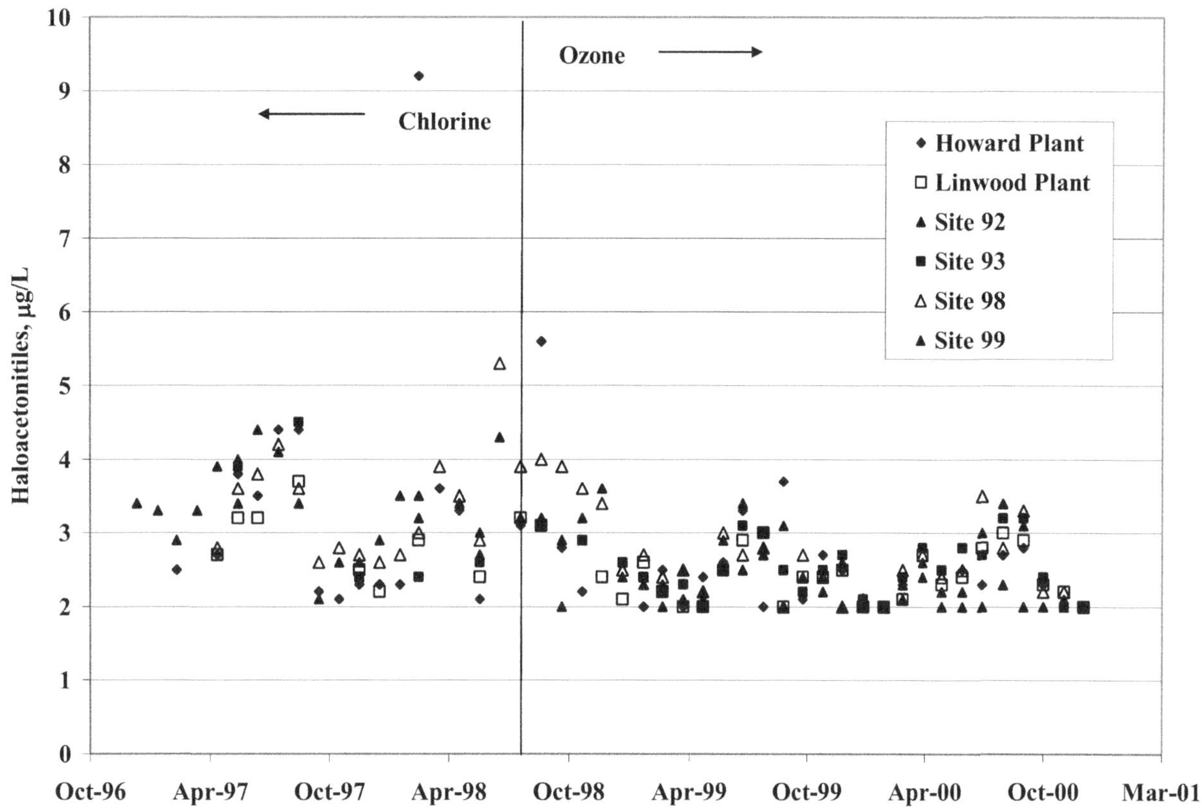

Figure 4.5 Milwaukee, WI haloacetonitrile data

Corrosion, Metal Release, etc.

Milwaukee did not observe any changes in water quality that would indicate an impact on corrosion or metal release. The addition of ozone was not incompatible with the selected corrosion control treatment as shown by the 90th percentile lead levels since the addition of ozone to the treatment process in Figure 4.6. Milwaukee Water Works began corrosion control chemical treatment with phosphoric acid beginning in the fall of 1996. The target dose remains at 2.0 mg/L.

Aesthetic Quality

Milwaukee did not observe any changes in water quality that would indicate an impact on aesthetic quality. The city has taken a number of steps to improve water quality in the distribution system and has observed decreasing customer complaints each year. They attribute this to improved flushing and distribution system operations. Figure 4.7 presents customer complaint data for the years from 1998 through 2003.

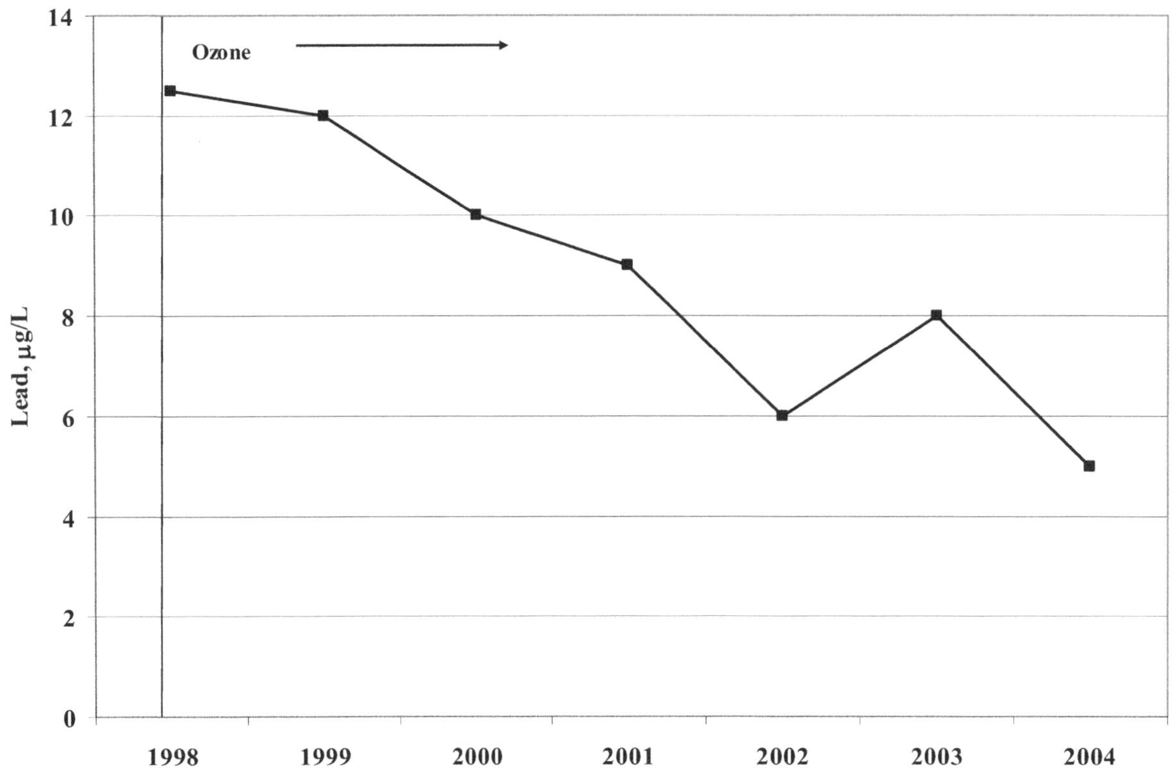

Figure 4.6 Milwaukee 90th percentile lead levels

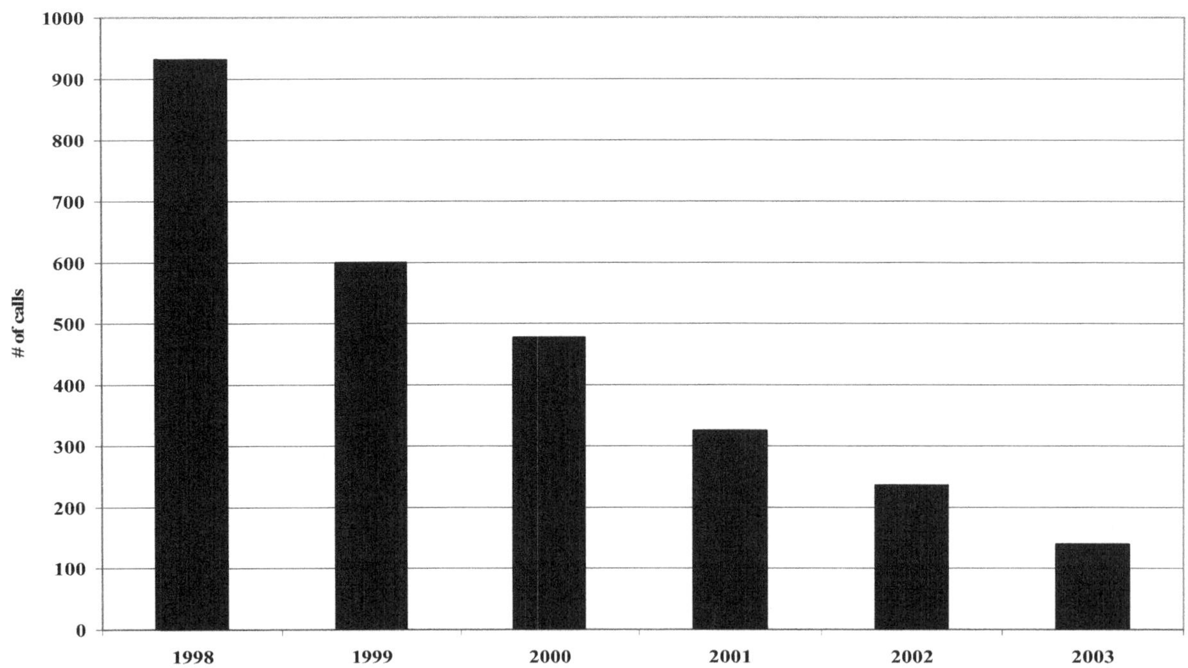

Figure 4.7 Milwaukee customer complaint data

Greater Vancouver Regional District, British Columbia, Canada

The Greater Vancouver Regional District (GVRD) utilizes water from three surface water supplies: Seymour, Capilano, and Coquitlam Reservoirs. All three of these sources are unfiltered. The Coquitlam Reservoir is treated at a single treatment plant to provide an average 125.6 mgd to approximately 710,000 customers in five municipalities in the region. GVRD switched from chlorine to ozone for primary disinfection of the Coquitlam source in 2000 and continues to use chlorine for secondary disinfection. In addition, GVRD adds soda ash for pH adjustment at the plant to promote ozone dissipation. Treatment for pH adjustment was initiated in 2000, increasing the pH from an average of 5.5 to 6.8.

During 2000 and 2001, GVRD's ozonation system was run intermittently due to start-up and source operations issues. Stable distribution system conditions were not achieved until 2002. Thus, the graphs and tables in this case study do not contain data from the years 2000 and 2001, and instead contain data from 2002 to 2003 to describe long-term, post-ozonation conditions.

Chemical Quality – DBPs

Data from two years of sampling before and after the switch to ozone were analyzed to identify changes in DBP levels. Tables 4.3, 4.4 and 4.5 and Figures 4.8, 4.9 and 4.10 present mean concentrations of TTHMs and HAA5s before and after the switch to ozone. Mean concentrations were calculated only for sites where multiple samples were collected before and after ozonation. As shown in Table 4.3 and Figure 4.8, significant changes in TTHM concentrations are not readily apparent; however, two sites experienced substantial decreases in TTHMs of 23% and 37%. Similar increases in TTHM levels were observed at other locations.

In addition, statistics were evaluated for data from all of the sites to determine whether ozonation resulted in a significant difference in overall TTHMs. At a 95% confidence level, there is no significant difference between the mean TTHM levels or the variance of the TTHM data sets before and after ozonation.

Table 4.3
Greater Vancouver average TTHM concentrations before and after ozonation

Source	Average TTHM concentration before ozone (μg/L)	Average TTHM concentration after ozone (μg/L)	Percent change
City A - Site 5	15	13	-9%
City B - Site 11	12	13	6%
City B - Site 9	17	20	19%
City D - Site 2	20	21	6%
City D - Site 5	23	18	-23%
City E - Site 10	17	19	15%
City E - Site 4	35	22	-38%
City E - Site 5	18	23	31%
City E - Site 8	22	20	-9%
City G - Site 2	24	28	20%

More dramatic results were observed in the HAA5 data (Table 4.4 and Figure 4.9). Decreases in HAA5 levels were observed at five of the six locations sampled before and after ozonation. Decreases ranged from 3% to 48% after the switch to ozone.

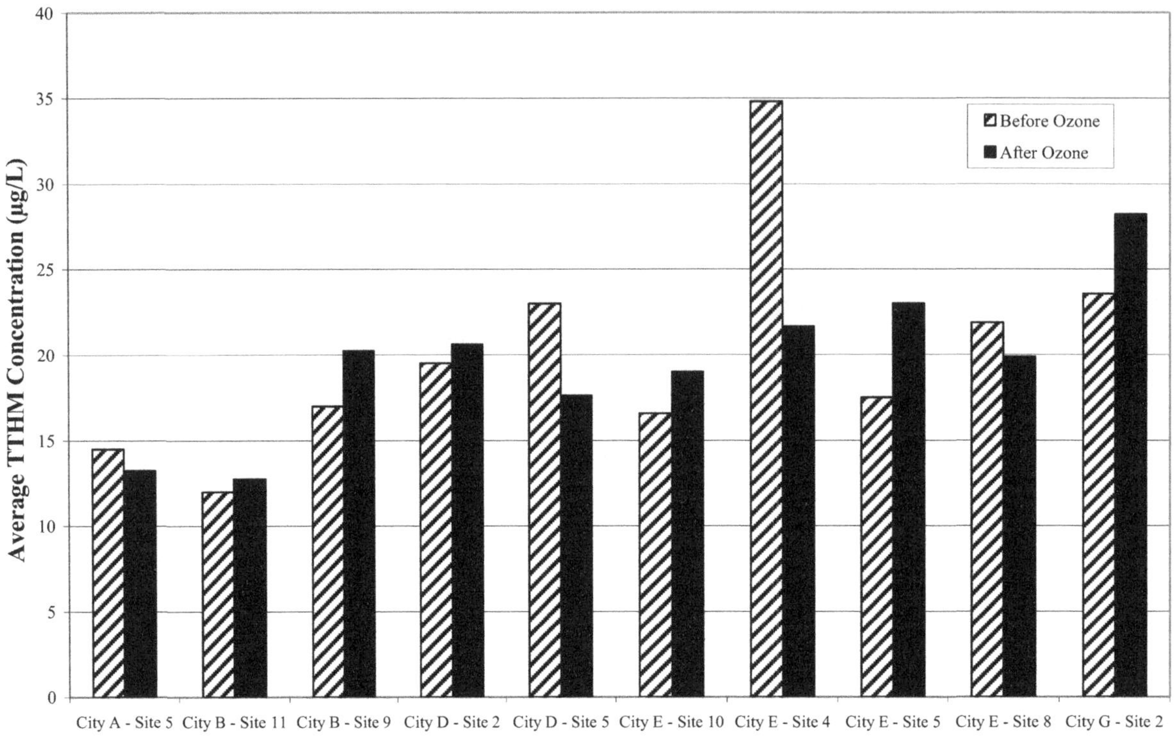

Figure 4.8 GVRD average TTHMs before and after ozonation

Table 4.4
Greater Vancouver average HAA5 levels before and after ozonation

Source	Average HAA5 concentration before ozone (µg/L)	Average HAA5 concentration after ozone (µg/L)	Percent change
City B - Site 11	47	24	-48%
City B - Site 9	50	32	-36%
City E - Site 10	52	30	-43%
City E - Site 4	37	45	22%
City E - Site 8	31	30	-3%
City G - Site 2	55	52	-6%

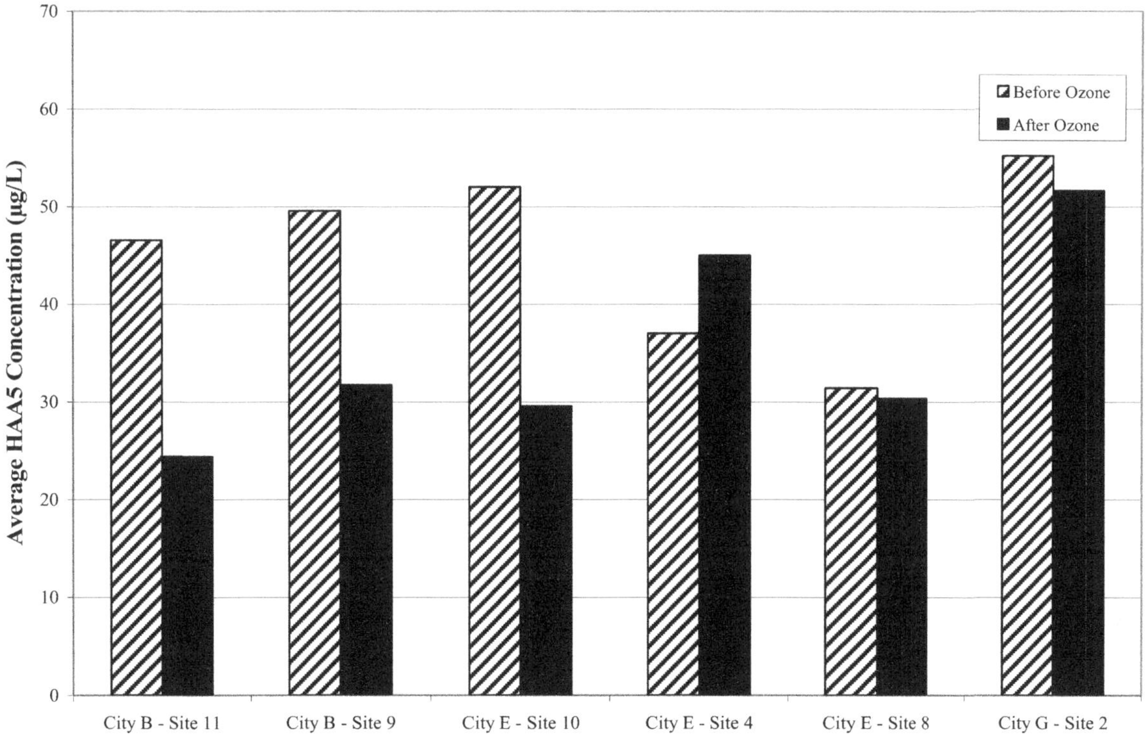

Figure 4.9 GVRD HAA5 before and after ozonation

However, a statistical evaluation of HAA5 data from all of the sites to determine whether ozonation resulted in a significant difference in overall HAA5 levels, found that at a 95% confidence level, there is no significant difference between the mean values or the variances of the HAA5 data sets before and after ozonation. A closer examination of specific concentrations of HAAs revealed that changes in dichloroacetic acid concentrations comprised the bulk of the change noted in HAA5 levels (Table 4.5 and Figure 4.10). Decreases in concentration of this species ranged from 31% to 47% after the switch to ozone.

Table 4.5
Greater Vancouver average dichloroacetic acid levels before and after ozonation

Source	Average dichloroacetic acid concentration before ozone (μg/L)	Average dichloroacetic acid concentration after ozone (μg/L)	Percent change
City B - Site 11	23	15	-36%
City B - Site 9	25	13	-46%
City E - Site 10	25	13	-47%
City E - Site 4	13	16	20%
City E - Site 8	13	9	-33%
City G - Site 2	25	17	-31%

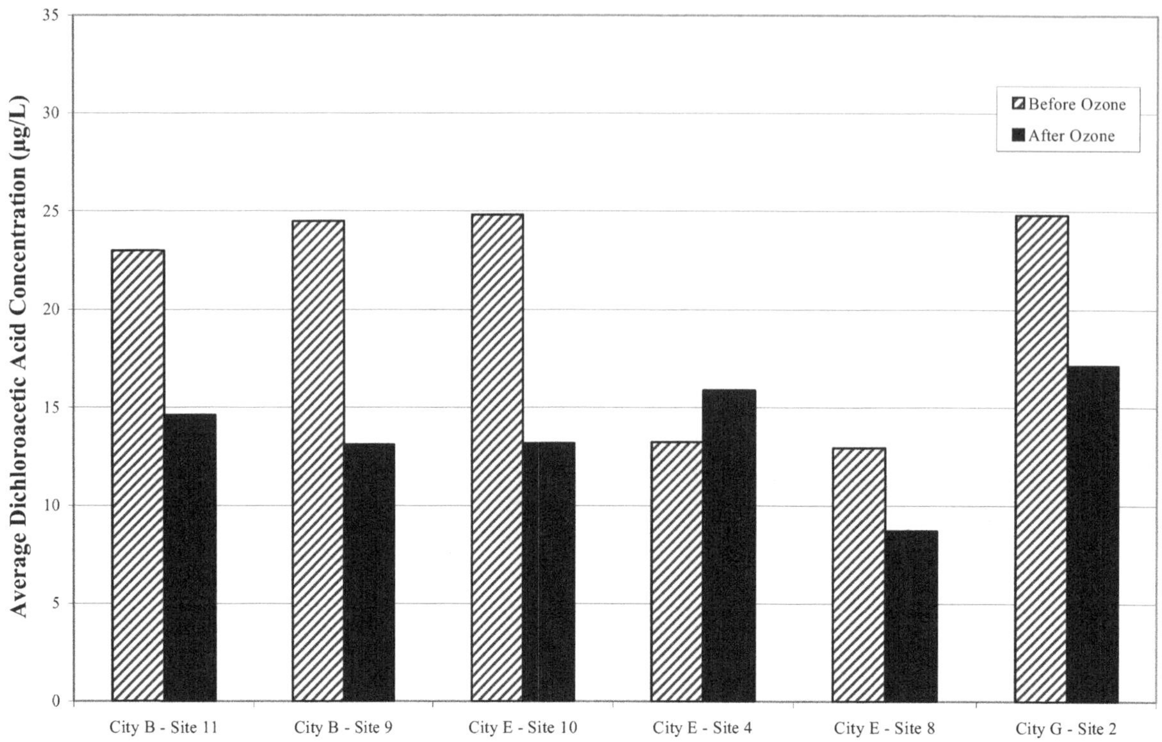

Figure 4.10 GVRD average dichloroacetic acid levels before and after ozonation

In addition, a statistical analysis was conducted to determine if dichloroacetic acid data from all of the sites demonstrated a significant difference after ozonation. This evaluation confirmed that at a 95% confidence level, there is a significant difference between the mean dichloroacetic acid levels and the variances before and after ozonation.

Canadian drinking water quality standards and regulations differ considerably from U.S. regulations. The *Guidelines for Canadian Drinking Water Quality* set a limit of 100 µg/L for TTHMs and no limit for HAA5 is provided; in contrast, Maximum Contaminant Levels of 80 µg/L TTHM and 60 µg/L HAA5 are required under the U.S. Stage 1 and Stage 2 Disinfectants and Disinfection By-Products Rules. The DBP data presented here indicate that GVRD is in compliance with both Canadian guidelines and American standards, and that the switch to ozone has likely enhanced the utility's ability to maintain compliance with DBP limits.

Microbial Quality - Coliform and HPCs

GVRD collected total coliform and HPC samples at 46 sites in 1999 and 55 sites in 2003. The majority of these sample sites showed no coliforms during the four years of data analyzed.

Table 4.6 and Figures 4.11 and 4.12 present a summary of the coliform positives at sites sampled before and after the switch to ozone. City D was not included in the figure as no positive coliform samples were recorded in that city in the four years of data analyzed. Similarly, no positive samples were recorded in City F before, and in City G after ozonation. Decreases were observed in the percent of positive samples in five of the six cities, despite increases in the number of sample sites associated with population increases in each city.

Table 4.6
Greater Vancouver total coliform positive samples before and after ozonation

City	Sample size before ozone	Percent positive before ozone	Sample size after ozone	Percent positive after ozone
City A	449	0.89%	590	0.51%
City B	1,458	0.48%	1,699	0.41%
City C	992	0.20%	1,606	0.19%
City E	1,187	0.34%	1,340	0.07%
City F	566	0.00%	2,178	0.14%
City G	1,245	0.24%	1,333	0.00%

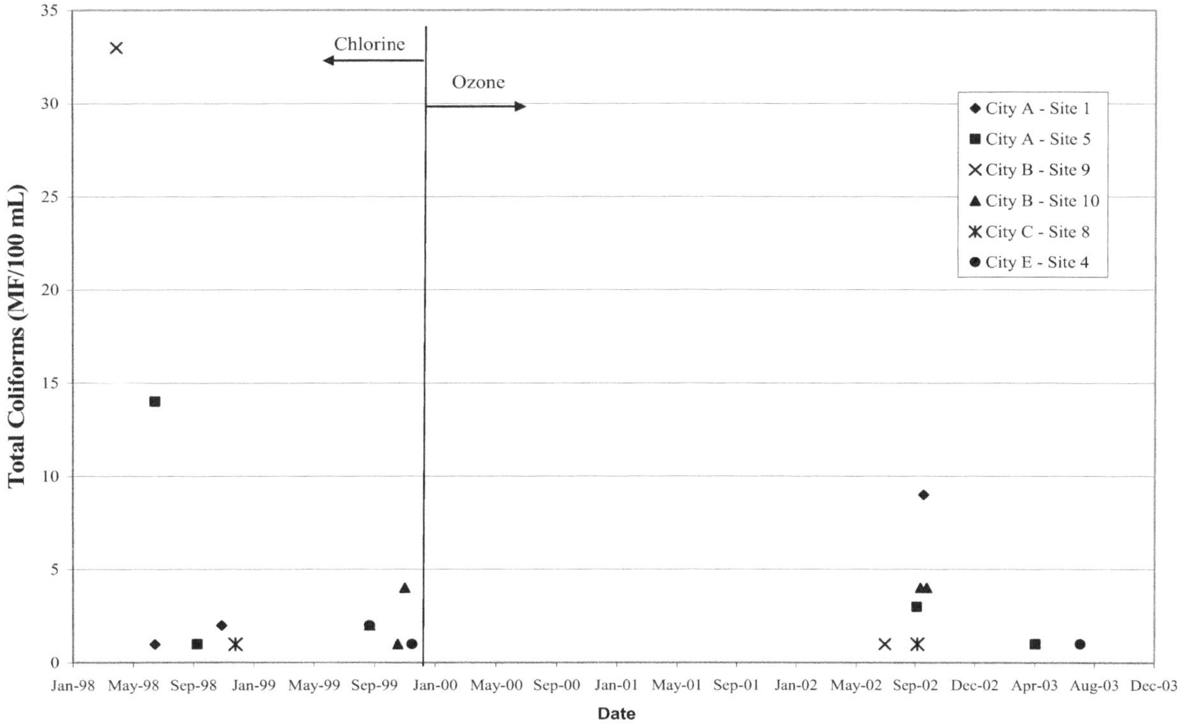

Figure 4.11 GVRD total coliform data summary

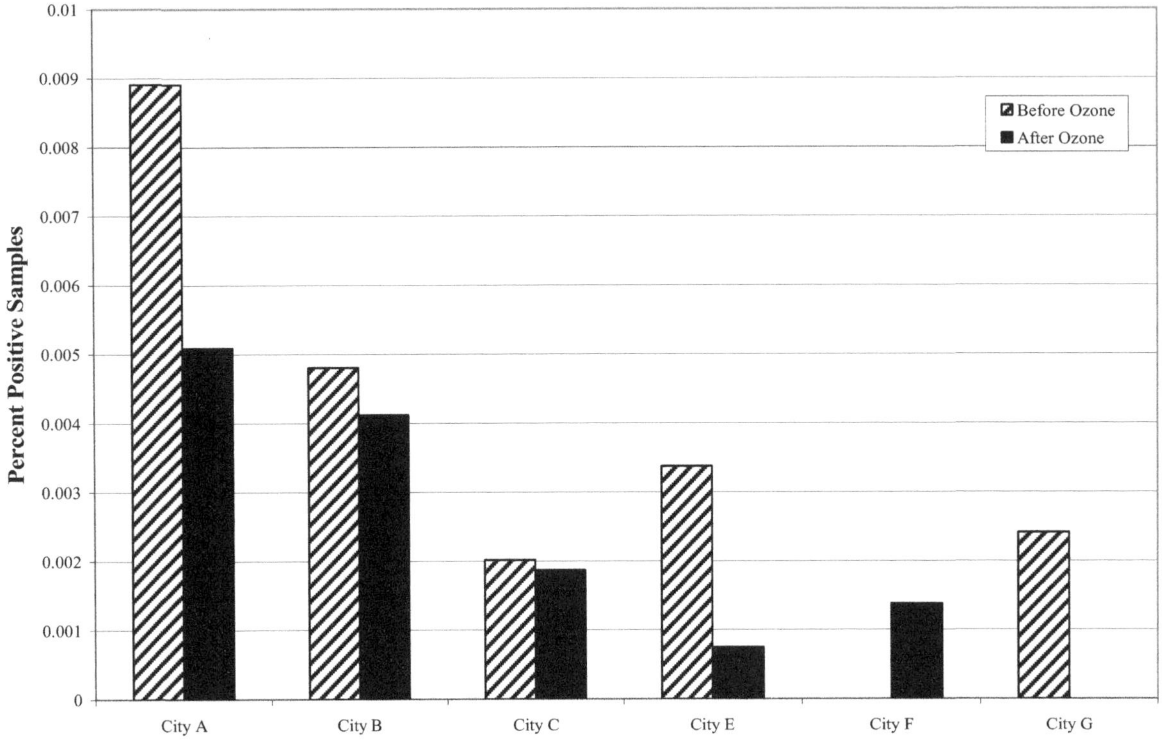

Figure 4.12 GVRD percent positive coliform samples before and after ozonation

GVRD conducts HPC analysis using R2A media with a 5-day incubation period at 28° C. This same method has been used for all of the data presented in this case study. Graphing the many data prints, it is difficult to assess changes in HPC levels.

Therefore percentile distributions to identify changes in HPC levels before and after ozonation were developed. Figure 4.13 and Table 4.7 present the percentile distribution of HPC levels for City B. This analysis shows decreases in HPC levels after the switch to ozone at several percentiles. For example, as shown in Table 4.7, 50% of samples before ozone were less than 32 cfu/mL; after the switch to ozone, 50% of samples in City B were less than 4 cfu/mL. Similarly, the 90th percentile value decreased from 3,800 to 290 cfu/mL after ozonation. Figure 4.13 presents this data in graphical form. A statistical analysis of City B's HPC data demonstrated that the means and variances of the two data sets differ significantly at the 95% confidence level.

Table 4.7
Percentile distribution of HPC in City B

Percentile	HPC Level Before Ozone	HPC Level After Ozone
1	0	0
10	0	0
25	2	0
50	32	4
75	400	48
90	3,800	290
99	22,000	5,700

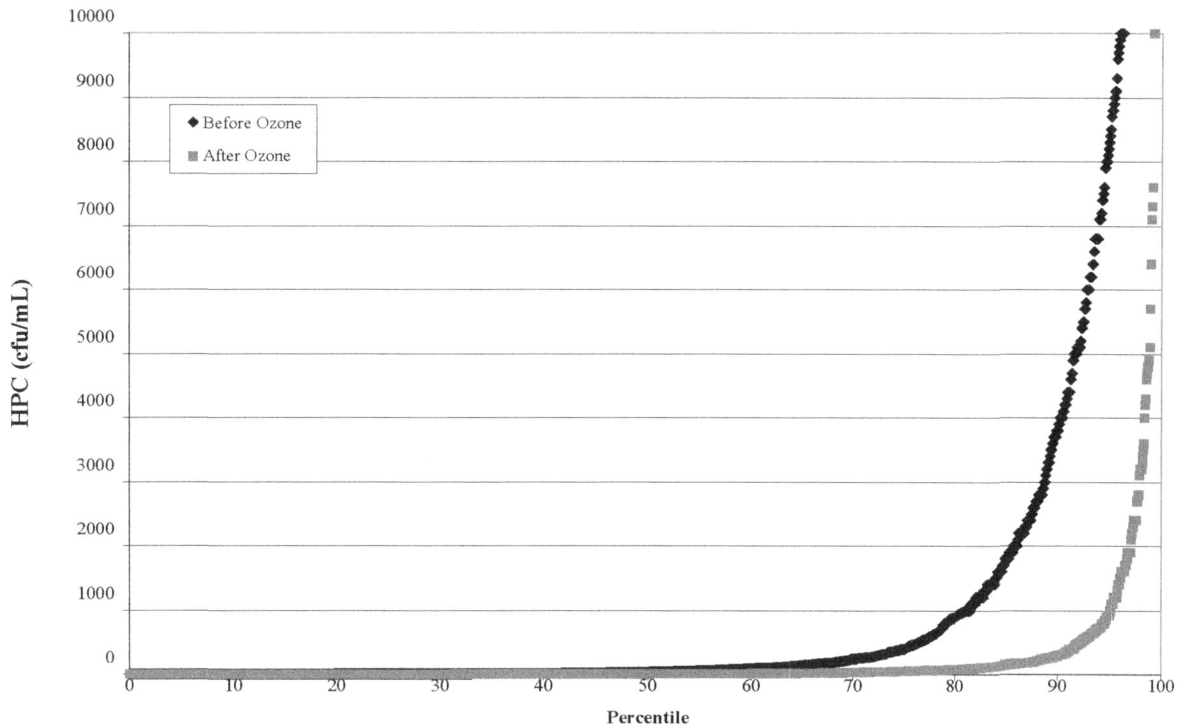

Figure 4.13 GVRD percentile distribution of HPC samples – City B

Figures 4.14 and 4.15 present the percentile distributions of HPC samples collected in City A and City C. Similar decreases in HPC levels were observed in these data. It is interesting to note that decreases in microbiological activity are not typically associated with the use of ozonation without subsequent filtration. Further information was gathered to determine if this could be the result of a change in operations and maintenance practices in GVRD's system. GVRD's transmission system has never been flushed due to the large diameter of the pipes and typical flow velocities are high enough to consider the pipes self-cleaning. Some of GVRD's member utilities began flushing their systems in 2000. However, there was no concerted effort on the part of all the systems receiving water from the Coquitlam source to alter distribution practices for the purpose of controlling microbial growth. With regard to treatment practices, prior to ozonation, GVRD was limited as to the amount of chlorine (gaseous) that could be added to the finished water due to resultant impacts on finished water pH. Prior to ozonation, the finished water pH was approximately 5.5. As part of the ozonation treatment process, pH adjustment to 6.8 was incorporated to facilitate ozone destruction. This pH increase also likely improved corrosion conditions within the distribution system, and allowed for increases in the disinfectant residual (from a maximum of 1.1 mg/L prior to ozonation to a maximum of 1.3 mg/L after ozonation) without negatively impacting pH. It is likely that the combination of all of these practices allowed the GVRD to avoid potential negative microbial impacts associated with ozonating an unfiltered surface water supply. In 2006, GVRD indicated that nearly all of the member utilities had recently begun receiving ozonated water and no additional water quality impacts have been reported.

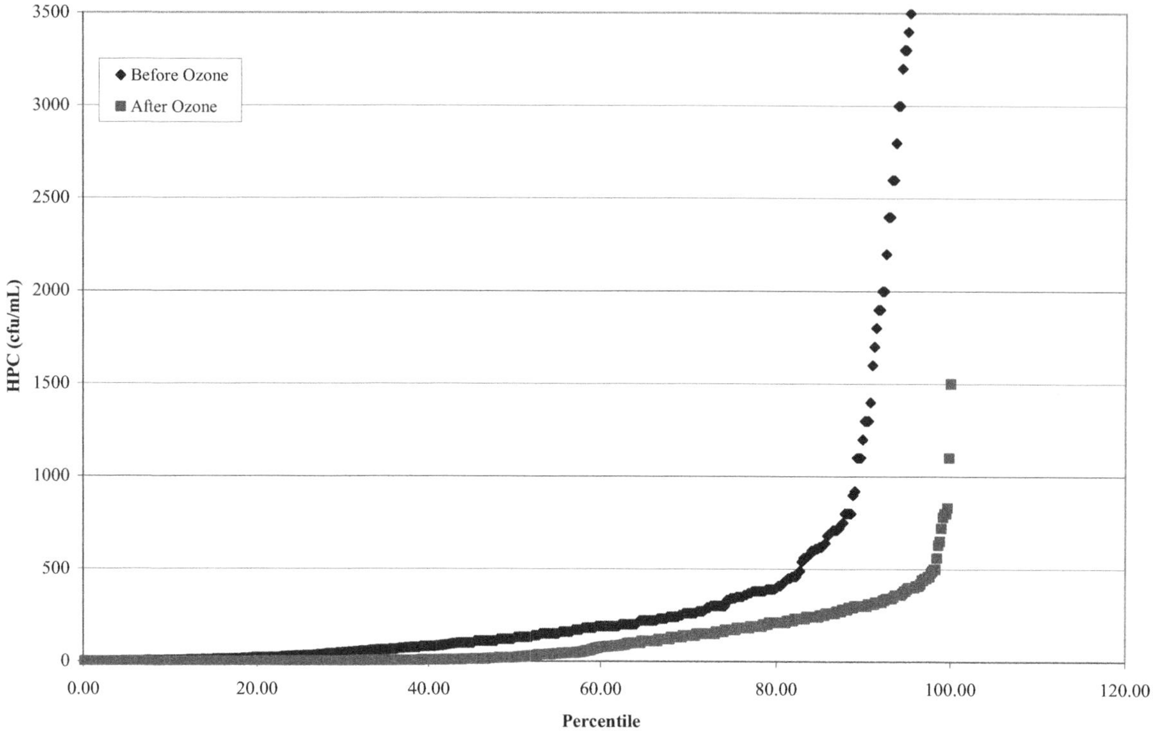

Figure 4.14 GVRD percentile distribution of HPC samples – City A

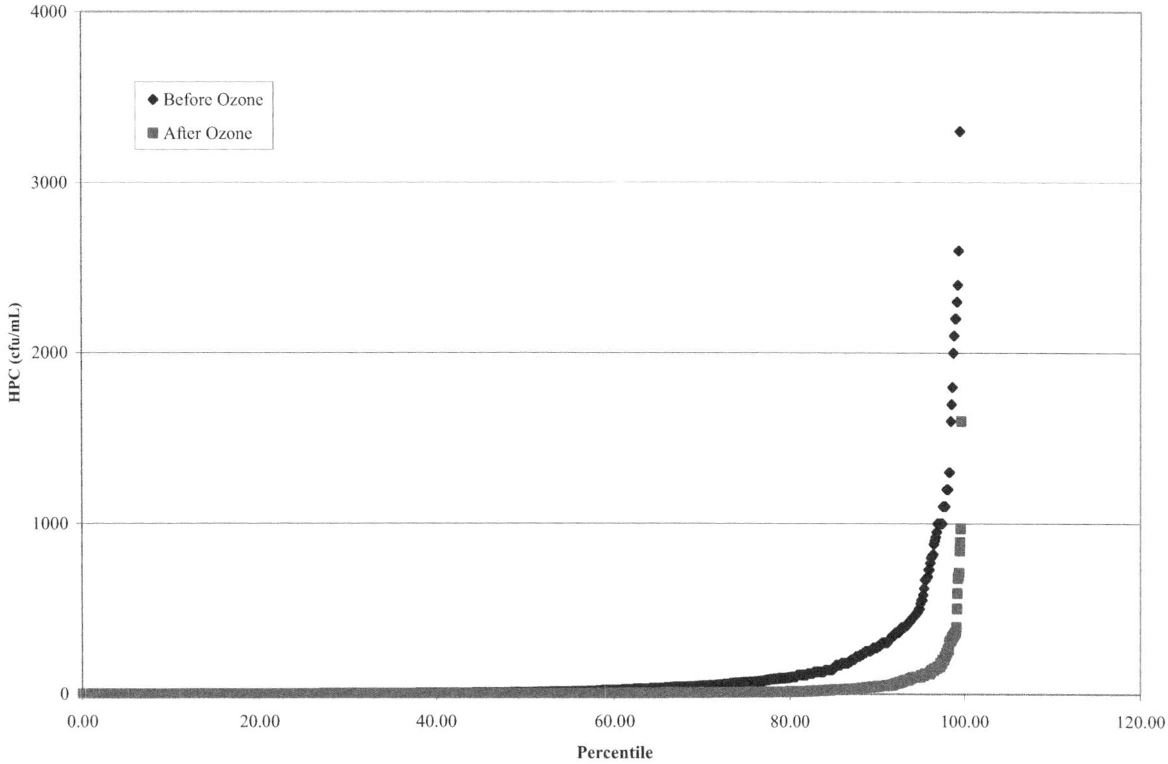

Figure 4.15 GVRD percentile distribution of HPC samples – City C

Metals Uptake and Corrosion – Iron, Manganese, and pH

GVRD has limited data on iron, manganese, and pH collected from member utility distribution systems. Iron and manganese samples were not collected in distribution systems prior to ozonation. Figure 4.16 below presents total and dissolved iron levels collected after treatment at Coquitlam Reservoir. In the distribution systems, total iron was sampled during the last two quarters of 2003 in five systems and not sampled prior to ozonation. Four of the five systems have average total iron levels of 0.05 mg/L, while one system has an average total iron level of 0.11 mg/L during the same period.

With respect to manganese, no samples were collected in member utilities' distribution systems until after ozonation. A review of samples collected at sites during the last two quarters of 2003 indicates that average manganese levels range from 0.002 to 0.004 mg/L, except in City A, which had an average of 0.02 mg/L, related to a relatively high sample collected during July 2003.

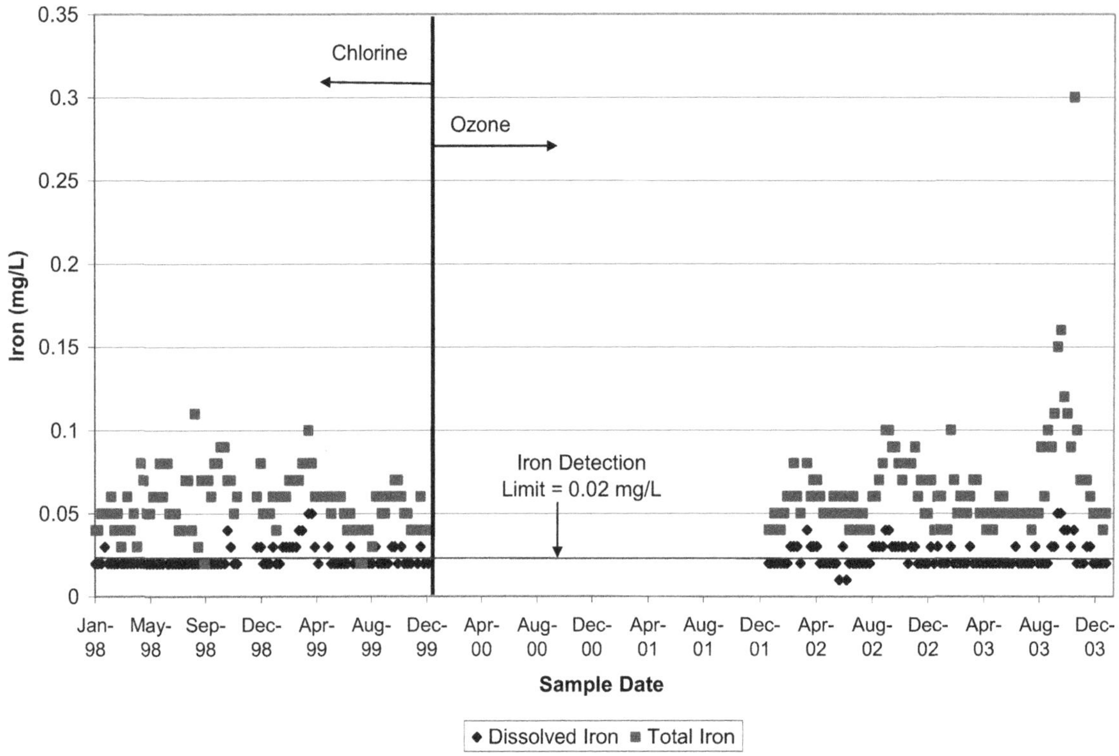

Figure 4.16 Total and dissolved iron measured after treatment at Coquitlam Reservoir

OBSERVATIONS

Based on the findings of the issue papers and the information provided by the participating utilities, the following observations can be made regarding a utility switching from chlorine to ozone for secondary disinfection:

Impacts on Microbial Quality

- Overall, microbial quality could be reduced based on higher HPC levels associated with more degradable matter in the bulk water in the distribution system. Treatment following ozonation to remove the degradable matter could reduce this possibility.
- HPC analyses should be based on use of R2A agar, since standard agar is too insensitive to detect differences before and after a conversion.
- No adverse impacts on coliform levels were observed.

Impacts on Chemical Quality – DBPs

- Overall, a reduction in disinfection by-product levels was observed with the use of ozone compared to the use of chlorine.
- Lower TTHM formation potential was observed – with both average and peak THM values being reduced.

- Lower HAA5 levels were observed.

Impacts on Chemical Quality - Corrosion, Metal Release, Color, and Particle Characteristics

- There is insufficient information to make any conclusions regarding corrosion and metal release.

Impacts on Aesthetic Quality - Taste and Odor

- Reductions in taste and odor complaints were observed.

CHAPTER 5
CHLORINE TO CHLORINE DIOXIDE

INTRODUCTION

Only two of the participating utilities reported a disinfectant change from chlorine to chlorine dioxide. The reported water quality impacts from the changes at both utilities are presented in this chapter.

The Gulf Coast Water Authority (GCWA) reported a disinfectant change from chlorine to chlorine dioxide and to chloramines. GCWA wholesales water to 13 municipalities in Texas between Houston and Galveston, serving approximately 185,000 people. The source of supply for the GCWA's treatment plant is the Brazos River, which has moderate to high levels of TOC, hardness, alkalinity, and bromide. The rated capacity of the plant is 50 mgd and includes coagulation, filtration and disinfection.

Prior to 1983, the GCWA plant used free chlorine for both primary and secondary disinfection. From late 1983 to late 1985, GCWA made several modifications to their disinfection process for several reasons:

- Improved disinfection
- Reduction in chlorinated DBPs
- Reduction in taste and odor

The disinfection changes that were investigated included various combinations of chlorine, chlorine dioxide and chloramines as summarized in Table 5.1.

Table 5.1
Disinfection changes for Gulf Coast Water Authority

Phase	Period	Preoxidant/ Primary Disinfectant	Post-Disinfectant/ Secondary Disinfectant
1	Prior to 11/83	Chlorine	Chlorine
2	11/83	Chloramine	Chloramine
3	5/84 – 2/85	Chlorine dioxide	Chlorine
4	3/85 – 4/85	Chlorine dioxide	Chlorine dioxide
5	5/85 – 11/85	Chlorine dioxide	Chlorine dioxide/chlorine
6	11/85-present	Chlorine dioxide	Chlorine dioxide/chloramines

The County of San Luis Obispo Lopez project plant withdraws and treats water from the Lopex Lake and Terminal. Lopez Lake water is alkaline with a hardness above 200 mg/L, moderate levels of TOC, and manganese. Conventional filtration treatment is provided with a switch from pre-chlorination to preliminary oxidation using chlorine dioxide in August 2002. The majority of the treated water is sold to seven municipal customers that serve a population of 50,000.

IMPACTS ON MICROBIAL QUALITY

Potential Impacts

Chorine dioxide is an oxidant with high disinfection capability against *Cryptosporidium*. It also may be better in controlling biofilms compared to other disinfectants because it may have less of an impact on iron corrosion as indicated in Issue Paper No.1 in Appendix B.

Positive Impacts

- Because of its oxidizing capability, chlorine dioxide should have the ability to effectively disinfect the bulk water, and in most cases this is seen. For example, strains of *Mycobacterium avium* were most susceptible to ozone and chlorine dioxide at CT levels orders of magnitude lower than that for chlorine (Taylor et. al, 2000). Botzenhart et. al, (1993) demonstrated that *Legionella pneumophila* was more susceptible to chlorine dioxide than *Escherichia coli*.
- Chlorite, a degradation product of chlorine dioxide, has also been seen to have bacteriostatic properties with a positive influence on microbial counts in drinking water (Masschelein, 1991). McGuire et. al, (1999) have shown that chlorite can reduce nitrification in distribution systems at concentrations below the current MCL.
- Recently completed work by Gagnon et. al, showed that chlorine dioxide had the greatest disinfection efficiency in a model reactor system, followed by free chlorine and monochloramine. Per unit of disinfectant, chlorine dioxide was 3-5 fold more effective on suspended cells than chlorine and 6-9 fold more effective than monochloramine. For biofilms, the results generally were a 2-4 and 5-6 fold increase in efficacy.

Adverse Impacts

- This disinfectant, like others, may have the ability to select for or against organisms of regulatory concern in the distribution system but this is presently unknown.
- Chlorine dioxide, like other oxidizing disinfectants, may change the nature of organic matter found in water and render it more easily degradable by bacteria in the distribution system.

Reported Impacts

While using free chlorine for both primary and secondary disinfection prior to 1983, GCWA found bacterial quality in the distribution system to be excellent, with coliform bacteria being absent in routine monitoring. HPC was not analyzed at that time. During the investigations of the various disinfection strategies, GCWA experienced changes in the bacterial quality and bacterial type in the distribution system.

In November 1983, GCWA implemented chloramines as both a pre-oxidant and post-disinfectant. A target residual of at least 2 mg/L at the customers' connections was established. Bacteriological analyses in distribution system samples yielded confluent growth in several plate cultures. In May 1984, chlorine dioxide was implemented as a pre-oxidant. Coliforms were not found in any distribution system samples and confluent growth was not observed in any standard plate counts.

GCWA received authorization from the state regulatory agency in March 1985 to use chlorine dioxide as both a pre-oxidant and post-disinfectant. No other residual disinfectant was used during this phase. This authorization was granted with the stipulation that GCWA maintain a chlorine dioxide residual of less 1.0 mg/L in the distribution system and increase microbiological water quality monitoring (e.g. plate counts). Total coliforms were not found in any samples from March through April 1985. Intermittent, elevated bacterial counts, however, were measured at three different points in the distribution system; 1) at the GCWA clearwell, 2: in Texas City, approximately two (2) miles from the plant, and 3) in Bacliff, approximately five (5) miles from the plant. In addition, it was difficult for GCWA to maintain a chlorine dioxide residual of 1.0 mg/L at the customer connections while at the same time providing a measurable chlorine dioxide residual throughout the wholesale customers' distribution systems. The chlorine dioxide decay was relatively rapid.

Figure 5.1 illustrates a comparison of average heterotrophic plate counts obtained at each monitoring location starting with the implementation of chlorine dioxide as the primary disinfectant under Phase 3. It is evident that during Phase 3, the use of chlorine dioxide increased standard plate counts above the desired maximum level of 500 cfu/mL in distribution system samples and that the use of chlorine dioxide for both pre- and post- disinfection exacerbated heterotrophic plate count numbers. The addition of free chlorine to the chlorine dioxide post-disinfection controlled microbial growth (Phase 5). Conversion of the final disinfection scheme to chlorine dioxide and chloramines further improved control of heterotrophic bacteria.

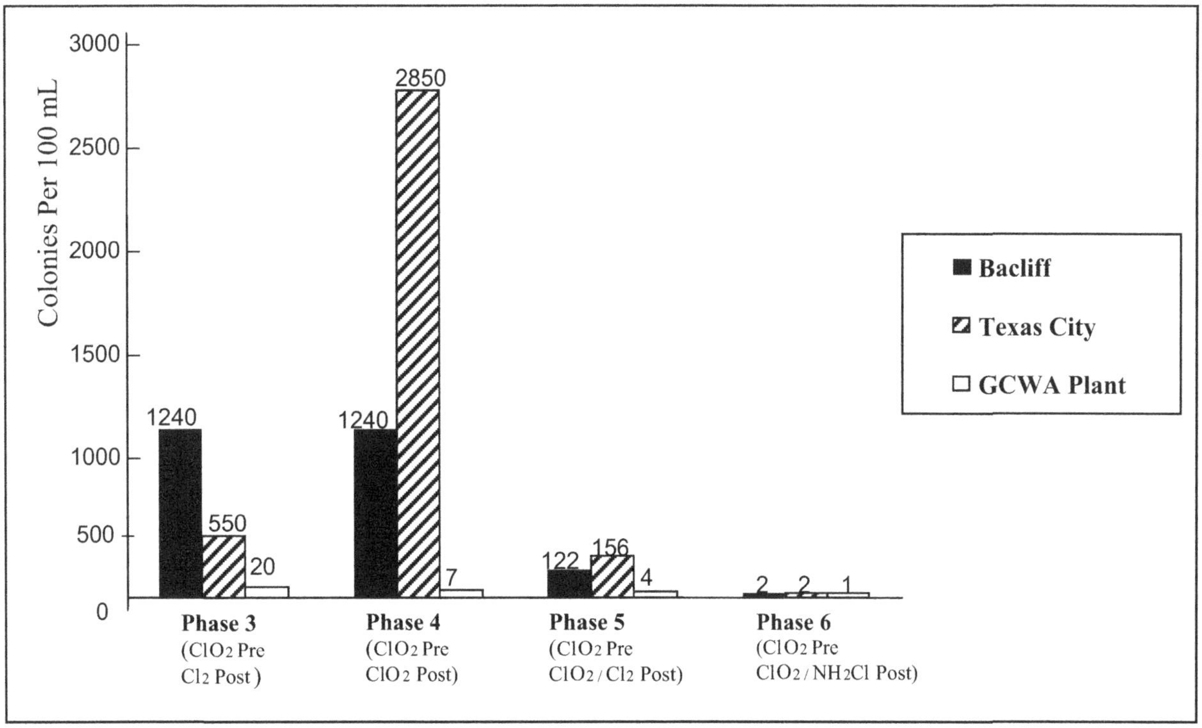

Figure 5.1 GCWA bacterial counts vs. disinfection alternatives

The distribution of bacterial species also changed upon the onset of chlorine dioxide disinfection. Bacterial evaluation indicated that a shift from orange and yellow gram (-) rods to white gram (+) rods, similar to slime-forming *Bacillus sp.*, was occurring in the plant effluent. Figure 5.2 illustrates a comparison of bacterial species distribution for the phase in which chlorine dioxide was used for preoxidation or post-disinfection. Figure 5.2 demonstrates that the microbial population that was found in the distribution system is a function of the disinfectant. What the data clearly demonstrate is that the population changes with disinfectant type and that both number and organism type are affected.

The intermittent, elevated bacterial counts and the occurrence of unidentified white colonies of gram (+) rods raised questions for GCWA regarding the bacteriological quality of the finished water. Therefore, chlorine was used (Phase 5), in addition to chlorine dioxide, as a post-disinfectant until the bacteria could be identified. The improvement in plate count was almost immediate. The shift in bacterial species distribution continued as the plate counts decreased, so that over 95% were either yellow gram (-) rods and/or white gram (+) rods.

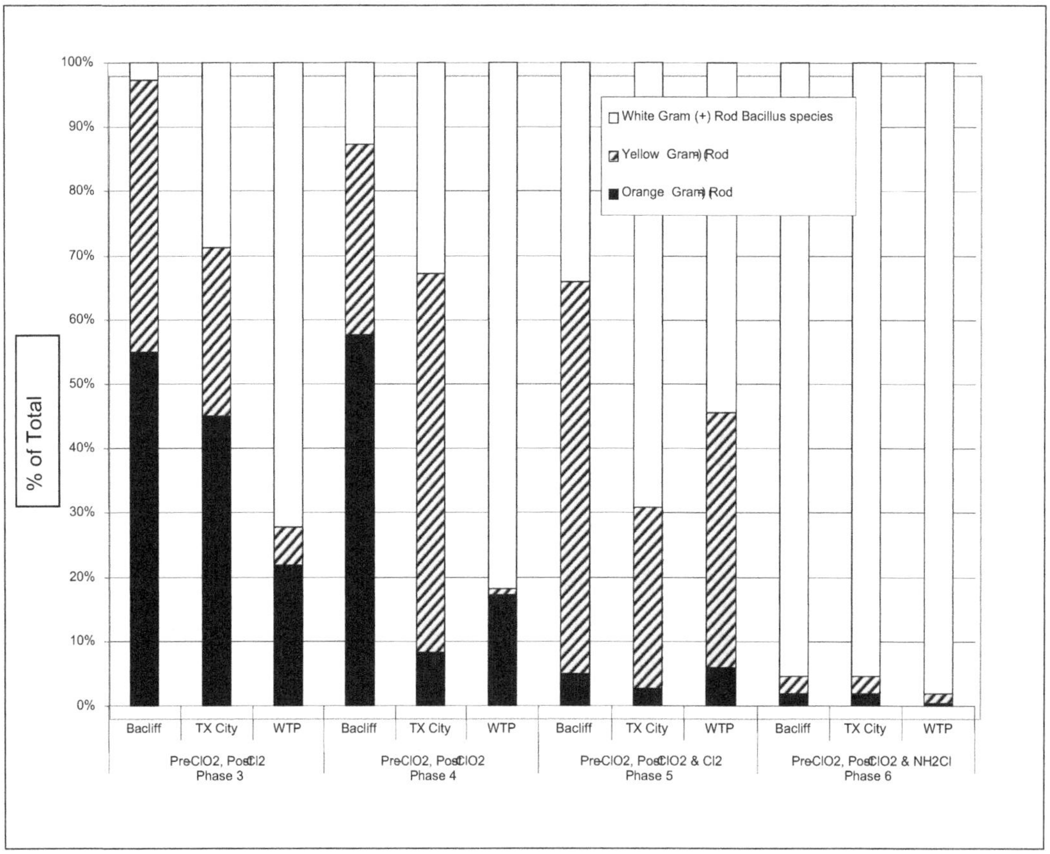

Figure 5.2 GCWA bacterial species distribution vs. disinfection alternatives

In December 1985, GCWA implemented chloramines for post disinfection (Phase 6). Chloramine residuals at customer connections have been consistently between 2.0 and 2.5 mg/L. This oxidation and disinfection strategy has remained to the present time. After this modification, bacteriological counts in the distribution system improved further. Between that time and the present, plate counts have ranged between 1 and 30 cfu/mL. Of those colonies identified, over 95% were the white, gram (+) *Bacillus sp.* variety. It is apparent that these organisms have preferential survivability when chlorine dioxide is used.

In retrospect, GCWA concluded that the implementation of chlorine dioxide possibly resulted in a "cleaning" of distribution pipe interiors, resulting in the release of bacteria embedded in biofilms. Upon further review of bacteriological sampling, plate counts gradually increased with time further into the system. After the counts peaked in a given location, they were reduced to lower levels than measured prior to the implementation of chlorine dioxide. GCWA believes that the strong disinfection capability of chlorine dioxide, together with a good biofilm control disinfectant such as chloramines, has allowed plate counts to continually remain low since December 1985. Up to the time that this work was completed, and since that time, GCWA had not conducted a significant amount of work relating the type of organisms as a function of disinfectant type, or a relationship between gram (+) and gram (-) organisms and other water quality issues, such as taste and odor. The observations by GCWA may be indicating a change in culturability of the organisms in the bulk water and may not represent actual changes in population in the distribution.

The County of San Luis Obispo reported no detectable changes in heterotrophic plate count numbers or coliform incidence after the conversion to chlorine dioxide, possibly because they used free chlorine followed by chloramines for secondary disinfection both before and after the conversion to chlorine dioxide for primary oxidation/disinfection.

IMPACTS ON CHEMICAL QUALITY - DBPs

Potential Impacts

The use of chlorine dioxide in place of chlorine as either a primary or secondary disinfectant should result in minimal formation of halogenated DBPs and other halogenated organics (TOX). DBPs of concern are chlorite and chlorate.

Positive Impacts

- Most utilities using chlorine dioxide as an oxidant or disinfectant experience essentially no production of TTHMs or HAA5s, and little formation of TOX as a result of chlorine dioxide addition.
- Most utilities using chlorine dioxide as an oxidant or disinfectant experience a reduction in the formation of halogenated DBPs upon subsequent chlorination. The extent of this decrease is dependent on the dose of chlorine dioxide applied. Chlorine dioxide does not participate in substitution reactions like free chlorine so that halogenated by-products are not formed to any appreciable extent; chlorine dioxide does not react with NOM to produce TTHMs and HAA5s, and chlorine dioxide reacts very slowly with bromide so that brominated by-products are not formed.
- A reduction in the chlorine demand of the water has been seen following pre-treatment with chlorine dioxide (Werdehoff and Singer 1987). The extent of the reduction is dependent on the dose of chlorine dioxide applied.
- Well-operated modern chlorine dioxide generators are capable of producing chlorine dioxide with little chlorite, chlorate, or free chlorine contamination.

- Application of chlorine to water pre-treated with chlorine dioxide results in production of some new chlorine dioxide which can have a positive influence as a disinfectant, and some chlorate which may be a negative outcome (Singer and O'Neil 1987).
- Application of monochloramine as a secondary disinfectant to water pre-treated with chlorine dioxide does not appear to produce any chlorine dioxide or chlorate, and does not appear to affect the residual chlorite concentration. The residual chlorite, in turn, has been shown to control nitrification.
- Few utilities encounter biofilm problems as a result of production of biodegradable organic material from chlorine dioxide treatment. This may be a result of the relatively low doses of chlorine dioxide typically used.

Adverse Impacts

- It has been found to be difficult to maintain a persistent chlorine dioxide residual in most distribution systems due to residual organic carbon in the treated water and to the wall demand of the pipes. In some cases, utilities erroneously believed that they were measuring a true chlorine dioxide residual, but this turned out to be an artifact of the analytical method they were using and positive coliform counts were observed (Singer 1988).
- About 50-70% of the applied chlorine dioxide is reduced to chlorite which has an MCL of 1.0 mg/L. This limits the chlorine dioxide dose to about 1.5 mg/L, and it is difficult to maintain a stable residual with ClO2 in most distribution systems.
- Chlorate is also a common oxyhalide species encountered when chlorine dioxide is used; there is currently no MCL for chlorate.
- Chlorine dioxide produces oxidation by-products similar to those produced by ozonation. Other oxidants used in conjunction with chlorine dioxide can lead to a variety of oxyhalide by-products; for example, if chlorine dioxide is applied to pre-chlorinated waters containing a free chlorine residual, chlorate will be formed. Likewise, if ozone is applied to waters that have been pre-treated with chlorine dioxide, ozone will oxidize the residual chlorite to chlorate.

Reported Impacts

One of the reasons GCWA and the County of San Luis Obispo converted to chlorine dioxide was the high natural organic content in the raw water and resultant high THM levels. Both the Brazos River and Lopez Lake are challenging source water in terms of controlling chlorinated DBPs. The sources are moderate to high in TOC (4 to 8 mg/L) and the Brazos is also is high in bromide (up to 0.3 mg/L). Consequently, THM formation under conditions of free chlorine in the distribution system can range up to 350 μg/L. It was these concerns that initially led GCWA to a modified disinfection strategy using chlorine dioxide in 1984. The disinfection strategy has been consistent with pre-disinfection using chlorine dioxide and post disinfection using chlorine dioxide and chloramines since 1985. However, GCWA has continued to make treatment improvements since that time. In 1992, the plant expanded and the filter media were changed from anthracite coal to granular activated carbon to further improve organic removal. In 2000, the plant was expanded from 25 mgd to 50 mgd, and a different chlorine dioxide generation system was installed. With these improvements, GCWA has been able to maintain THMs between 30 and 50 μg/L, as demonstrated on Figure 5.3.

GCWA analyzes for TTHMs in their finished water on a quarterly basis, but wholesales their water to 13 municipal customers. THM values in the customers' distribution systems range between 50 and 70 μg/L, with a running annual average of about 55 μg/L. In a 35-utility survey conducted by Metropolitan Water District of Southern California (Krasner et. al, 1989b), TTHMs were sampled quarterly at GCWA for one year. The values for TTHM ranged between 35 and 58 μg/L at the customer connections, supporting the values measured in customer distribution systems.

For the County of San Luis Obispo, average TTHM concentrations went from 86 μg/L το 62 μg/L when the utility replaced pre-chlorinzation with pre-chlorine dioxide addition.

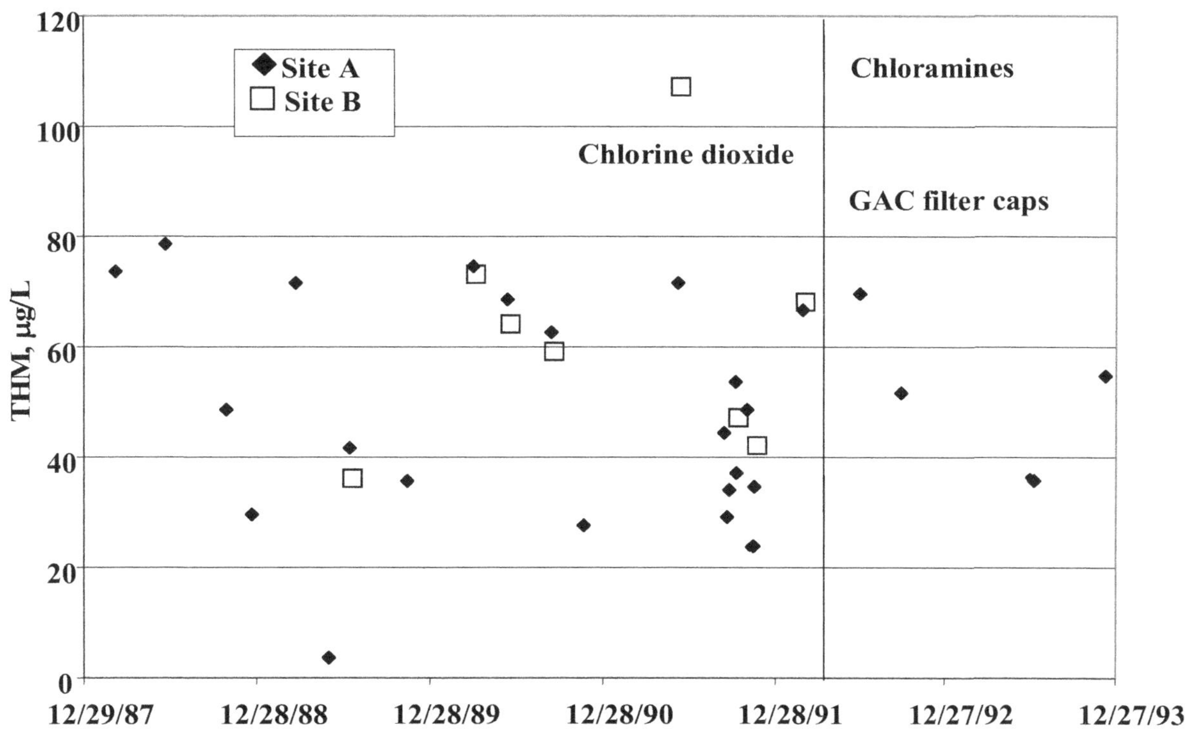

Figure 5.3 Gulf Coast Water Authority TTHM data

In the 35-utility study, HAA5 concentrations were measured between 13 and 16 μg/L. Since 2005, HAA5 concentrations generally have been between 10 and 30 μg/L. HAA5 will not likely represent a compliance problem, given the relatively high source water bromide concentration resulting in the preferential formation of brominated HAAs. This makes compliance with HAA5 less of an issue than TTHM compliance. Future regulatory efforts, which may focus upon HAAs not included in HAA5, may cause HAA formation to be more constraining.

IMPACTS ON CHEMICAL QUALITY - CORROSION, METAL RELEASE, COLOR, AND PARTICLE CHARACTERISTICS

Potential Impacts

Very limited data are available on the long-term effects of chlorine dioxide on corrosion, metal release, color and particle characteristics.

Positive Impacts

- Work by some researchers has attempted to compare chlorine dioxide to other disinfectants with regard to iron release (Eisnor et. al, 2001). In the study, the researchers found that chlorine dioxide resulted in the lowest corrosion rate and iron release compared to chlorine and monochloramine.

Adverse Impacts

- Unknown at this time.

Reported Impacts

No data were available from GCWA or the Country of San Luis Obispo to document any impacts on corrosion or metal release.

IMPACTS ON AESTHETIC QUALITY - TASTE AND ODOR

Potential Impacts

The use of chlorine dioxide can result in a reduction of taste and odors in the delivered water. Its use also can generate certain types of taste and odors.

Positive Impacts

- Chlorine dioxide is widely used as an effective oxidant for purposes of iron and manganese oxidation, taste and odor destruction, and oxidation of color.
- The earliest uses of chlorine dioxide during water treatment included were for reduction in odors caused by phenols, chlorophenols, and algal products (Aston, 1947; Mounsey and Hagar, 1946, Walker and Lee, 1983).
- Numerous examples of successful uses of chlorine dioxide for reducing halogentated DBP formation have been published (Andrews et. al, 2001; Hulsey et. al, 2000; Routt, 1989). One can infer from this information that the formation of odorous halogenated DBPs would be minimized, thus improving the organoleptic properties of the treated water.
- Chlorine dioxide was effective for controlling algae in open reservoirs by the Los Angeles Department of Water and Power distribution system (Stolarik and Liu, 2000). Research has shown that some, but not all, sulfur-containing compounds that cause odors described as "fishy/swampy/grassy" can be effectively eliminated by chlorine dioxide treatment (Glaze et. al, 1990; Krasner et. al, 2004).

Adverse Impacts

- Chlorine dioxide is generally not effective in removing many biological products that cause odor tastes and odors in drinking water, especially aldehydes and alcohols. Glaze et. al, (1990) reported only marginal removals of 2,4-decadienal (fishy/swampy odor) and geosmin and 2-MIB (earthy and musty odors, respectively) from Colorado River water. Krasner et. al, (2004) reported that chlorine dioxide was relatively ineffective for eliminating other compounds that cause "fishy/swampy/grass" odors and the compound most commonly associated with cucumber odor (trans, 2-cis, 6-nonadienal). Dietrich et. al, (1995) used FPA to evaluate the effectiveness of low dosages of chlorine dioxide for reducing odors caused by a variety of compounds in water. They found that the oxidant was not effective in removing odors caused by β-cyclocitral (tobacco/grape odor), 2t,6c-nonadienal (cucumber odor), and phenethyl alcohol (floral/rose) but did eliminate the odor of linolenic acid (watermelon). Chlorine dioxide generated grassy odors from two organic acids (linoleic and linolenic).
- Hoehn et. al, (1990) reported that many utilities that use chlorine dioxide frequently receive customer complaints of odors that resemble kerosene or cat urine odors, especially if the final disinfectant is free chlorine. Their investigation showed that trace amounts of chlorine dioxide are regenerated in the distribution system by reactions between chlorite ion and free chlorine. When a customer's tap is opened, the chlorine dioxide is released and reacts with organic compounds already present in the air. In another study, Hoehn et. al, (2002) found that chlorine dioxide concentrations in the distribution system less than 0.1 mg/L were sufficient to cause customer complaints of kerosene- and cat-urine odors.

Reported Impacts

The other major role of chlorine dioxide for GCWA is the control of taste-and-odors in the finished water. The warm temperatures in Texas, coupled with the organic and nutrient content of the source water, promote algae growth. The reservoirs used to store water between the river and the treatment plant provides ample residence time to foster growth. While chlorine dioxide together with powdered activated carbon (PAC) provided superior control compared to previous oxidation and disinfection strategies, GCWA also upgraded their filters in 1991 to add granular activated carbon as a filtering medium. This allows a more proactive approach to absorbing taste-and-odor causing compounds, compared to PAC. Further, GCWA baffled their source water reservoirs in 1994, effectively eliminated "dead zones" and promoting a flow-through regime. These enhancements collectively have reduced taste-and-odor complaints, and a significant algal bloom has not been measured during the critical period (summer) since 1994.

The County of San Luis Obispo reported no changes in taste or odor, nor any increases in customer complaints as a result of the change to chlorine dioxide.

OBSERVATIONS

Based on the findings of the issue papers and the information provided by the participating utilities, the following observations can be made regarding a utility switching from chlorine to chlorine dioxide for either primary or secondary disinfection:

Impacts on Microbial Quality

- Overall, with use of chlorine dioxide as a secondary disinfectant, there could be a degradation in microbial quality based on higher HPC levels resulting from more degradable matter in the bulk water in the distribution system.
- The use of chlorine dioxide as a primary disinfectant should provide better microbial quality in the distribution system based on lower HPC levels.
- HPC analyses should be based on use of R2A agar, as the use of standard agar will yield levels that are too low to detect a difference between before and after the conversion.
- There should be no adverse impact on coliform levels, and coliform levels may be reduced.

Impacts on Chemical Quality - DBPs

- Overall, disinfection by-product levels should be reduced with the use of chlorine dioxide compared to the use of chlorine.
- Lower TTHM formation potential should be expected – with both average and peak TTHM values being reduced.
- Lower HAA5 levels should be expected.

Impacts on Chemical Quality - Corrosion, Metal Release, Color, and Particle Characteristics

- There is insufficient information to make any observations regarding corrosion and metal release.

Impacts on Aesthetic Quality - Taste and Odor

- There should be a reduction in taste and odor complaints.

CHAPTER 6
CHLORINE TO UV

INTRODUCTION

The use of ultraviolet (UV) radiation for primary disinfection has received much attention over the past 5 to 10 years. The initial interest in UV resulted from its ability to inactivate *Cryptosporidium parvum oocysts.* More recently, more stringent disinfection by-product regulations have resulted in the need for water utilities to critically evaluate their disinfection practices to reduce DBPs in the distribution system. UV has been investigated because it does not produce DBPs of concern. However, it does not produce a residual, and so there is still a need for a secondary disinfectant such as free chlorine or chloramines. The impacts (microbial, chemical, and aesthetic) of switching from chlorine to UV as a primary disinfectant are presented in this chapter. The potential impacts are based on the findings of the issue papers and the reported impacts are based on the findings of the utility survey. In addition, the results of some recent AwwaRF projects on UV are presented to provide more information on distribution system water quality.

Of the utilities that participated in the project, five (5) reported switching from chlorine to UV for primary disinfection. Aqua NJ installed UV to meet the filtration guidance waiver. Chlorine still is used for disinfection, although UV provides some additional disinfection. These utilities are listed in Table 6.1 along with the prior and current disinfection strategies.

Table 6.1
Participating utilities that converted from chlorine to UV

Utility	Capacity mgd	Source	Prior Disinfection			Current Disinfection			Date Changed
			Preox	Primary	Secondary	Preox	Primary	Secondary	
SMCMUA, NJ	20.0	GW/SW	None	Cl_2	Cl_2	none	UV	Cl_2	2003
UWNY, NY	40.0	GW/Reservoir	None	Cl_2	Cl_2	none	UV	Cl_2	2002
Aqua NJ	13.0	GW	None	Cl_2	Cl_2	none	Cl_2	Cl_2/UV	2005
Flagstaff, AZ	8.0	Lake	Cl_2	Cl_2	Cl_2	Cl_2	UV	Cl_2	2003
Clayton Co. Water Authority, GA	10.0	Reservoir – Freeman Plant	None	Cl_2	Cl_2	none	UV	Cl_2	2003
Freeman	10.0	Reservoir	None	Cl_2	Cl_2	none	UV	Cl_2	2003
W.J. Hooper	20.0	Reservoir	None	Cl_2	Cl_2	none	UV	Cl_2	2003
J.W. Smith	12.0	Reservoir	None	Cl_2	Cl_2	none	UV	Cl_2	2003

SMCMUA – Southeast Morris County Municipal Utilities Authority
UWNY – United Water New York

It should be noted that for SMCMUA and UWNY, their capacities shown in Table 6.1 represent total production capacity. In both cases, UV was installed on one or two of many wells. The capacities of these wells ranged from 0.5 to 1.0 mgd. In both these cases, UV was installed because of periodic positive coliform occurrences in the raw water from the respective wells. In both cases, the installation of UV eliminated the coliform problem. However, because the UV treated wells pump into distribution systems that have many other points of entry, it was impossible to determine definitively whether the UV had any impact on the water quality in the distribution system. No negative changes in distribution system water quality were reported by these utilities.

IMPACTS ON MICROBIAL QUALITY

Potential Impacts

One of the concerns about the use of UV on microbial quality in the distribution system is the impact on biodegradability of the organic matter in the water and the effects on biofilms.

Positive Impacts

- As with ozone, UV light has greater efficacy against cysts and oocysts than free chlorine. UV can inactivate cysts at low water temperatures.
- UV light can be effective at reducing bacterial numbers and inactivating pathogenic organisms (Harris et. al, 1987; Parrota and Bekdash, 1998; Wolfe, 1990). In a recently published report (Lehtola et. al, 2003), 90% of bacteria in drinking water were inactivated at doses below 50 mWs/cm^2.
- One expectation of UV treatment would be that the water entering the distribution system would carry a much lower microbial load and therefore biofilm growth may be decreased. In one case, (Lund and Omerod, 1995) reported less biofilm accumulation in a distribution system receiving UV-treated water compared to ozonated water, which the investigators attributed to inhibitory hydroxyl radicals produced during UV treatment.
- There have been two published reports on the interactions between UV and chorine, chloramines, and/or chlorine dioxide. When UV treated water was subsequently treated with chorine or chloramines, there was no difference between the performance of the two secondary disinfectants, but biofilm growth was reduced compared to the UV-treated only water (Snicer et. al, 1996). In another study where water was subsequently treated with chlorine, chlorine dioxide or chloramine, synergistic effects between UV and the secondary disinfectant were noted (Dykstra et. al, 2002).

Adverse Impacts

- It has been hypothesized that UV treatment may increase the bioavailability of natural organic matter in water. For example, Kieber et. al, (1990) demonstrated that natural UV irradiation of humic substances resulted in the production of low molecular weight carbonyl groups. If this occurred, then one might expect to see the same potential for regrowth that often accompanies ozonation. One investigation with well established biofilms that were subjected to UV treated water showed increased biofilm growth (Lund and Hongve, 1994). In another case, a one log increase in biofilm numbers was seen post-UV treatment (Snicer et. al, 1996). However, this phenomenon is not universal.

Reported Impacts

None of the participating utilities reported a change in the microbial quality of the water following UV disinfection. There were no reported increases in HPC or coliform levels that might indicate an impact on the biodegradability of organic matter in the water or on biofilm control.

In the 2002 AwwaRF report entitled "Impact of UV Disinfection on Biological Stability", Dykstra et. al, present the results of bench-scale studies employing annular reactors to determine possible synergistic effects of UV and chlorine, chlorine dioxide, and chloramines. They evaluated the impacts on HPC in both the bulk water and in biofilms. The authors state the following conclusions:

- The use of UV alone did not control the growth of HPC in the bulk water or the biofilm.
- The use of UV and a secondary, chemical disinfectant (chlorine, chlorine dioxide, or chloramines) enhanced microbial inactivation in the bulk phase compared to the use of chemical disinfectants alone. There was no significant change in the biofilm.
- Both chlorine and chlorine dioxide performed better at controlling microbial growth when used in conjunction with UV compared to without UV.
- Chlorine dioxide performed better than chlorine on an equivalent CT basis for control of suspended bacteria, with or without the use of UV.

In another AwwaRF project (Andrews et. al, 2005), bench-scale tests were conducted in association with East Bay Municipal Utility District in California using annular reactors to determine the impact of UV on raw water as it travels through 90 miles of aqueduct. Tests were conducted using chlorine and chlorine dioxide, with and without pre-treatment using UV. The results indicated the following:

- The use of UV in combination with either chlorine or chlorine dioxide provided better microbial control in the aqueduct compared to the use of chemical disinfectants alone based on lower HPC levels.
- UV/chlorine dioxide performed better than UV/chlorine based on controlling HPCs.
- The use of UV ahead of either chlorine or chlorine dioxide did not reduce the required chemical dose to maintain a target residual when using chlorine or chlorine dioxide.

IMPACTS ON CHEMICAL QUALITY – DBPs

Potential Impacts

There are no anticipated DBPs of concern with the use of UV. If it is used as part of an advanced oxidation process, biodegradable oxidation by-products can be expected, and bromate formation is possible in bromide-containing waters.

Positive Impacts

- There are no anticipated DBPs of concern, unless UV irradiation is used in conjunction with a chemical disinfectant/oxidant as part of an advanced oxidation process. In the latter case, biodegradable oxidation by-products can be expected, and bromate formation is possible in bromide-containing waters.
- UV irradiation should not produce TTHMs, HAA5s, or any other halogenated organic DBPs. Likewise, it should not produce bromate or chlorite like some of the other alternative primary disinfectants. There is limited plant-scale experience to demonstrate the fact that UV irradiation does not produce TTHMs, HAA5s, other halogenated organic DBPs, chlorite, or bromate.
- UV irradiation should not increase or increase the DBP formation potential of a water. DBP precursors remain in the water and are subject to subsequent reaction with secondary disinfectants added to the water later in the treatment train.

Adverse Impacts

- At the doses typically used for disinfection, UV irradiation should be incapable of oxidizing DBP precursors. As a result, DBP precursors remain in the water and are subject to subsequent reaction with secondary disinfectants added to the water later in the treatment train.
- Comparison of different types of UV designs (e.g. monochromatic, polychromatic, low pressure, medium pressure, pulsed UV) need to be evaluated in a variety of waters with different chemical characteristics to verify whether or not UV produces DBPs of concern under practical dosage conditions.

Reported Impacts

There were no reported changes in DBP levels in the distribution systems of the participating utilities as a result of the use of UV. In the AwwaRF project conducted by Rand et. al, it was reported that, on average, both TTHM and HAA5 levels were lower with UV/chlorine compared to chlorine alone.

IMPACTS ON CHEMICAL QUALITY - CORROSION, METAL RELEASE, COLOR, AND PARTICLE CHARACTERISTICS

Potential Impacts

Although no major impacts are expected with the use of UV on this area of chemical quality, there is very little data to support this.

Positive Impacts

- Of all the possible disinfectants used, UV disinfection is considered least likely to have an impact on distribution system water quality, as it leaves no residual.
- Does not change redox potential of the treated water – may be best option for groundwaters under reducing conditions.

Adverse Impacts

- Nitrate has been shown to reduce to nitrite when exposed to certain UV "doses". Other constituents may be reduced which may affect their fate in the distribution system.

Reported Impacts

There were no reported impacts on corrosion or metal release as a result of the use of UV for primary disinfection by any of the participating utilities.

AESTHETIC QUALITY -T&O

Potential Impacts

There may be some benefit in the reduction of taste and odors, but little data are available to support this.

Positive Impacts

- Glaze et. al, (1990) reduced concentrations of odorous compounds in Colorado River water by treating it for 20 min with a combination of UV and hydrogen peroxide (0.56 watts/L and 0.2 mg/L·min, respectively). They reported the following reductions in compounds spiked into the raw water:

  - 1- heptanal, 74 percent
  - dimethyltrisulfide > 99 percent
  - 2,4-decadiental, 73 percent
  - geosmin, 40 percent
  - 2-MIB, 29 percent.

- Liu et. al, (2002) saw no significant changes in chlorinated DBP concentrations when water was treated first with UV at dosages < 500 mJ/cm^2 then with chlorine. When UV was applied at higher dosages and supplemented with high dosages of hydrogen peroxide (100 mg/L), chlorinated DBP concentrations actually decreased, but aldehydes and carboxylic acids were formed.

Adverse Impacts

- Linden et. al, (2002) showed that neither LP nor MP UV was effective for destroying MIB and geosmin. When used in combination with hydrogen peroxide (2 mg/L and 7 mg/L), MP UV was more effective than LP UV for degrading the two odor compounds. UV dosages required to achieve complete elimination of the MIB and geosmin were between 2,000 mJ/gm^2 and 5,000 mJ/gm^2. Romain et. al, (2003) evaluated geosmin destruction in pilot-scale and bench-scale studies using UV irradiation in combination with hydrogen peroxide and an aluminum-based catalyst, similar to what is known as "Fenton's Reaction."

- Glaze et. al, (1990) found that concentrations of an odorous aldehyde (1-heptanal) increased by 27 percent following treatment of Colorado River water for 20 min with a combination of UV and hydrogen peroxide (0.56 watts/L and 0.2 mg/L·min, respectively).

Reported Impacts

There were no reported impacts on aesthetic water quality as a result of the use of UV by the participating utilities.

OBSERVATIONS

Based on the findings of the issue papers and the information provided by the participating utilities, the following conclusions can be made regarding a utility switching from chlorine to UV for primary disinfection:

Impacts on Microbial Quality

- The use of UV in combination with a chemical disinfectant should provide better microbial control in the distribution system based on lower HPC levels in the bulk water.
- The use of UV in combination with a chemical disinfectant is unlikely to result in the use of a lower disinfectant dose to maintain a target residual in the distribution system.

Impacts on Chemical Quality - DBPs

- The use of UV in combination with chlorine may result in lower TTHM and HAA5 levels in the distribution system.

Impacts on Chemical Quality - Corrosion, Metal Release, Color, and Particle Characteristics

- No data were available to draw conclusions in this area. However, none of the participating utilities reported any negative impacts.

Impacts on Aesthetic Quality - Taste and Odor

- No data were available to draw conclusions in this area. However, none of the participating utilities reported any negative impacts.

CHAPTER 7
IDENTIFICATION OF FUTURE RESEARCH NEEDS

The results of this project have led to an identification of future research that is required to better understand the long-term impacts of changing disinfectants on distribution system water quality. Data gaps and research needs were identified for each of the major changes in disinfection strategy:

- Chlorine to chloramines
- Chlorine to ozone
- Chlorine to chlorine dioxide
- Chlorine to UV

There are many data gaps regarding the use of booster chlorination, and research needs in this area also were identified.

This information was developed as the issue papers were prepared and during the workshop that was held at the beginning of the project. This chapter presents the data gaps and research needs identified as part of the project.

CHLORINE TO CHLORAMINES

Data Gaps and Research Needs

Microbial Quality

- Need for a study where chlorine and chloramine levels are varied to determine at what concentrations equivalent results for biofilm and suspended cell control are seen (if they are seen). Study should include corroding surface, inert surface, and corrosion control.
- Better data/more complete studies on the survival of opportunistic pathogens in the presence of both disinfectants.
- More information on the impact of these disinfectants on microbial water quality in health care facilities, other buildings.
- A better understanding of conditions leading to nitrification, predicting when and where it will occur, and approaches for control.

Chemical Quality - DBPs

- What is the nature of the TOX produced by combined chlorine?
- What is the linkage, if any, between chloramination, nitrification, and NDMA production?
- Are there other nitrosamines that may be formed in distribution systems in addition to NDMA and are they associated with chloramination practice?
- How widespread is the occurrence of the newly identified halogenated DBPs of potential toxicological concern?

- From an analytical perspective, how do you quench samples containing monochloramine without altering the by-products present? Most analytical schemes for measuring DBPs involve quenching the residual chlorine with ammonium sulfate. Quenching a sample containing combined chlorine with ammonium sulfate does not quench the combined chlorine and allows for some continuing formation of halogenated organics.

Chemical Quality – Corrosion, Metal Release, Color, and Particle Characteristics

- Studies are needed to quantify the extent and rate of color reduction by chloramines and the role pH plays in the process. Highly colored natural waters should be used instead of laboratory-prepared waters containing commercial humic substances.
- Need studies of redox potential, iron and turbidity at various locations within distribution system before and after change to chloramines to look at red water formation and particle release.
- Need studies of nitrification effects on oxidant and dissolved oxygen losses and alkalinity decreases.
- Need more data about formation of amine complexes with plumbing materials.

Aesthetic Quality – T&O

- Not much is known about the impacts of biofilms on taste-and-odors and other aesthetic properties of water in distribution system systems where chloramines are used. Biofilms harvested from actual distribution systems should be examined for actinomycetes and fungi that produce odors. The microbiological examination should be supplemented with some type of odor evaluation procedure such as triangle tests and Flavor Profile Analysis.
- Customer complaint records before and after conversion of a system to chloramines should be examined to determine if complaints decreased, and efforts should be made throughout the drinking water industry to involve water treatment plant personnel in the investigation of taste and odor complaints.
- Little is known about the formation and persistence of highly odorous iodonated and brominated compounds during chloramine reactions with bromide ion and chloride ion.
- The role of bromide ion in producing odorous DBPs in chloraminated water needs further investigation.

CHLORINE TO OZONE

Data Gaps and Research Needs

Microbial Quality

- Better methods for assessing the biodegradable fraction of NOM produced by ozonation or other disinfectants; improved ability to predict microbial growth based on measurable water quality parameters.

- An understanding of how utilities may use corrosion control to mitigate potential increases in microbial growth due to ozonated NOM.

Chemical Quality – DBPs

- The nature and characteristics of the oxidation by-products of ozonation need to be identified and quantified.
- New bromate control technologies, especially the application of chlorine and ammonia developed by the Southern Nevada Water Authority and expanded upon by Buffle et. al, (2003) need to be demonstrated on a plant-scale basis at a number of different water utilities.
- A more comprehensive survey should be conducted to assess the occurrence of the newly identified halogenated DBPs of potential toxicological concern.

Chemical Quality – Corrosion, Metal Release, Color, and Particle Characteristics

- Affects of ozone application on particle and natural organic surface charge should be further explored.
- Interaction between ozone and high pH treatment on iron release/ corrosion control needs more research.

Aesthetic Quality – T&O

- Utilities should examine customer service records before and after ozonation was included in the treatment train and determine if the change affected the frequency of complaints, especially taste-and-odor complaints. Correlating the complaint records with the institution of ozonation may be difficult if the final disinfectant is also switched from chlorine to chloramines.
- Additional research is needed to determine the odor- and flavor threshold concentrations of odorous DBPs, especially brominated DBPs, formed during ozonation of natural waters.
- Additional research is also needed to determine the efficacy of ozone at different dosages and in waters with varying NOM concentrations for eliminating odorous aldehydes commonly produced during ozonation. Special attention should be given to the relationship between ozone dose and percent aldehyde removals following ozonation of waters containing high NOM concentrations.
- Color reduction is an ancillary benefit of ozone that is usually not emphasized during pilot-scale and full-scale evaluations of ozone. More data are needed to clarify the conditions (e.g. pH, ozone dose, contact time) that produce acceptable treatment in highly colored waters.

CHLORINE TO CHLORINE DIOXIDE

Data Gaps and Research Needs

Microbial Quality

- Changes in microbial ecology created by chlorine dioxide; is there a selection for opportunistic pathogens in distribution systems and health care facilities.
- Further exploration of the possible synergistic interactions between ozone and UV with chlorine dioxide.
- Better insight on behavior of chlorine dioxide in a wider variety of water qualities and seasonal effects.
- Interaction with adsorbed natural organic matter at the pipe wall and impact on corrosion is unknown; both would have an affect on the growth of biofilms on pipe surfaces.
- Better understanding of the effects of a change from free chlorine or monochloramine to chlorine dioxide on biofilms and microbial populations in controlled settings.

Chemical Quality - DBPs

- A survey of water treatment plants using chlorine dioxide should be conducted to determine levels of biodegradable organic material (BDOC, AOC), and other chlorine dioxide by-products of the type suggested by Richardson et. al, (1994).

Chemical Quality - Corrosion, Metal Release, Color, and Particle Characteristics

- Information and data with regard to the impact of chlorine dioxide on corrosion and particle transformations are very limited.
- Not much information regarding color reduction by chlorine dioxide was found in the available literature. Additional studies are needed to evaluate the chlorine dioxide reduction of color caused by NOM, especially fulvic acids.

Aesthetic Quality – T&O

- More data are needed that identify odorous organic DBPs produced by chlorine dioxide treatment of NOM containing iodide and bromide ions.
- Research by Krasner and his associates indicate that chlorine dioxide is not very effective for reducing tastes and odors caused by aldehydes and other non-humic components of NOM in natural waters.
- Additional studies are needed to identify the chemicals that cause the kerosene- and cat-urine odors that are produced when chlorine dioxide reacts with substances released from new carpeting.

CHLORINE TO UV

Data Gaps and Research Needs

Microbial Quality

- More information/controlled studies on the synergistic effects of UV treatment followed by chlorination/chloramination/chlorine dioxide application.
- Greater understanding of the impact of UV light on the degradable fraction of NOM.
- For small systems and point of use/point of entry UV units, a better understanding of the conditions that lead to decreased efficacy over time against suspended cells.
- Controlled studies comparing the efficacy of realistic UV wavelengths, including 254 nm, for inactivation of organisms.

Chemical Quality - DBPs

- Plant-scale experience is needed to see how effectively UV irradiation performs on a long-term basis with respect to fouling, how it can be integrated with oxidative pre-treatment strategies for waters that are in need of an oxidant for iron and manganese oxidation and taste and odor destruction, how effectively it controls DBP formation when it is integrated with various secondary disinfection practices, and how effectively it does indeed inactivate pathogens of concern.
- Comparison of different types of UV designs (e.g. monochromatic, polychromatic, pulsed UV) need to be evaluated in a variety of waters with different chemical characteristics.
- A methodology is needed to guarantee that a given piece of UV irradiation equipment (lamps and reactor) is performing up to its specifications and continues to perform effectively over the lifetime of the design.

Chemical Quality - Corrosion, Metal Release, Color, and Particle Characteristics

- One area that has had little study is the ability of ultraviolet illumination to act as a reductant for some constituents.
- One other area where UV disinfection may have an impact is on groundwater supplies where UV illumination would replace chlorine as the disinfectant. A system that changes to UV in place of free chlorine will not see the same redox potential and may experience a change in metal behavior.

Aesthetic Quality – T&O

- Little is known about the impact of UV or UV/H_2O_2 on the aesthetic properties of drinking water, and the goals of UV treatment are not typically concerned with such issues. However, improvements in taste-and-odor and other aesthetic properties of water could easily be evaluated during UV studies designed for other purposes. Studies like those reported by Glaze et. al, (1990) could serve as a model for evaluating UV and UV/ H_2O_2 effects on other model odor compounds.

- No studies of color reduction by UV/H_2O_2 treatment were found in the literature. The subject should be investigated in pilot-plant-scale studies with a variety of high DOC natural waters.

CHLORINE TO BOOSTER CHLORINATION

Data Gaps and Research Needs

Microbial Quality

- Studies illustrating the response of model or actual distribution systems with established biofilms to the sudden or gradual increase in free chlorine residuals.
- Better insight on the interaction between free chlorine, corrosion, and microbial water quality.

Chemical Quality – DBPs

- There is limited experience in the waterworks industry and few, if any, publications specifically involving the impact of booster chlorination stations on DBP levels, and how the use of such stations has impacted the regulatory compliance picture for these utilities, both for the Stage 1 and Stage 2 DBP Rules.
- Distribution system models should be used to predict how different booster system scenarios might impact the concentrations of TTHMs and HAA5s in a variety of water distribution systems, and how these scenarios impact regulatory compliance for different regulatory paradigms such as the Stage 1 and Stage 2 DBP Rules. These models should then be verified with real data from utilities employing booster chlorination stations in their distribution systems.

Chemical Quality - Corrosion, Metal Release, Color, and Particle Characteristics

- Studies illustrating the response of model or actual distribution systems with established surface scales or pipe surfaces to the sudden or gradual increase in free chlorine residuals.

Aesthetic Quality – T&O

- The major deficiency in the current state of knowledge regarding the impacts of booster chlorination or chloramination is the lack of data on customer complaints associated with the practice. Customer complaint records should be reviewed to determine the nature of the complaint, the chlorine or chloramine concentrations in the distribution system at the point where the complaint was registered, and the association between the complaint and the addition of chlorine or chloramine at the booster station. Other data such as pH, temperature, HPC, etc. should also be noted if they exist.

APPENDIX A
BIBLIOGRAPHY

ID:	**1**
DISINFECTANT:	Chloramines
KEY WORDS:	Biofilm, organics, distribution systems, inhibition, pipes, iron, corrosion, chlorine, chloramines
TITLE:	Interactions Between Pipe Materials, Disinfectants, Corrosion Inhibitors, Organics, and Distribution Biofilms
AUTHORS:	Abernathy, Calvin; Camper, Anne
YEAR:	1997
SOURCE:	1997 Water Quality Technology Conference: AWWA
ABSTRACT?:	Y
FULL ARTICLE?:	N
COMMENTS:	This paper discusses the interactions among pipe materials, disinfectants, corrosion inhibitors, organics, and distribution system biofilms in order to provide information on how treatment changes may affect biofilms.
ID:	**2**
DISINFECTANT:	Chloramines
KEY WORDS:	Distribution systems, chloramines, nitifircation, water quality monitoring, reservoirs, cleaning, flushing, regulations, water demand, California, Operation and Maintenance, Design
TITLE:	Operation and Maintenance of Water Distribution Systems During and After Conversion to Chloramine Disinfectant
AUTHORS:	Acker, Courney; Kraska, Dave
YEAR:	2001
SOURCE:	2001 Annual Conference Proceedings; AWWA
ABSTRACT?:	Y
FULL ARTICLE?:	Y
COMMENTS:	This paper presents useful operation, maintenance and system design concepts for systems that have or are considering a conversion to chloramine disinfection.
ID:	**3**
DISINFECTANT:	Chloramines
KEY WORDS:	
TITLE:	Nitrification
AUTHORS:	AWWA
YEAR:	2002
SOURCE:	USEPA
ABSTRACT?:	N
FULL ARTICLE?:	Y
COMMENTS:	One of the white papers prepared for development of TCR revisions.
ID:	**4**
DISINFECTANT:	Chloramines
KEY WORDS:	Contaminant Candidate List, Microbial, DBPs, Microbiology, disinfection, treatment
TITLE:	Inactivation of Waterborne Emerging Pathogens by Selected Disinfectants
AUTHORS:	AwwaRF
YEAR:	2002
SOURCE:	AwwaRF
ABSTRACT?:	Y
FULL ARTICLE?:	N
COMMENTS:	This project determines the efficiencies of conventional and newly developing disinfectants for emerging waterborne pathogens.
ID:	**5**

DISINFECTANT:	Chloramines
KEY WORDS:	Cryptosporidium, microbial, DBPs, ozone
TITLE:	Synergistic Effects of Multiple Disinfectants
AUTHORS:	AwwaRF
YEAR:	2000
SOURCE:	AwwaRF
ABSTRACT?:	Y
FULL ARTICLE?:	N
COMMENTS:	Examines and quantifies the presence and significance of synergistic effects through the use of disinfectant strategies.
ID:	**6**
DISINFECTANT:	Chloramines
KEY WORDS:	Microbial, DBPs Assessment of Seasonal Chlorination Practices and Impacts to Chloraminating Utilities
TITLE:	
AUTHORS:	AwwaRF
YEAR:	to be published
SOURCE:	AwwaRF
ABSTRACT?:	Y
FULL ARTICLE?:	N
COMMENTS:	Will identify and assess the impacts of seasonal or periodic conversion of chloraminated systems to chlorine.
ID:	**7**
DISINFECTANT:	Chloramines
KEY WORDS:	Filtration, distribution systems, chlorine, ozonation, reactors, biomass, biological filters, residual chlorine, chloramines
TITLE:	Effects of Biological and Conventional Treatment on Biological Stability of Distribution Systems
AUTHORS:	Baribeau, H.; Hacker, P.; Mofidi, A.; Coffeey, B.; De Leon, R.
YEAR:	1998
SOURCE:	1998 Symposium Proceedings; Protecting Water Quality in the Distribution System: What is the Role of disinfection Residuals?
ABSTRACT?:	Y
FULL ARTICLE?:	N
COMMENTS:	This study compares the biological stability of water in simulated distribution systems following three treatment scenarios: ozonation followed by biofiltration; ozonation followed by non-biological filtration, and now ozone followed by non-biological filtration. Chloramination was used as a primary disinfectant.
ID:	**8**
DISINFECTANT:	Chloramines
KEY WORDS:	Distribution systems, treatment plants, surveys, wastewater treatment, degradation
TITLE:	Occurrence of NDMA in Drinking Water: A North American Survey 2001-2002
AUTHORS:	Barrett, S.; Hwang, C.; Guo, Y.; Andrews, S.; Valentine, R.
YEAR:	2003
SOURCE:	2003 Annual Conference Proceedings: AWWA
ABSTRACT?:	Y
FULL ARTICLE?:	Y
COMMENTS:	This paper presents the results of a survey of NDMA occurrence in treated waters of 21 North American drinking water treatment plants.
ID:	**9**
DISINFECTANT:	Chloramines
KEY WORDS:	Water quality, bacteria, coliforms, chloramines, distribution systems, flushing
TITLE:	Methods of Plant Operation and Distribution Control to Meet Microbial Goals: A Fairfax County Water Authority Perspective

AUTHORS:	Bonacquisti, T
YEAR:	1993
SOURCE:	1993 Distribution System Symposium Proceedings
ABSTRACT?:	Y
FULL ARTICLE?:	N
COMMENTS:	The paper describes the Fairfax County Water Authority's methods for control by outlining several key programs which have proven to be an effective means of distribution system microbial control over the years. The Authority employs monochloramine as a secondary disinfectant the majority of the year.
ID:	**10**
DISINFECTANT:	Chloramines
KEY WORDS:	Water treatment, chloramines, distribution system, ammonia, corrosion, pH, natrual organic matter, coagulation, CT values, Lakes, pilot plants
TITLE:	Ammonia Release from Chloramine Decay: Implications for the Prevention of Nitrification Episodes
AUTHORS:	Bone, Christopher; Harrington, Gregory; Oldenburg, Patrick; Noguera, Daniel
YEAR:	1999
SOURCE:	1999 Annual Conference Proceedings: AWWA
ABSTRACT?:	Y
FULL ARTICLE?:	Y
COMMENTS:	Corrosion appears as a key word. This research used Lake Mendota (Madison Wisconsin) water that was treated as necessary with a pilot-plant.
ID:	**11**
DISINFECTANT:	Chloramines
KEY WORDS:	
TITLE:	Formation and Decay of THMs and HAAs in Distribution Systems: Results of Five Case Studies
AUTHORS:	Boulos, L.; Singer, P.; Schlesinger, S.; Gullick, R.; Baribeau, H.; Nichols, C.; Williams, S.
YEAR:	2003
SOURCE:	2003 Water Quality Technology Conference Proceedings: AWWA
ABSTRACT?:	N
FULL ARTICLE?:	Y
COMMENTS:	This paper presents case studies for five utilities on DBPs formation in the distribution system. One of the utilities uses chloramine on a seasonal basis as a disinfectant.
ID:	**12**
DISINFECTANT:	Chloramines
KEY WORDS:	
TITLE:	Choosing and Optimal Chloramine Residual Goal for Philadelphia's Tap Water
AUTHORS:	Burlingame, G.; Mackey, E.; Choi, J.
YEAR:	2003
SOURCE:	2003 Water Quality Technology Conference Proceedings: AWWA
ABSTRACT?:	N
FULL ARTICLE?:	Y
COMMENTS:	This paper reviews the various factors involved with setting a goal for the maintenance of a chloramine residual for Philadelphia's tap water. These factors include public perception and DBP control.
ID:	**13**
DISINFECTANT:	Chloramines
KEY WORDS:	
TITLE:	Factors Influencing Microbial Water Quality in Distribution Systems
AUTHORS:	Camper, A.
YEAR:	2003
SOURCE:	2003 Annual Conference Proceedings: AWWA
ABSTRACT?:	N

FULL ARTICLE?:	Y
COMMENTS:	This paper presents an overview of the methods for assessing microbial water quality and the factors associated with biofilm proliferation.
ID:	**14**
DISINFECTANT:	Chloramines
KEY WORDS:	
TITLE:	A New Awareness of Copper Pipe Failures in Water Distribution Systems
AUTHORS:	Cantor, A.; Bushman, J.; Glodoski, M.
YEAR:	2003
SOURCE:	2003 Water Quality Technology Conference Proceedings: AWWA
ABSTRACT?:	N
FULL ARTICLE?:	Y
COMMENTS:	This paper looks into the causes of copper pipe failures and determines that microbiologically influenced corrosion is the likely cause. The study points out that treatment was switched to ozonation followed by chloramination in 1998.
ID:	**15**
DISINFECTANT:	Chloramines
KEY WORDS:	
TITLE:	Charaterizing the Effect of Chlorine and Chloramines on the Formation of Biofilm in a Simulated Drinking Water Distribution System
AUTHORS:	Clark, R.; Sivaganesan, M.
YEAR:	1999
SOURCE:	USEPA/600/R-01/024: USEPA
ABSTRACT?:	N
FULL ARTICLE?:	Y
COMMENTS:	This is a study conducted by USEPA and University of Nancy (France) to examine the control of microorganisms in treated water and at the pipe wall.
ID:	**16**
DISINFECTANT:	Chloramines
KEY WORDS:	Chlorine, chloramines, microorganisms, water quality, performance
TITLE:	Effect of Changing from Chlorination to Chloramination on Microbiological Quality
AUTHORS:	Cuncliffe, D.; Christy, P.; Robinson, B.; Walters, R.
YEAR:	1990
SOURCE:	Water (17)1: 28-30: Aaustrailian Water and Wastewater Association
ABSTRACT?:	Y
FULL ARTICLE?:	N
COMMENTS:	The results of this study show that the use of chloramination improved the microbiological quality of water supplies.
ID:	**17**
DISINFECTANT:	Chloramines
KEY WORDS:	Chemicals, utilities, CA, laboratories, distribution systems, analysis, water treatment, DBPs, standard methods, regulations
TITLE:	N-Nitrosodimethylamine (NDMA) in Surface Water
AUTHORS:	Davis, M.; Barrett, S.; Hwant, C.; Guo, Y.; Liang, S.
YEAR:	2000
SOURCE:	2000 Water Quality Technology Conference Proceedings: AWWA
ABSTRACT?:	Y
FULL ARTICLE?:	Y
COMMENTS:	This paper describes Metropolitan Water District's efforts after NDMA was detected in their distribution system.
ID:	**18**
DISINFECTANT:	Chloramines
KEY WORDS:	Interagency Agreements, Chloramines, Disinfection by products, Trihalomethanes, Design, Microbiology, Regulations, Blending, Residual chlorine, Health effects, Planning,

	Management, Public information, Chemical handling and feeding, chemistry, Operation and Maintenance, Taste and odor, nitrification, fish, training
TITLE:	Practical Aspects of Implementing Chloramines
AUTHORS:	Dennis, J.P.; Rauscher, D.C.; Foust, D.A
YEAR:	1991
SOURCE:	1991 Annual Conference Proceedings; AWWA
ABSTRACT?:	Y
FULL ARTICLE?:	Y
COMMENTS:	This paper summarizes Metropolitan Water District's 1985 conversion from chlorine to chloramine disinfection. Covers practical aspects of implementing chloramine disinfection and dealing with problems such as the blending of chloraminated ahd chlorinated water, public notification, material and equipment failures, investigative and corrective measures in support of kidney dialysis facilities and others, taste and odor episodes, loss of residuals as a result of nitrification, and assistance to member agencies.
ID:	**19**
DISINFECTANT:	Chloramines
KEY WORDS:	Chlorine dioxide, corrosion, chlorine, distribution system, trihalomethanes, disinfection, pipes, modeling, iron, water quality, red water, oxidants, chloramines
TITLE:	Impact of Chlorine Dioxide on Corrosion in a Model Distribution System
AUTHORS:	Eisnor, J.D.; Gagnon, G.A.; Volk, C.; Chauret, C.P.; Andrews, R.C.
YEAR:	2001
SOURCE:	2001 Proceedings Water Quality Technology Conference: AWWA
ABSTRACT?:	Y
FULL ARTICLE?:	Y
COMMENTS:	The objective of this study is to study the impacts of free chlorine, monochloramine, chlorine dioxide and chlorite on internal corrosion and water quality in a pilot-scale pipe loop system.
ID:	**20**
DISINFECTANT:	Chloramines
KEY WORDS:	Water treatment, chemical, disinfection, chlorination, water chemistry
TITLE:	Boston MDC Experience with Chloramine Disinfection
AUTHORS:	Elwood, J.; Foss, G.; Sheridan, P.; Couture, T.
YEAR:	1985
SOURCE:	1985 Conference Proceedings "Environmental Engineering" edited by J. Shaughnessy
ABSTRACT?:	N
FULL ARTICLE?:	N
COMMENTS:	None
ID:	**21**
DISINFECTANT:	Chloramines
KEY WORDS:	Disinfectant residuals, chloramines, trihalomethanes, Wisconsin, ultraviolet radiation, chlorine, chlorine demand
TITLE:	Effects of UV Irradiation on Disinfectant Residuals and the THM Formation in Chlorinated and Chloraminated Drinking Water
AUTHORS:	Fleming, K.; Harrington, G.
YEAR:	2001
SOURCE:	2001 Microbial/Disinfectio By-Products Health Effects Symposium
ABSTRACT?:	Y
FULL ARTICLE?:	N
COMMENTS:	The effects of UV irradiation on disinfectant residuals and the formation of four THMS in chlorinated and chloraminated drinking waters, produced by three Wisconsin water utilities, were studied.
ID:	**22**
DISINFECTANT:	Chloramines
KEY WORDS:	

TITLE:	A Survey on Chloramine Treatment
AUTHORS:	Baribeau, H.; Flores, A.; Kirmeyer, G.J.
YEAR:	2002
SOURCE:	2002 Water Quality Technology Conference Proceedings: AWWA
ABSTRACT?:	Y
FULL ARTICLE?:	Y
COMMENTS:	The Chloramine Treatment Survey presented in this paper is part of a larger project with goals to develop practical chloramination guidelines for use in the Utility Operation Toolbox.
ID:	**23**
DISINFECTANT:	Chloramines
KEY WORDS:	Disinfection, utilities, chlorine, trihalomethanes, haloacetic acids, residual chlorine, treatment plants, disinfection by-products
TITLE:	Evaluation of "Low-Cost" Strategies for the Control of Trihalomethanes and Haloacetic Acids
AUTHORS:	Franchi, A.; Singer, P.; Chowdhury, Z.; Carter, J.; Grace, N.
YEAR:	2002
SOURCE:	2002 Water Quality Technology Conference Proceedings: AWWA
ABSTRACT?:	Y
FULL ARTICLE?:	Y
COMMENTS:	This paper presents and compares the results and implications of bench-scale testing commissioned by four different utilities in the Eastern US.
ID:	**24**
DISINFECTANT:	Chloramines
KEY WORDS:	Distribution systems, disinfection, contamination, biofilm, regulations, compliance,
TITLE:	Disinfectant Type and Levels in Distribution Systems
AUTHORS:	Gagnon, Graham; Skinner, Dawne
YEAR:	2002
SOURCE:	2002 Distribution and Plant Operations Conference Proceedings; AWWA
ABSTRACT?:	Y
FULL ARTICLE?:	N
COMMENTS:	Provides an overview of the secondary disinfectants chlorine, chloramine, and chlorine dioxide. Also evaluates the physical and chemical factors that affect the management of disinfectant residual.
ID:	**25**
DISINFECTANT:	Chloramines
KEY WORDS:	Water treatment, DBPs, research, chlorination, GAC, BAC, alternative disinfectants, biofilm
TITLE:	Microbial Aspects of Water Treatment Processes: A Progress Report
AUTHORS:	Geldreich, E.; Lykins, B.; Clark, R.
YEAR:	1990
SOURCE:	1990 Proceedings of AWWA Seminar on Current Research Activities in Support of USEPA's Regulatory Agenda
ABSTRACT?:	Y
FULL ARTICLE?:	N
COMMENTS:	The paper focuses on the microbial quality of treated and distribution water. The investigation looks at studies that have been performed on alternative disinfectants.
ID:	**26**
DISINFECTANT:	Chloramines
KEY WORDS:	T&O, Oxidants, Ozone
TITLE:	Evaluating Oxidants for the Removal of Model Taste and Odor Coumpounds from a Municipal Water Supply
AUTHORS:	Glaze, W.; Schep, R.; Chauncey, W.; Rugh, E.; Zarnoch, J.; Aieta, E.; Tate, C.; McGuire, M.

YEAR:	1990
SOURCE:	JAWWA: (82) 5: 79-84
ABSTRACT?:	Y
FULL ARTICLE?:	Y
COMMENTS:	Nine oxidizing agents were evaluated for removing six T&O compounds spiked into Colorado River water.
ID:	**27**
DISINFECTANT:	Chloramines
KEY WORDS:	
TITLE:	Managing Conversion to Chloramines
AUTHORS:	Harms, L.; Owen, C.
YEAR:	2003
SOURCE:	2003 Water Quality Technology Conference Proceedings: AWWA
ABSTRACT?:	N
FULL ARTICLE?:	Y
COMMENTS:	Tampa Bay Water Authority conducted 13 site visits to better understand experiences of other utilities that have converted to chloramines.
ID:	**28**
DISINFECTANT:	Chloramines
KEY WORDS:	Reservoirs, chloramines, algae, chlorine, chemicals, LA CA
TITLE:	Converting a 117 MG Open Finished Water Reservoir to Chloramine
AUTHORS:	Hirai, S.; Karimi, A; Male, M.; Orzechowski, W.; Trussell, R.; Wetstein, D.
YEAR:	2002
SOURCE:	2002 Water Quality Technology Conference Proceedings: AWWA
ABSTRACT?:	Y
FULL ARTICLE?:	Y
COMMENTS:	This study evaluated the ability of chloramines to control algae growth in the reservoir.
ID:	**30**
DISINFECTANT:	Chloramines
KEY WORDS:	DBPs, chlorine dioxide, giardia, cryptosporidium, CT values, chlorine, chloramines, pH, temperature, dosage, disinfection residuals
TITLE:	Chlorine Dioxide-Related By-Products When Disinfecting to Giardia and Cryptosporidium CT Criteria
AUTHORS:	Hofmann, R.; Ye, Q.; Andrews, R.
YEAR:	1997
SOURCE:	1997 Annual Conference Proceedings: AWWA
ABSTRACT?:	Y
FULL ARTICLE?:	N
COMMENTS:	The primary objective of this research was to investigate the required chlorine dioxide dosages and DBP formation when disinfecting to different levels of Giardia and Cryptosporidium control using chlorine dioxide as a primary disinfectant and when using chlorine or monochloramine as a secondary disinfectant.
ID:	**31**
DISINFECTANT:	Chloramines
KEY WORDS:	Amino acids, aldehydes, chloramines, chlorine, T&O
TITLE:	Odorous Aldehydes Produced by Disinfectant Reaction With Common Amino Acids
AUTHORS:	Hrudey, S.; Gac, A.; Daignault, S.
YEAR:	1988
SOURCE:	1988 Annual Conference Proceedings: AWWA
ABSTRACT?:	Y
FULL ARTICLE?:	N
COMMENTS:	This research suggests that chlorine or chloramine-mediated conversion of four specific amino acids into aldehydes possessing low threshold odor levels will occur under come conditions.

ID:	**32**
DISINFECTANT:	Chloramines
KEY WORDS:	Chlorine, Natural Organic Matter, Trihalomethanes, Disinfection Byproducts, chloramines, halogenated organics, ozonation
TITLE:	DBP Yields of Polar NOM Fractions From Low Humic Waters
AUTHORS:	Hwang, Cordelia; Sclimenti, Michael; Bruchet, Auguste; Croue, Jean-Philippe; Amy, Gary
YEAR:	2001
SOURCE:	2001 Water Quality Technology Conference Proceedings: AWWA
ABSTRACT?:	Y
FULL ARTICLE?:	Y
COMMENTS:	In order to better understand the source of common halogenated DBPs, the NOM from four low humic water was evaluated for HAAs and THMs after conventional treatment and ozonation/no ozonation and chloramination. Gives treatment applications.
ID:	**33**
DISINFECTANT:	Chloramines
KEY WORDS:	Disinfection, chloramines, distribution system, chlorine, planning
TITLE:	Decision Process on Whether to Switch from Free Chlorine to Chloramines
AUTHORS:	Kirmeyer, G.; LeChevallier, M.; Baribeau, H.
YEAR:	2002
SOURCE:	2002 Water Quality Technology Conference Proceedings: AWWA
ABSTRACT?:	Y
FULL ARTICLE?:	Y
COMMENTS:	This paper outlines 10 logical steps to be followed in making the decision on whether to switch from free chlorine to chloramines for final disinfection.
ID:	**34**
DISINFECTANT:	Chloramines
KEY WORDS:	
TITLE:	Optimizing Chloramine Treatment: Second Edition
AUTHORS:	Kirmeyer, GJ; Martel, K; Thompson, G; Radder, L; Klement, W.; LeChevallier, M.; Baribeau, H.; Flores, A.
YEAR:	2003 (DRAFT)
SOURCE:	2003 AWWARF Manual (currently unpublished)
ABSTRACT?:	N
FULL ARTICLE?:	N
COMMENTS:	This report covers DBP formation, nitrification, corrosivity and materials deterioration, T&O, maintaining residual, and blending.
ID:	**35**
DISINFECTANT:	Chloramines
KEY WORDS:	Chlorination, water treatment, trihalomethanes, alternative disinfectants, Netherlands
TITLE:	Chlorination By-Products: Production and Control
AUTHORS:	KIWA
YEAR:	1986
SOURCE:	1986: KIWA (available through AWWA)
ABSTRACT?:	Y
FULL ARTICLE?:	N
COMMENTS:	Describes an analytical investigation of the presence of THMSs in the drinking water of 14 treatment plants in the Netherlands. Discusses use of alternative disinfectants.
ID:	**36**
DISINFECTANT:	Chloramines
KEY WORDS:	Chloramination, chlorine, chloramines, trihalomethanes, bromide, regulations, utilities, giardia, haloacetic acids, chlorination, DBPs
TITLE:	DBP Formation Resulting from Short-Term Contact with Chlorine, Followed by Long-Term Contact with Chloramines

AUTHORS:	Krasner, S.; Mofidi, A.; Liang, S.
YEAR:	2003
SOURCE:	2003 Annual Conference Proceedings: AWWA
ABSTRACT?:	Y
FULL ARTICLE?:	Y
COMMENTS:	This paper describes THM formation in bench-scale tests with short free-chlorine contact times and post-chloramination.
ID:	**37**
DISINFECTANT:	Chloramines
KEY WORDS:	
TITLE:	DBP Formation From Chlorine Dioxide/Chlorine/Chloramine Disinfection
AUTHORS:	Krasner, S.; Yates, R.; Liang, S.
YEAR:	2003
SOURCE:	2003 Water Quality Technology Conference Proceedings: AWWA
ABSTRACT?:	N
FULL ARTICLE?:	Y
COMMENTS:	The purpose of this paper was to study the formation of DBPs from pre-disinfection with chlorine dioxide, intermediate chlorination, and post-chloramination.
ID:	**38**
DISINFECTANT:	Chloramines
KEY WORDS:	Pipe material, biofilm biomass, quantification, drinking water EPA
TITLE:	Effect of Distribution System Materials and Water Quality on Heterotrophic Plate Counts and Biofilm Proliferation
AUTHORS:	Le Puil, M.; Young, C.; Randall, A.; Taylor, J.
YEAR:	2003
SOURCE:	2003 Annual Conference Proceedings: AWWA
ABSTRACT?:	N
FULL ARTICLE?:	Y
COMMENTS:	This study used chloramines as a disinfectant in a model distribution system.
ID:	**39**
DISINFECTANT:	Chloramines
KEY WORDS:	Distribution systems, biofilm, disinfection, chlorine, pipes, corrosion, ozone, chlorine dioxide, residual chlorine
TITLE:	Disinfecting Biofilms in a Model Distribution System
AUTHORS:	LeChevallier, M.; Lowry, C.; Lee, R.
YEAR:	1990
SOURCE:	JAWWA: 82 (7): 87-99
ABSTRACT?:	Y
FULL ARTICLE?:	N
COMMENTS:	This study examined the disinfection efficiency of free chlorine and monochloramine for controlling biofilm organisms in a model pipe system.
ID:	**40**
DISINFECTANT:	Chloramines
KEY WORDS:	Iron, pipes, corrosion, disinfection, biofilm, water mains, chloride, sulfates, chloramines
TITLE:	Examining the Relationship Between Iron Corrosion and the Disinfection of Biofilm Bacteria
AUTHORS:	LeChevallier, M.; Lowry, C.; Lee, R.; Gibbon, D.
YEAR:	1993
SOURCE:	JAWWA, 85(7)
ABSTRACT?:	Y
FULL ARTICLE?:	Y
COMMENTS:	The results of this study show that even low levels of corrosion could interfere with free chlorine disinfection. High corrosion rates affected monochloramine disinfection.
ID:	**41**

DISINFECTANT:	Chloramines
KEY WORDS:	Disinfection byproducts, distribution systems, treatment plants, chlorine, chloramines, ozone, trihalomethanes, weather, chlorinated organics
TITLE:	Monthly Variation of Disinfection By-Products in Water From Treatment Plants Using Three Different Disinfection Processes
AUTHORS:	Lebel, G.; Benoit, F.; Williams, D.
YEAR:	1994
SOURCE:	Planning for Tomorrow: Proceedings of the Sixth National Conference on Drinking Water: AWWA
ABSTRACT?:	Y
FULL ARTICLE?:	N
COMMENTS:	This study assesses the temporal and spatial variation in DBP levels over a one-year period for three facilities each of which utilizes chlorine- chloramine, chlorine-chlorine, ozone-chlorine.
ID:	**42**
DISINFECTANT:	Chloramines
KEY WORDS:	
TITLE:	Comparison of Treatments Efficiency for Legionella Control in Domestic Water Systems
AUTHORS:	Loret, J.; Robert, S.; Thomas, V.; Levi, Y.; Cooper, A.; Myers, E.
YEAR:	2003
SOURCE:	2003 Water Quality Technology Conference Proceedings
ABSTRACT?:	N
FULL ARTICLE?:	Y
COMMENTS:	In this study, CIRSEE developed a domestic water supply simulation unit to study the Legionella inactivation efficacy of different disinfectants.
ID:	**43**
DISINFECTANT:	Chloramines
KEY WORDS:	Potable water, disinfectants, halogen organic compounds, activated carbon, filtration
TITLE:	Alternative Disinfectants for Drinking Water Treatment
AUTHORS:	Lykins, B.; Koffskey, W.; Patterson, K.
YEAR:	1994
SOURCE:	Journal of Environmental Engineering: 120 (4): 754-758
ABSTRACT?:	Y
FULL ARTICLE?:	N
COMMENTS:	This paper discusses a one-year study in Louisiana of the chemical, microbiological, and mutagenic effects of using alternative drinking water disinfectants.
ID:	**44**
DISINFECTANT:	Chloramines
KEY WORDS:	Chloramines, disinfection, water quality monitroing, bacteria, THMs, residual chlorine, coliforms, Florida
TITLE:	Selected Effects on Distribution Water Quality as a Result of Conversion to Chloramines
AUTHORS:	MacLeod, B.; Zimmerman, J.
YEAR:	1986
SOURCE:	1986 Water Quality Technology Conference Proceedings: AWWA
ABSTRACT?:	Y
FULL ARTICLE?:	N
COMMENTS:	This paper discusses the effectiveness of conversion to chloramines in meeting a utility's water quality goals. Was presented three years after conversion.
ID:	**45**
DISINFECTANT:	Chloramines
KEY WORDS:	
TITLE:	Entry Point Treatment for Consecutive Distribution System Disinfectant By-Product Control
AUTHORS:	Mahon, J.

YEAR:	2003
SOURCE:	2003 Water Quality Technology Conference Proceedings: AWWA
ABSTRACT?:	N
FULL ARTICLE?:	Y A utility providing water to a consecutive distribution system studied impacts on DBPs of implementing point of entry chloramination.
COMMENTS:	
ID:	**46**
DISINFECTANT:	Chloramines
KEY WORDS:	Groundwater, disinfection, disinfection by-products, alternative disinfectants, UV radiation, UV disinfection, regulations, compliance
TITLE:	Evaluation of By-Products Produced by Treatment of Groundwater with Ultraviolet Irradiation
AUTHORS:	Malley, J.; Shaw, J.; Ropp, J.
YEAR:	1995
SOURCE:	AWWARF
ABSTRACT?:	Y
FULL ARTICLE?:	N
COMMENTS:	Performed UV irradiation on 20 groundwater and 10surface water supplies. Looked at DBP formation and compared with chlorine and chloramines used as post disinfectants.
ID:	**47**
DISINFECTANT:	Chloramines
KEY WORDS:	Ozone, biological filters, granular activated carbon, trihalomethanes, organics, organic carbon, contactors, pilot pltans, chloramines, chlorine
TITLE:	Removal of THM and HAA Precursors Using Ozone and Biologically Active GAC
AUTHORS:	McClellan, J.; Reckhow, D.; Tobiason, J.
YEAR:	1996
SOURCE:	1996 Annual Conference Proceedings: AWWA
ABSTRACT?:	Y
FULL ARTICLE?:	N
COMMENTS:	This study includes results from a long-term pilot study of a post-filtration ozone/biologically active GAC process. Compares DBP formation chloramines disinfection with free chlorine only.
ID:	**48**
DISINFECTANT:	Chloramines
KEY WORDS:	Disinfection, trihalomethanes, chlorine, costs, chlorination, chloramination, chlorine dioxide, Philadelphia
TITLE:	The Evolution of a Modified Disinfection Strategy to Reduce Trihalomethane Formation
AUTHORS:	McKeon, W.; Muldowney, J.; Aptowica, B.
YEAR:	1986
SOURCE:	1986 Annual Conference Proceedings: AWWA
ABSTRACT?:	Y
FULL ARTICLE?:	N
COMMENTS:	This paper reviews the experiences at Philadelphia's Baxter Plant over the past 10 years in modifying chlorination practices and problems, including chloramination of finished water and utilization of chlorine dioxide at the raw water basin.
ID:	**49**
DISINFECTANT:	Chloramines
KEY WORDS:	Distribution systems, chlorination, chloramination, ozonation, water quality, public information, trihalomethanes, color, bacteria, taste and odor, treatment plants
TITLE:	Observations of Reconditioning a Chlorinated Water Distribution System to Chloramines
AUTHORS:	McMorran, Carl; Moriarty, Kathy
YEAR:	1997
SOURCE:	1997 Annual Conference Proceedings: AWWA
ABSTRACT?:	Y

FULL ARTICLE?:	N
COMMENTS:	Discusses the conversion of Bangor ME Water District supply from an unfiltered, chlorinated supply to an unfiltered, ozonated, chloraminated system.
ID:	**50**
DISINFECTANT:	Chloramines
KEY WORDS:	Chlorine, Ontario CA, Chloramines, Regulations, iron, direct filutration, performance, disinfection, water supply, surface water, analysis, blending, lakes, costs, water quality, distribution systems, ammonia, cast iron pipe, color, water mains, residual chlorine
TITLE:	Success! Switching from Chloramines to Chlorine in the Distribution System: Results From a One-Year Full-Scale Trial
AUTHORS:	Muylwyk, Q.; MacDonald, J.; Klawunn, M.
YEAR:	2001
SOURCE:	2001 Proceedings Water Quality Technology Conference: AWWA
ABSTRACT?:	Y
FULL ARTICLE?:	Y
COMMENTS:	The system used chloramines for 10 years and then converted to chlorine because of the need to blend with a chlorinated drinking water. Maybe this would provide some background/comparison of water quality conditions.
ID:	**51**
DISINFECTANT:	Chloramines
KEY WORDS:	Chloramination, California, chloramines, trihalomethanes, chlorine, regulations, testing, haloacetic acids, organic carbon, DBPs, coagulation, distribution systems
TITLE:	Formation of Nitrosamines as By-Products of Chloramination
AUTHORS:	Najm, I.; Ma, J.
YEAR:	2002
SOURCE:	2002 Annual Conference Proceedings: AWWA
ABSTRACT?:	Y
FULL ARTICLE?:	Y
COMMENTS:	This paper studies NDMA and other nitrosamines formation for Elsinore Valley Municipal District, which is considering a conversion from free chlorine to chloramine.
ID:	**52**
DISINFECTANT:	Chloramines
KEY WORDS:	Distribution systems, water quality, corrosion control, chloramination, coliforms, bacteria, disinfection by-products
TITLE:	Implementation of Chloramination and Corrosion Control to Limit Microbial Activity in the Distribution System
AUTHORS:	Norton, Cheryl; LeChevallier, Mark; Boswell, Malcolm; Hanchak, John; Robinson, Jeff
YEAR:	1995
SOURCE:	1995 Water Quality Technology Conference: AWWA
ABSTRACT?:	Y
FULL ARTICLE?:	Y
COMMENTS:	The water quality characteristics examined included coliforms, heterotrophic plate count bacteria, disinfectant residuals, free ammonia, pH, temperature, disinfection by-products, assimilable organic carbon, and corrosion rates.
ID:	**53**
DISINFECTANT:	Chloramines
KEY WORDS:	Distribution systems, chlorine, chloramines, Information Collection Rule, haloacetic acids, trihalomethanes, DBPs
TITLE:	Information Collection Rule Data Analysis, Chapter 7 - Distribution System DBP Results and SDS Performance
AUTHORS:	Obolensky, A.; Frey, M.
YEAR:	2003
SOURCE:	AWWARF
ABSTRACT?:	Y

FULL ARTICLE?:	N
COMMENTS:	This chapter presents analyses of DBP data collected under the ICR. The analyses focus on differences between DBP concentrations in finished water and in distribution systems and on variations in DBP concentrations across individual distribution systems.
ID:	**54**
DISINFECTANT:	Chloramines
KEY WORDS:	Chloramines, chlorine, disinfection, Tampa FL, Public information, permits, planning
TITLE:	Overcoming Problems Associated with Switching from Free Chlorine to Chloramine Disinfection
AUTHORS:	Owen, C.
YEAR:	2002
SOURCE:	2002 Distribution and Plant Operations Conference Proceedings; AWWA
ABSTRACT?:	Y
FULL ARTICLE?:	N
COMMENTS:	The 3-year process of consensus building, educating the public, coordinating with consecutive systems and obtaining permits for chloramines conversion is outlined in this paper.
ID:	**55**
DISINFECTANT:	Chloramines
KEY WORDS:	Disinfection, disinfection by-products, chloramines, taste and odor, biofilm, dosage, optimization, chemical handling and feeding, pH, mixing, heterotrophic plate count
TITLE:	Chloramine Optimization Studies
AUTHORS:	Phillipee, D; Price, S.; Zafer, M.
YEAR:	1992
SOURCE:	1992 Proceedings of the Water Quality Technology Conference - Pt 1; AWWA
ABSTRACT?:	Y
FULL ARTICLE?:	Y
COMMENTS:	This paper presents data from chloramine optimization studies conducted on a surface water source.
ID:	**56**
DISINFECTANT:	Chloramines
KEY WORDS:	Disinfection, oxidation, iron, manganese, volatile organic chemicals, disinfection byproducts, taste and odor, color removal, ozone, chlorine, chloramines, chlorine dioxide
TITLE:	Disinfection Alternatives
AUTHORS:	Reckhow, D.
YEAR:	1996
SOURCE:	1996 Annual Conference Proceedings: AWWA
ABSTRACT?:	Y
FULL ARTICLE?:	N
COMMENTS:	The purpose of this paper is to highlight some of the important characteristics of the available oxidants and disinfectants.
ID:	**57**
DISINFECTANT:	Chloramines
KEY WORDS:	Chloramines, disinfection and disinfectants, water supply engineering US, corrosion and anticorrosives, water purification chlorinations
TITLE:	Chloramine Effects on Distribution System Materials
AUTHORS:	Reiber, S.
YEAR:	1993
SOURCE:	AwwaRF
ABSTRACT?:	N
FULL ARTICLE?:	N
COMMENTS:	
ID:	**58**
DISINFECTANT:	Chloramines

KEY WORDS:	
TITLE:	Disinfectant Efficacy in Distribution Systems: A Field Scale Evaluation
AUTHORS:	Rutledge, S.; Gagnon, G.; Baribeau, H.; Boulos, L.; Chauret, C.; Andrews, S.; Dumancic, R.
YEAR:	2003
SOURCE:	2003 Water Quality Technology Conference Proceedings: AWWA
ABSTRACT?:	N
FULL ARTICLE?:	Y
COMMENTS:	The purpose of this study is to operate a pipe loop system for a minimum of six months to correlate the interaction of various factors, such as pipe material and water quality, that affect the efficacy of secondary disinfection. The study included use of a chloraminated supply.
ID:	**59**
DISINFECTANT:	Chloramines
KEY WORDS:	Sampling, regulations, distribution systems, disinfection residuals, chloramines, water quality, monitoring
TITLE:	Modification of Total Coliform Sampling Locations in Response to Conversion to Chloramines
AUTHORS:	Sadowski, S.; Hiltebrand, D.; Antoun, E.
YEAR:	1996
SOURCE:	1996 Annual Conference Proceedings: AWWA
ABSTRACT?:	Y
FULL ARTICLE?:	N
COMMENTS:	This paper discusses current sampling regulations and presents Virginia Beach's criteria for the selection of sampling site locations.
ID:	**60**
DISINFECTANT:	Chloramines
KEY WORDS:	Ozone, surface water, organics, color, THMs, chloramines, chlorine, disinfection, residual chlorine, turbidity
TITLE:	Ozonation at Belle Glade, Florida: A Case History
AUTHORS:	Singer, P.; Robinson, K.; Elefritz, R.
YEAR:	1990
SOURCE:	Ozone: Science & Engineering: 12 (2): 199-215: Lewis Publishers
ABSTRACT?:	N
FULL ARTICLE?:	N
COMMENTS:	The authors describe experiences of implementing ozonation and chloramination. Implementation occurred in 1984 and 1987, respectively. This appears to provide some long-term water quality information.
ID:	**61**
DISINFECTANT:	Chloramines
KEY WORDS:	Disinfection Byproducts, history, chemistry, modeling, natural organic matter, epidemiology, toxicity, risks, regulations, chloramines, ozone, chlorine dioxide, ultraviolet disinfection, coagulation, softening, adsoprtion, oxidation, membranes, organic carbon, costs, Europe
TITLE:	Formation and Control of Disinfection By-Products in Drinking Water
AUTHORS:	Singer, Philip
YEAR:	1999
SOURCE:	American Water Works Association
ABSTRACT?:	Y
FULL ARTICLE?:	N
COMMENTS:	This book includes discussion of alternative disinfectants on DBP formation.
ID:	**62**
DISINFECTANT:	Chloramines
KEY WORDS:	Nitrification, pH, utilities, Ann Arbor MI, chloramines, chlorine, regulations, water

	quality, chloramination, chlorination, DBPs
TITLE:	Effectiveness of High pH in Controlling Nitrification
AUTHORS:	Skadsen, J.
YEAR:	2002
SOURCE:	JAWWA: (94) 7: 73-83.
ABSTRACT?:	Y
FULL ARTICLE?:	Y
COMMENTS:	This study focused on using pH to control nitrification. I think DBPs are mentioned because this prompted an investigation into chloramines. This study was also selected because Ann Arbor uses chloramination.
ID:	**63**
DISINFECTANT:	Chloramines
KEY WORDS:	Distribution systems, water quality, residual chlorine, chloramines, nitrification, sampling, flushing, ammonia
TITLE:	Excessive Loss of Chloramine Residual: A Case Study
AUTHORS:	Smith, C.; Smith, J.; Milosky, B.
YEAR:	1996
SOURCE:	1996 Annual Conference Proceedings: AWWA: Water Quality
ABSTRACT?:	Y
FULL ARTICLE?:	N
COMMENTS:	This report describes the evaluation of water quality and provides recommendation. The utility operates a booster chlorination facility and uses chloramines for disinfection.
ID:	**64**
DISINFECTANT:	Chloramines
KEY WORDS:	Ozonation, coagulation, halogens, DBPs, chlorination, chloramines
TITLE:	The Impact of Ozonation and Coagulation on Disinfection By-Product Formation
AUTHORS:	Smith, M.; Cowman, G.; Singer, P.
YEAR:	1993
SOURCE:	1993 Annual Conference Proceedings: AWWA
ABSTRACT?:	Y
FULL ARTICLE?:	Y
COMMENTS:	The objective of this research was to evaluate the impact of ozonation on the formation of halogenated DBPs produced by subsequent chlorination and chloramination with and without coagulation.
ID:	**65**
DISINFECTANT:	Chloramines
KEY WORDS:	Ultraviolet Disinfection, dosage, groundwater, water quality, chloramines, chlorine, pilot plants, viruses, copper pipe, PVC pipe, cast iron pipe, performance, fouling, costs, biofilm
TITLE:	Evaluation of Ultraviolet Technology in Drinking Water Treatment
AUTHORS:	Snicer, G.; Malley, J.; Margolin, A.; Hogan, S.
YEAR:	1996
SOURCE:	1996 Water Quality Technology Conference: AWWA
ABSTRACT?:	Y
FULL ARTICLE?:	N
COMMENTS:	The major objectives of the study were to: evaluate the long-term operational parameters and problems associated with UV disinfection; examine the effects of groundwater quality on UV effectiveness; evaluate chlorine and chloramines as possible secondary disinfectants and determine their effectiveness on copper, iron, and PVC pipe.
ID:	**66**
DISINFECTANT:	Chloramines
KEY WORDS:	Lead, nitrification, disinfection, corrosion control, pH, alkalinity, copper, chloramines, iron, carbon dioxide, calcium carbonate
TITLE:	Corrosion Control and Chloramination, Discolored Water and Nitrification
AUTHORS:	Sung, Windsor

YEAR:	2002
SOURCE:	2002 Water Quality Technology Conference Proceedings: AWWA
ABSTRACT?:	Y
FULL ARTICLE?:	Y
COMMENTS:	This paper focuses on the interactions between the effects of increased buffering capacity on iron release and the interaction of iron release with chloramine and pH stability, as well as chloramine stability and nitrification.
ID:	**67**
DISINFECTANT:	Chloramines
KEY WORDS:	Chlorine dioxide, residual chlorine, distribution systems
TITLE:	Chlorite and Chlorate Residuals in the Distribution System
AUTHORS:	Thompson, A.; Matthews, N.; Mittl, A.; Owen, D.
YEAR:	1988
SOURCE:	1988 Proceedings Water Quality Technology Conference; Advances in Water Analysis and Treatment: AWWA
ABSTRACT?:	Y
FULL ARTICLE?:	Y
COMMENTS:	This paper addresses the formation of chlorite and chlorate in the treatment process and distribution system after addition of chlorine dioxide as a pre-oxidant and co-disinfectant with chloramines.
ID:	**68**
DISINFECTANT:	Chloramines
KEY WORDS:	Chloramines, distribution systems, chlorine, disinfection byproducts, pipes, corrosion
TITLE:	Chloramine Decay in Drinking Water Distribtuion Systems
AUTHORS:	Valentine, R., Ozekin, K.; Vikesland, P.
YEAR:	1998
SOURCE:	1998 Symposium Proceedings; Protecting Water Quality in the Distribution System: What is the Role of disinfection Residuals?
ABSTRACT?:	Y
FULL ARTICLE?:	N
COMMENTS:	This paper was selected because it provides info on interactions between chloramines and pipe materials and the formation of DBPs.
ID:	**69**
DISINFECTANT:	Chloramines
KEY WORDS:	DBP, distribution systems, chlorine, chloramines, ductile iron pipe, organic carbon, inorganic carbon, reactors, hydrodynamics
TITLE:	Characterization of Disinfectant Decay and DBP Formation in the Presence of Water Distribution System Deposits
AUTHORS:	Valentine, R.; Videsland, P.; Angerman, B.; Hackett, S.
YEAR:	1999
SOURCE:	1999 Proceedings of the Water Quality Technology Conference: AWWA
ABSTRACT?:	Y
FULL ARTICLE?:	Y
COMMENTS:	This paper presents a study on the consumption of free chlorine and chloramines and formation of DBPs in the presence of pipe deposits collected from cities with ductile iron pipe distribution systems.
ID:	**70**
DISINFECTANT:	Chloramines
KEY WORDS:	Modeling, iron, distribution systems, chlorine, kinetics, DBPs, flow, pH, Iowa, oxidation, nitrogen, hydrodynamics, chloramines, trihalomethanes, disinfection residuals
TITLE:	The Role of Pipe-Water Interface in DBP Formation and Disinfectant Loss
AUTHORS:	Valentine, R.; Videsland, P.; Angerman, B.; Hackett, S.
YEAR:	2000
SOURCE:	AwwaRF Report

ABSTRACT?:	Y
FULL ARTICLE?:	N
COMMENTS:	The overall objective of this project was to study the processes and reactions, which account for disinfectant loss and DBP formation.
ID:	**71**
DISINFECTANT:	Chloramines
KEY WORDS:	T & O, oxidation, chlorine dioxide, chloramines, PAC, Alberta CA, North America
TITLE:	Oxidative Control of Taste and Odour in Drinking Water: The Edmonton Experience
AUTHORS:	Walker, G.; Lee, F.
YEAR:	1983
SOURCE:	35th Annual Convention, Western Canada Water and Sewage Conference
ABSTRACT?:	Y
FULL ARTICLE?:	N
COMMENTS:	This paper can be obtained through AWWA.
ID:	**72**
DISINFECTANT:	Chloramines
KEY WORDS:	Chlorine, choramination, nutrients, reactors, biofilm, disinfection, ozone, biofilters, statistical analysis
TITLE:	The Effects of Ozone and Chloramination on Distribution System Regrowth
AUTHORS:	West, T.; Daniel, P.; Labonte, J.; DeGraca, A.; Boozarpour, M.; Chemlick, D.; Hou, S.
YEAR:	1998
SOURCE:	1998 Annual Conference Proceedings
ABSTRACT?:	Y
FULL ARTICLE?:	Y
COMMENTS:	This paper discusses a study in which the primary objectives were to answer two key questions: how will a change from free chlorine to chloramine influence bacterial regrowth; and will chloramine control increase regrowth potential associated with an increase in nutrient levels due to ozonation of an unfiltered water?
ID:	**73**
DISINFECTANT:	Chloramines
KEY WORDS:	Polymers, distribution systems, treatment plants, ammonia, backwash, filters, monitoring, nitrification, chloride
TITLE:	Minization of N-Nitrosodimethylamine (NDMA) in EBMUD Distribution Systems
AUTHORS:	Wilczak, A.; Assadi-Rad, A.; Lai, H.; Smith, J.; Hoover, L; Lazzelle, L.;
YEAR:	2003
SOURCE:	2003 Annual Conference Proceedings: AWWA
ABSTRACT?:	Y
FULL ARTICLE?:	Y
COMMENTS:	This paper investigates the formation of NDMA in EMBUD's distribution system, which has a chloramine residual.
ID:	**74**
DISINFECTANT:	Chloramines
KEY WORDS:	Oxidation, organic carbon, biological filters, chloramines, disinfection byproducts, decomposition, testing, ozone, hydrogen peroxide, chlorine dioxide ultrafiltration
TITLE:	Investigations of Oxidation, TOC Removal and Biological Filtration Effects on Chloramine Stability and DBP Formation at EBMUD CA
AUTHORS:	Wilczak, A.; Hoover, l.; Song, D.; Lai, H.; Kachur, S.; Long, C.
YEAR:	1999
SOURCE:	1999 Water Quality Technology Conference
ABSTRACT?:	Y
FULL ARTICLE?:	Y
COMMENTS:	This paper discusses EBMUD's first year of chloramines conversion.
ID:	**76**
DISINFECTANT:	Chloramines

KEY WORDS:	
TITLE:	Strategies for Combining Free Ammonia and Boosting Chloramines in Distribution Systems - Survey of Utility Practice
AUTHORS:	Wilczak, A.; Smith, C.; Cohen, Y.; Martin, P.
YEAR:	2003
SOURCE:	2003 Water Quality Technology Conference Proceedings: AWWA
ABSTRACT?:	N
FULL ARTICLE?:	Y
COMMENTS:	This survey was conducted to obtain data from actual full-scale chloramine boosting applications, review process control and evaluate the success of these applications.
ID:	**77**
DISINFECTANT:	Chloramines
KEY WORDS:	Chloramines, Distribution systems, chloramination, nitrification
TITLE:	Chloramine Decay Rate: Factors and Research Needs
AUTHORS:	Wilczak, Andrzej
YEAR:	2001
SOURCE:	2001 Annual Conference Proceedings; AWWA
ABSTRACT?:	Y
FULL ARTICLE?:	Y
COMMENTS:	This paper attempts to summarize the published factors impacting chloramine decay rate with some practical experiences.
ID:	**78**
DISINFECTANT:	Chloramines
KEY WORDS:	DBPs, chlorination, chloramines, ozone, weather, treatment plants, distribution systems, THMs, surveys, chlorinated organics
TITLE:	A Survey of Disinfection By-Products in Canadian Drinking Water
AUTHORS:	Williams, D.; Lebel, G.; Benoit, F.
YEAR:	1994
SOURCE:	Planning for Tomorrow: Proceedings of the Sixth National Conference on Drinking Water: AWWA
ABSTRACT?:	Y
FULL ARTICLE?:	N
COMMENTS:	This study was designed to determine the concentrations of halogenated DBPs in Canadian drinking water supplies where chlorine was used at some stage in the treatment process.
ID:	**79**
DISINFECTANT:	Chloramines
KEY WORDS:	temperature, monitoring, disinfectant residuals, distribution systems, organic carbon, residual chlorine, heterotrophic plate count, microbiological analysies, water quality, bacteria, chloramines, St. Paul Minnesota
TITLE:	An Evaluation of Biostability and Microbiological Quality in a Full-Scale Drinking Water Distribution System
AUTHORS:	Zhang, M.; Hozalski, R.; Semmens, M.; Schuler, D.
YEAR:	2000
SOURCE:	2000 Water Quality Technology Conference Proceedings
ABSTRACT?:	Y
FULL ARTICLE?:	Y
COMMENTS:	The biostability and microbiological quality in the full-scale chloraminated drinking water distribution system of St. Paul Minnesota were evaluated.
ID:	**81**
DISINFECTANT:	Chlorine Dioxide
KEY WORDS:	Odor, chlorine dioxide, organics, analysis, laboratories, disinfection, aldehydes
TITLE:	Investigation into Carpet Odors With ClO2
AUTHORS:	Dluzniewski, T.

YEAR:	2001
SOURCE:	Proceedings of the Fourth International Symposium: Chlorine Dioxide: The State of Science, Regulatory, Environmental Issues, and Case Histories
ABSTRACT?:	Y
FULL ARTICLE?:	N
COMMENTS:	This paper describes how some customers experienced disagreeable odors after having new carpet installed in their homes and when they ran water in their showers or sinks.
ID:	**83**
DISINFECTANT:	Chlorine Dioxide
KEY WORDS:	Oxidation, organic carbon, biological filters, chloramines, disinfection byproducts, decomposition, testing, ozone, hydrogen peroxide, chlorine dioxide ultrafiltration
TITLE:	Investigations of Oxidation, TOC Removal and Biological Filtration Effects on Chloramine Stability and DBP Formation at EBMUD CA
AUTHORS:	Wilczak, A.; Hoover, l.; Song, D.; Lai, H.; Kachur, S.; Long, C.
YEAR:	1999
SOURCE:	1999 Water Quality Technology Conference
ABSTRACT?:	Y
FULL ARTICLE?:	Y
COMMENTS:	This paper discusses EBMUD's first year of chloramines conversion. Included in chlorine dioxide list because it is one of the keywords.
ID:	**84**
DISINFECTANT:	Chlorine Dioxide
KEY WORDS:	Treatment plants, dosage, organics, odor, chlorination, utilities, chlorine, distribution systems, GAC, contactors, Virginia
TITLE:	ClO2 and By-Product Persistence in a Drinking Water System
AUTHORS:	Hoehn R.; Ellenberger, C.; Gallagher, D.; Wiseman, E.; Benninger, R.; Rosenblatt, A.
YEAR:	2003
SOURCE:	JAWWA: 95(4): 141-150: AWWA
ABSTRACT?:	Y
FULL ARTICLE?:	N
COMMENTS:	This study, conducted at the Spring Hollow Water Treatment Facility in Virginia was the first full-scale evaluation of a chlorine gas-solid sodium chlorite generation system.
ID:	**85**
DISINFECTANT:	Chlorine Dioxide
KEY WORDS:	Chlorine dioxide, residual chlorine, distribution systems
TITLE:	Chlorite and Chlorate Residuals in the Distribution System
AUTHORS:	Thompson, A.; Matthews, N.; Mittl, A.; Owen, D.
YEAR:	1988
SOURCE:	1988 Proceedings Water Quality Technology Conference; Advances in Water Analysis and Treatment: AWWA
ABSTRACT?:	Y
FULL ARTICLE?:	Y
COMMENTS:	This paper addresses the formation of chlorite and chlorate in the treatment process and distribution system after addition of chlorine dioxide as a pre-oxidant and co-disinfectant with chloramines.
ID:	**86**
DISINFECTANT:	Chlorine Dioxide
KEY WORDS:	Chlorine dioxide, wastewater treatment, disinfection, regulations, zebra mussels, corrosion, Europe, drinking water, Netherlands, Germany, Italy, France
TITLE:	Chlorine Dioxide: A European Perspective
AUTHORS:	Dernat, M.; Leburgue De Oliveira, A.; Gautier, J.; Pouillet, M.
YEAR:	1995
SOURCE:	Proceedings Chlorine Dioxide: Drinking Water, Process Water, and Wastewater Issues: Third International Symposium

ABSTRACT?:	Y
FULL ARTICLE?:	N
COMMENTS:	This paper presents the regulatory aspects and the consequences for the treatment of water destined for human consumption, the increase of use in industrial water treatment, and also discusses two case studies of wastewater disinfection.
ID:	**87**
DISINFECTANT:	Chlorine Dioxide
KEY WORDS:	Chlorine dioxide, corrosion, chlorine, distribution system, trihalomethanes, disinfection, pipes, modeling, iron, water quality, red water, oxidants, chloramines
TITLE:	Impact of Chlorine Dioxide on Corrosion in a Model Distribution System
AUTHORS:	Eisnor, J.D.; Gagnon, G.A.; Volk, C.; Chauret, C.P.; Andrews, R.C.
YEAR:	2001
SOURCE:	2001 Proceedings Water Quality Technology Conference: AWWA
ABSTRACT?:	Y
FULL ARTICLE?:	Y
COMMENTS:	The objective of this study is to study the impacts of free chlorine, monochloramine, chlorine dioxide and chlorite on internal corrosion and water quality in a pilot-scale pipe loop system.
ID:	**88**
DISINFECTANT:	Chlorine Dioxide
KEY WORDS:	Chlorination, water treatment, trihalomethanes, alternative disinfectants, Netherlands
TITLE:	Chlorination By-Products: Production and Control
AUTHORS:	KIWA
YEAR:	1986
SOURCE:	1986: KIWA (available through AWWA)
ABSTRACT?:	Y
FULL ARTICLE?:	N
COMMENTS:	Describes an analytical investigation of the presence of THMSs in the drinking water of 14 treatment plants in the Netherlands. Discusses use of alternative disinfectants.
ID:	**89**
DISINFECTANT:	Chlorine Dioxide
KEY WORDS:	Disinfection Byproducts, history, chemistry, modeling, natural organic matter, epidemiology, toxicity, risks, regulations, chloramines, ozone, chlorine dioxide, ultraviolet disinfection, coagulation, softening, adsoprtion, oxidation, membranes, organic carbon, costs, Europe
TITLE:	Formation and Control of Disinfection By-Products in Drinking Water
AUTHORS:	Singer, Philip (ed.); Hoehn, R.; Gates, D.
YEAR:	1999
SOURCE:	American Water Works Association
ABSTRACT?:	Y
FULL ARTICLE?:	N
COMMENTS:	This book includes discussion of alternative disinfectants on DBP formation.
ID:	**90**
DISINFECTANT:	Chlorine Dioxide
KEY WORDS:	DBPs, chlorine dioxide, giardia, cryptosporidium, CT values, chlorine, chloramines, pH, temperature, dosage, disinfection residuals
TITLE:	Chlorine Dioxide-Related By-Products When Disinfecting to Giardia and Cryptosporidium CT Criteria
AUTHORS:	Hofmann, R.; Ye, Q.; Andrews, R.
YEAR:	1997
SOURCE:	1997 Annual Conference Proceedings: AWWA
ABSTRACT?:	Y
FULL ARTICLE?:	N
COMMENTS:	The primary objective of this research was to investigate the required chlorine dioxide

	dosages and DBP formation when disinfecting to different levels of Giardia and Cryptosporidium control using chlorine dioxide as a primary disinfectant and when using chlorine or monochloramine as a secondary disinfectant.
ID:	**91**
DISINFECTANT:	Chlorine Dioxide
KEY WORDS:	Turbidity, chlorine, filters, sedimentation, performance, optimization, trihalomethanes, chlorination, oxidants, DBPs
TITLE:	Optimizing Preoxidation for DBPs Control and Its Impact on Filter Performance
AUTHORS:	Arias, M.; Caufield, J.; Summers, R.
YEAR:	2002
SOURCE:	2002 Annual Conference Proceedings: AWWA
ABSTRACT?:	Y
FULL ARTICLE?:	Y
COMMENTS:	The objectives of this study are to: assess and evaluate the conditions under which optimizing preoxidation is effective in controlling DBPs and determine the conditions under which preoxidation impacts filter performance. Chlorine dioxide was one of the preoxidants used in the study.
ID:	**92**
DISINFECTANT:	Chlorine Dioxide
KEY WORDS:	Chlorine dioxide, bromide, Israel, chemicals chlorine, surface water, organics, chromotography, DBPs, gas chromotography
TITLE:	Chlorine Dioxide DBPs: Overview and Results of Recent Work
AUTHORS:	Richardson, S.; Thruston, A.; Rav-Acha, C.; Groisman, L.; Popilevsky, I.; Glezer, V.
YEAR:	2001
SOURCE:	Proceedings of the Fourth International Symposium: Chlorine Dioxide: The State of Science, Regulatory, Environmental Issues, and Case Histories
ABSTRACT?:	Y
FULL ARTICLE?:	N
COMMENTS:	An overview of studies performed on DBPs and chlorine dioxide.
ID:	**93**
DISINFECTANT:	Chlorine Dioxide
KEY WORDS:	Compliance, monitoring, chlorine dioxide, chemicals, risks, treatment plants, process control, testing, equipment, ion chromatography, oxidants, DBPs, distribution systems, regulations
TITLE:	Practical Suggestions for Meeting USEPA Compliance Monitoring Requirements and In-Plant Operational Control of Chlorine Dioxide in Drinking Water
AUTHORS:	Gates, D.
YEAR:	1999
SOURCE:	Ozone: Science & Engineering: (21) 5: 433-445.
ABSTRACT?:	Y
FULL ARTICLE?:	N
COMMENTS:	An overview of Current USEPA compliance objectives in the use and control of chlorine dioxide in drinking water treatment.
ID:	**94**
DISINFECTANT:	Chlorine Dioxide
KEY WORDS:	Reservoirs, LA CA, chlorine dioxide, DBPs, distribution systems, algae, trihalomethanes, costs, monitoring, sampling
TITLE:	Chlorine Dioxide for Algae and DBP control: The Los Angeles Experience
AUTHORS:	Stolarik, G.; Liu, P.
YEAR:	2000
SOURCE:	2000 Water Quality Technology Conference Proceedings: AWWA
ABSTRACT?:	Y
FULL ARTICLE?:	Y
COMMENTS:	Chlorine dioxide was evaluated for its potential to control algal growth in an uncovered

	distribution reservoir that provides finished water to a large service area in LA.
ID:	**95**
DISINFECTANT:	Chlorine Dioxide
KEY WORDS:	Oxidants, DBPs, chlorine, chlorine dioxide, halogenated organics, water quality, modeling
TITLE:	Evolution of Oxidants and Disinfection By-Products in Distribution Systems
AUTHORS:	Helene, B.; Prevost, M.; LaFrance, P.
YEAR:	1994
SOURCE:	1994 Annual Conference Proceedings: AWWA
ABSTRACT?:	Y
FULL ARTICLE?:	Y
COMMENTS:	Objectives of the study were to: monitor and model oxidant consumption in clear wells and distribution systems; monitor and model the formation of DBPs in clear wells and distribution systems; and distinguish the consumption of oxidants and the formation of DBPs caused by the water quality from that caused by the pipes.
ID:	**96**
DISINFECTANT:	Chlorine Dioxide
KEY WORDS:	Water treatment, DBPs, research, chlorination, GAC, BAC, alternative disinfectants, biofilm
TITLE:	Microbial Aspects of Water Treatment Processes: A Progress Report
AUTHORS:	Geldreich, E.; Lykins, B.; Clark, R.
YEAR:	1990
SOURCE:	1990 Proceedings of AWWA Seminar on Current Research Activities in Support of USEPA's Regulatory Agenda
ABSTRACT?:	Y
FULL ARTICLE?:	N
COMMENTS:	The paper focuses on the microbial quality of treated and distribution water. The investigation looks at studies that have been performed on alternative disinfectants.
ID:	**97**
DISINFECTANT:	Chlorine Dioxide
KEY WORDS:	Utilities, disinfection, surface water, South Carolina, small water systems, DBPs, halogenated organics, trihalomethanes, costs, chlorine dioxide
TITLE:	Evaluation of Alternative Disinfection Strategies for Small/Medium Surface Water Utilities
AUTHORS:	Rivers, P.; Carlson, K.; Gregory, D.
YEAR:	2001
SOURCE:	2001 Annual Conference Proceedings: AWWA
ABSTRACT?:	Y
FULL ARTICLE?:	Y
COMMENTS:	This paper evaluates several alternative disinfection strategies for 3 small to medium surface water systems in South Carolina.
ID:	**98**
DISINFECTANT:	Chlorine Dioxide
KEY WORDS:	Legionella, chlorine dioxide, distribution systems, disinfection, home water treatment units
TITLE:	Chlorine Dioxide: A Point of Entry Treatment Technology for the Control of Legionella in Sensitive Secondary Distribution Systems
AUTHORS:	Sidari, F.; Stout, J.; VanBriessen, J.; Bowman, A.; Grubb, D.; Neuner, A.; Yu, V.
YEAR:	2002
SOURCE:	2002 Water Quality Technology Conference Proceedings: AWWA
ABSTRACT?:	Y
FULL ARTICLE?:	Y
COMMENTS:	This paper summarizes a study of the efficacy of using chlorine dioxide as a POE treatment technology to address Legionella.
ID:	**99**

DISINFECTANT:	Chlorine Dioxide
KEY WORDS:	Disinfection, utilities, chlorine, trihalomethanes, haloacetic acids, residual chlorine, treatment plants, disinfection by-products
TITLE:	Evaluation of "Low-Cost" Strategies for the Control of Trihalomethanes and Haloacetic Acids
AUTHORS:	Franchi, A.; Singer, P.; Chowdhury, Z.; Carter, J.; Grace, N.
YEAR:	2002
SOURCE:	2002 Water Quality Technology Conference Proceedings: AWWA
ABSTRACT?:	Y
FULL ARTICLE?:	Y
COMMENTS:	This paper presents and compares the results and implications of bench-scale testing commissioned by four different utilities in the Eastern US.
ID:	**100**
DISINFECTANT:	Chlorine Dioxide
KEY WORDS:	Chlorine dioxide, bacteria, distribution systems, sampling, HPC, Modeling, organic carbon, utilities, France, water quality
TITLE:	Drinking Water Use of Chlorine Dioxide: A Case Study in France
AUTHORS:	Gatel, D.; Dernat, M.; Bazer-Bachi, A.
YEAR:	1995
SOURCE:	Proceedings Chlorine Dioxide Drinking Water Issues: Third International Symposium
ABSTRACT?:	Y
FULL ARTICLE?:	N
COMMENTS:	This paper describes the use of chlorine dioxide by Toulouse, France.
ID:	**101**
DISINFECTANT:	Chlorine Dioxide
KEY WORDS:	Treatment plants, filtration, distribution system, water quality, microbiological analyses
TITLE:	Comparison of Microbiological Data From Two Water Filtration Plants and Their Distribution Systems
AUTHORS:	Payment, P.; Gamache, F.; Paquette, G.
YEAR:	1989
SOURCE:	Water Science and Technology 21(3): 287-289
ABSTRACT?:	Y
FULL ARTICLE?:	N
COMMENTS:	As part of a larger, epidemiological study aiming to evaluate drinking water related health effects, two water filtration plants and their distribution systems were studied.
ID:	**102**
DISINFECTANT:	Chlorine Dioxide
KEY WORDS:	Disinfection, trihalomethanes, chlorine, costs, chlorination, chloramination, chlorine dioxide, Philadelphia
TITLE:	The Evolution of a Modified Disinfection Strategy to Reduce Trihalomethane Formation
AUTHORS:	McKeon, W.; Muldowney, J.; Aptowica, B.
YEAR:	1986
SOURCE:	1986 Annual Conference Proceedings: AWWA
ABSTRACT?:	Y
FULL ARTICLE?:	N
COMMENTS:	This paper reviews the experiences at Philadelphia's Baxter Plant over the past 10 years in modifying chlorination practices and problems, including chloramination of finished water and utilization of chlorine dioxide at the raw water basin.
ID:	**104**
DISINFECTANT:	Chlorine Dioxide
KEY WORDS:	Microorganisms, DBPs, distribution systems, biofilm, chlorine demand, pH, temperature, cold weater operations, optimization GAC, BAC, alternative disinfectants, chlorine dioxide, potassium permanganate, ozone
TITLE:	Microbiological Changes in Source Water Treatment: Reflections in Distribution Water

	Quality
AUTHORS:	Geldreich, E.;
YEAR:	1991
SOURCE:	1991 Annual Conference Proceedings: AWWA
ABSTRACT?:	Y
FULL ARTICLE?:	Y
COMMENTS:	The author uses case studies to discuss distribution system issues associated with treatment changes.
ID:	**105**
DISINFECTANT:	Chlorine Dioxide
KEY WORDS:	Organic carbon, molecular weight, oxidants, iron, iron removal, chlorine, alum, coagulation, chlorine dioxide, ozone, hydrogen peroxide, potassium permanganate, research
TITLE:	Impacts of Dissolved Organic Carbon on Iron Removal
AUTHORS:	Knocke, W.; Shorney, H.; Bellamy, J.
YEAR:	1993
SOURCE:	AwwaRF
ABSTRACT?:	Y
FULL ARTICLE?:	N
COMMENTS:	This research study addressed the need for a better understanding of the interactions between soluble ferrous iron, dissolved organic carbon, and alternative oxidants.
ID:	**106**
DISINFECTANT:	Chlorine dioxide
KEY WORDS:	Contaminant Candidate List, Microbial, DBPs, Microbiology, disinfection, treatment
TITLE:	Inactivation of Waterborne Emerging Pathogens by Selected Disinfectants
AUTHORS:	Jacangelo, J.; Patania, N.; Trussell, R.; Haas, C.; Gerba, C.
YEAR:	2002
SOURCE:	AwwaRF
ABSTRACT?:	Y
FULL ARTICLE?:	N
COMMENTS:	This project determines the efficiencies of conventional and newly developing disinfectants for emerging waterborne pathogens.
ID:	**107**
DISINFECTANT:	Chlorine dioxide
KEY WORDS:	Cryptosporidium, microbial, DBPs, ozone
TITLE:	Synergistic Effects of Multiple Disinfectants
AUTHORS:	Finch, G.; Liyanage, L.; Gyurek, L.; Bradbury, J.; Belosevic, M.
YEAR:	2000
SOURCE:	AwwaRF
ABSTRACT?:	Y
FULL ARTICLE?:	N
COMMENTS:	Examines and quantifies the presence and significance of synergistic effects through the use of disinfectant strategies.
ID:	**108**
DISINFECTANT:	Chlorine dioxide
KEY WORDS:	T & O
TITLE:	Taste-and-Odor Problems Associated with Chlorine Dioxide
AUTHORS:	AwwaRF
YEAR:	to be published
SOURCE:	AwwaRF
ABSTRACT?:	Y
FULL ARTICLE?:	N
COMMENTS:	Determines causes of and methods to control distribution system taste and odor problems related to chlorine dioxide use. Will be completed in June 2004.

ID:	**109**
DISINFECTANT:	Chlorine dioxide
KEY WORDS:	DBPs
TITLE:	Impact of Chlorine Dioxide on Transmission, Treatment, and Distribution Performance
AUTHORS:	AwwaRF
YEAR:	to be published
SOURCE:	AwwaRF
ABSTRACT?:	Y
FULL ARTICLE?:	N
COMMENTS:	Will compare the effectiveness of chlorine dioxide to that of chlorine in suppressing biofilm growth in drinking water while minimizing formation of DBPs. Will be completed in December 2005.
ID:	**110**
DISINFECTANT:	Chlorine Dioxide
KEY WORDS:	Chlorine dioxide, research, disinfection, DBPs
TITLE:	Practical Applications of Chlorine Dioxide: Field Data
AUTHORS:	Lykins, B.
YEAR:	1992
SOURCE:	Proceedings Chlorine Dioxide Drinking Water Issues: Second International Symposium
ABSTRACT?:	Y
FULL ARTICLE?:	N
COMMENTS:	Field studies have been initiated by the USEPA to evaluate the biocidal effectiveness of chlorine dioxide and control the total oxidants to acceptable levels.
ID:	**111**
DISINFECTANT:	Chlorine Dioxide
KEY WORDS:	DBPS, chlorine dioxide, design, equipment, chemical handling and feeding, resource recovery, oxidation, treatment plants, odor, dosage, biofilm, taste and odor
TITLE:	Disinfection Alternatives for Safe Drinking Water: Chapter 6, Chlorine Dioxide
AUTHORS:	Bryant, E.; Fulton, G.; Budd, G.; Hazen (ed.); Sawyer, (ed.)
YEAR:	1992
SOURCE:	Van Nostrand Reinhold
ABSTRACT?:	Y
FULL ARTICLE?:	N
COMMENTS:	A reference for chlorine dioxide as an alternative disinfectant.
ID:	**112**
DISINFECTANT:	Chlorine Dioxide
KEY WORDS:	Oxidants, utilities, oxidation, manganese oxide, surface water, kinetics, ozone, potassium permanganate, chlorine dioxide
TITLE:	Effect of Soluble Mn Concentration on Oxidation Kinetics
AUTHORS:	Dean, G.; Carlson, K.
YEAR:	2003
SOURCE:	JAWWA: (95) 1: 98-108
ABSTRACT?:	Y
FULL ARTICLE?:	N
COMMENTS:	The overall objective of this study was to evaluate chlorine dioxide, potassium permanganate, and ozone for the oxidation of low initial Mn2+ concentrations to Mn2+ of more than 10 micrograms/Liter.
ID:	**113**
DISINFECTANT:	Chlorine Dioxide
KEY WORDS:	Iron, manganese removal, standards, chemistry, methods, training, filtration, hydraulics, microbiology, particle counting, adsorption, pretreatment, ozone, chlorine dioxide, hyrdogen peroxide, testing, equipment, filter media, softening, flotation, cleaning
TITLE:	Iron and Manganese Removal Handbook
AUTHORS:	Sommerfeld, E.

YEAR:	1999
SOURCE:	AWWA
ABSTRACT?:	Y
FULL ARTICLE?:	N
COMMENTS:	A reference covering chlorination, ozonation and manganese and iron.
ID:	**114**
DISINFECTANT:	Chlorine Dioxide
KEY WORDS:	Treatment plants, optimization, chlorine dioxide, manganese, seasonal effects, performance, coagulation, particulates, trihalomethanes, CT values
TITLE:	Optimizing Plant Performance Using Chlorine Dioxide: A Case Study at Soldier Canyon Water Treatment Plant, Fort Collins, CO
AUTHORS:	Ullmann, T.; Reed, B.; Turner, J.
YEAR:	1997
SOURCE:	1997 Annual Conference Proceedings: AWWA
ABSTRACT?:	Y
FULL ARTICLE?:	N
COMMENTS:	This paper discusses the implementation of chlorine dioxide for control of manganese and describes the other benefits observed from three years of operating experience at the Soldier Canyon Water Treatment Plant.
ID:	**115**
DISINFECTANT:	Chlorine Dioxide
KEY WORDS:	Manganese, oxidation, chlorine dioxide, potassium permanganate, ozone, dosage, organic carbon
TITLE:	Oxidation of Dissolved Manganese in Natural Waters
AUTHORS:	Dean, G.; Carlson, K.
YEAR:	1996
SOURCE:	1996 Annual Conference Proceedings: AWWA
ABSTRACT?:	Y
FULL ARTICLE?:	N
COMMENTS:	The overall objective of this study was to evaluate chlorine dioxide, potassium permanganate, and ozone for the oxidation of low initial Mn2+ concentrations to Mn2+ of more than 10 micrograms/Liter.
ID:	**116**
DISINFECTANT:	Chlorine Dioxide
KEY WORDS:	Oxidants, iron removal, manganese removal, chlorine dioxide, potassium permanganate, ozone, chlorine, hydrogen peroxide, humic substances, fulvic acid, oxidation, pH, temperature
TITLE:	Alternative Oxidants for the Removal of Soluble Iron and Manganese
AUTHORS:	Knocke, W.; Van Benschoten, J.; Kearney, M.; Soborski, A.; Reckhow, D.
YEAR:	1990
SOURCE:	AWWARF
ABSTRACT?:	Y
FULL ARTICLE?:	N
COMMENTS:	This study examines the efficiency of various alternative oxidants, for use in removal of soluble iron and manganese.
ID:	**117**
DISINFECTANT:	Chlorine Dioxide
KEY WORDS:	Disinfection, oxidation, iron, manganese, volatile organic chemicals, disinfection byproducts, taste and odor, color removal, ozone, chlorine, chloramines, chlorine dioxide
TITLE:	Disinfection Alternatives
AUTHORS:	Reckhow, D.
YEAR:	1996
SOURCE:	1996 Annual Conference Proceedings: AWWA
ABSTRACT?:	Y

FULL ARTICLE?:	N
COMMENTS:	The purpose of this paper is to highlight some of the important characteristics of the available oxidants and disinfectants.
ID:	**118**
DISINFECTANT:	Chlorine Dioxide
KEY WORDS:	Reactors, distribution systems, chemicals, UV disinfection, reduction, chlorine, biofilm, chlorine dioxide, disinfection residuals
TITLE:	Impact of UV Disinfection on Biological Stability in Distribution Systems
AUTHORS:	Dykstra, T.; O'Leary, K.; Chauret, C.; Andrews, R.; Gagnon, G.;
YEAR:	2002
SOURCE:	2002 Water Quality Technology Conference Proceedings
ABSTRACT?:	Y
FULL ARTICLE?:	Y
COMMENTS:	This study evaluates the synergistic effects of UV coupled with secondary disinfectants using a model distribution system.
ID:	**119**
DISINFECTANT:	Chlorine Dioxide
KEY WORDS:	
TITLE:	Comparison of Treatments Efficiency for Legionella Control in Domestic Water Systems
AUTHORS:	Loret, J.; Robert, S.; Thomas, V.; Levi, Y.; Cooper, A.; Myers, E.
YEAR:	2003
SOURCE:	2003 Water Quality Technology Conference Proceedings
ABSTRACT?:	N
FULL ARTICLE?:	Y
COMMENTS:	In this study, CIRSEE developed a domestic water supply simulation unit to study the Legionella inactivation efficacy of different disinfectants.
ID:	**120**
DISINFECTANT:	Chlorine Dioxide
KEY WORDS:	Distribution systems, disinfection, contamination, biofilm, regulations, compliance,
TITLE:	Disinfectant Type and Levels in Distribution Systems
AUTHORS:	Gagnon, Graham; Skinner, Dawne
YEAR:	2002
SOURCE:	2002 Distribution and Plant Operations Conference Proceedings; AWWA
ABSTRACT?:	Y
FULL ARTICLE?:	N
COMMENTS:	Provides an overview of the secondary disinfectants chlorine, chloramine, and chlorine dioxide. Also evaluates the physical and chemical factors that affect the management of disinfectant residual.
ID:	**122**
DISINFECTANT:	Chlorine Dioxide
KEY WORDS:	Biofilm, pipes, disinfection, linings, hydrocarbons, distribution systems, water hammer, hydraulics, laboratories, coatings, chlorine dioxide, flow, chlorine, velocity, substrates, corrosion, contamination, coal
TITLE:	The Mobilization of Polycyclic Aromatic Hydrocarbons (PAHs) from the Coal-Tar Lining of Water Pipes
AUTHORS:	Maier, M.; Maier, D.; Lloyd, B.
YEAR:	1999
SOURCE:	Aqua (Journal of Water Services Research and Technology): 48 (6): 238-249: Blackwell Science Ltd.
ABSTRACT?:	Y
FULL ARTICLE?:	N
COMMENTS:	The goal of this project was to investigate the processes and mechanisms responsible for the occurrence of PAHs in drinking water distribution systems.
ID:	**123**

DISINFECTANT:	Chlorine Dioxide
KEY WORDS:	Iron removal, organic carbon, oxidants, coagulation, separation techniques, potassium permanganate, chlorine dioxide
TITLE:	Examining the Reactions between Soluble Iron, DOC, and Alternative Oxidants During Conventional Treatment
AUTHORS:	Knocke, W.; Shorney, H.; Bellamy, J.
YEAR:	1994
SOURCE:	JAWWA 86(1): 117 - 127: AWWA
ABSTRACT?:	Y
FULL ARTICLE?:	Y
COMMENTS:	A paper that concludes that speciation of iron in source and treated water samples is needed to determine whether oxidant addition will prove advantageous.
ID:	**124**
DISINFECTANT:	Chlorine Dioxide
KEY WORDS:	Chlorine dioxide, disinfection, regulations, health effects, chemical and physical analyses, disinfection residuals, oxidation, DBPs, iron, manganese, T&O, algae, nitrification, design, construction, O&M, Information Collection Rule
TITLE:	The Chlorine Dioxide Handbook: Water Disinfection Series
AUTHORS:	Gates, D.
YEAR:	1998
SOURCE:	AWWA
ABSTRACT?:	Y
FULL ARTICLE?:	N
COMMENTS:	This general handbook provides information about the use of chlorine dioxide in water treatment.
ID:	**125**
DISINFECTANT:	Chlorine Dioxide
KEY WORDS:	algae, reservoirs, trihalomethanes, chlorine dioxide
TITLE:	Chlorine Dioxide: An Attractive Alternative for Algae Control in Open, Finished Water Reservoirs
AUTHORS:	Warf, C
YEAR:	1999
SOURCE:	1999 Annual Conference Proceedings: AWWA
ABSTRACT?:	Y
FULL ARTICLE?:	N
COMMENTS:	This paper investigates the use of chlorine dioxide in finished water reservoirs for algae control.
ID:	**126**
DISINFECTANT:	Chlorine Dioxide
KEY WORDS:	Chlorine dioxide, surface water, disinfection, trihalomethanes, statistics, sodium chlorite
TITLE:	Chlorine Dioxide in Drinking Water: Views From a User
AUTHORS:	Moser, R.
YEAR:	1989
SOURCE:	Proceedings of the Workshop on Chlorine Dioxide: Scientific, Regulatory and Application Issues
ABSTRACT?:	Y
FULL ARTICLE?:	N
COMMENTS:	This presentation discusses the practical problems encountered in making the change to higher doses of chlorine dioxide for the purpose of controlling DBPs and microorganisms.
ID:	**127**
DISINFECTANT:	Chlorine Dioxide
KEY WORDS:	Chlorine dioxide, water systems, alternative disinfectants, taste and odor, water treatment treatment plants, disinfection, sodium chlorite, HPC
TITLE:	Practical Considerations for Applications of Chlorine Dioxide in a Municipal Water

	System
AUTHORS:	Thompson, A.
YEAR:	1989
SOURCE:	Proceedings of the Workshop on Chlorine Dioxide: Scientific, Regulatory and Application Issues
ABSTRACT?:	Y
FULL ARTICLE?:	N
COMMENTS:	This paper describes Galveston's use of chlorine dioxide and other disinfectants.
ID:	**128**
DISINFECTANT:	Chlorine Dioxide
KEY WORDS:	T & O, aeration, chlorination, chlorine dioxide, filtration, coagulation, activated carbon, ozonation, sedimentation, disinfection
TITLE:	Advances in Taste-and-Odor Treatment and Control: Treatment Trains
AUTHORS:	Hoehn, R.; Mallevialle, J.
YEAR:	1995
SOURCE:	AwwaRF
ABSTRACT?:	Y
FULL ARTICLE?:	N
COMMENTS:	This Chapter discusses the sequences of water treatment processes that can be used to combat taste and odor problems.
ID:	**129**
DISINFECTANT:	Chlorine Dioxide
KEY WORDS:	T & O, chlorine dioxide, pretreatment, water quality, DBPs, PAC
TITLE:	Taste-and-Odor Problems Associated With Chlorine Dioxide
AUTHORS:	Dietrich, A.; Hoehn, R.; Via, C.
YEAR:	1991
SOURCE:	AwwaRF
ABSTRACT?:	Y
FULL ARTICLE?:	N
COMMENTS:	This research investigates the cause(s) and control(s) for taste and odor problems associated with the use of chlorine dioxide in drinking water treatment.
ID:	**130**
DISINFECTANT:	Chlorine Dioxide
KEY WORDS:	Chlorine dioxide, health effects, DBPs
TITLE:	Drinking Water Disinfection Practices: Chlorine Dioxide in the Nineties
AUTHORS:	Gates, D.; Harrington, R.
YEAR:	1992
SOURCE:	Disinfection Dilemma: Microbiological Control Vs. By-Products: Proceedings of the Fifth National Conference on Drinking Water: AWWA
ABSTRACT?:	Y
FULL ARTICLE?:	N
COMMENTS:	This paper describes "new" studies on the use and benefits of using chloride dioxide in drinking water treatment.
ID:	**131**
DISINFECTANT:	Chlorine Dioxide
KEY WORDS:	T&O, Oxidants, Ozone
TITLE:	Evaluating Oxidants for the Removal of Model Taste and Odor Compounds from a Municipal Water Supply
AUTHORS:	Glaze, W.; Schep, R.; Chauncey, W.; Rugh, E.; Zarnoch, J.; Aieta, E.; Tate, C.; McGuire, M.
YEAR:	1990
SOURCE:	JAWWA: (82) 5: 79-84
ABSTRACT?:	Y
FULL ARTICLE?:	Y

COMMENTS:	Nine oxidizing agents were evaluated for removing six T&O compounds spiked into Colorado River water.
ID:	**132**
DISINFECTANT:	Chlorine Dioxide
KEY WORDS:	Chlorine dioxide, THMs, T & O
TITLE:	Chlorine Dioxide Trihalomethane Control and Distribution System Odor Problem: A Utility's Five-Year Experience
AUTHORS:	Routt, J.
YEAR:	1989
SOURCE:	Proceedings AWWA Seminar on Impact of Changing Disinfectants
ABSTRACT?:	Y
FULL ARTICLE?:	N
COMMENTS:	This paper provides a summary of the Kentucky-American Water Company's use of chlorine dioxide.
ID:	**133**
DISINFECTANT:	Chlorine Dioxide
KEY WORDS:	Water treatment, chlorine dioxide, disinfection, trihalomethanes, T&O, surface water, organic carbon, temperature, Laval, Quebec
TITLE:	Use of Chlorine Dioxide in the City of Laval
AUTHORS:	Duchesne, D.; LaFrance, P.
YEAR:	1987
SOURCE:	Sciences et Techniques de l'Eau: 20(1): 12-13, 15, 17, 19: Association Quebecoise desTechniques de l'Eau
ABSTRACT?:	Y
FULL ARTICLE?:	N
COMMENTS:	This paper is in French with both a French and English Summary
ID:	**134**
DISINFECTANT:	Chlorine Dioxide
KEY WORDS:	T & O, oxidation, chlorine dioxide, chloramines, PAC, Alberta CA, North America
TITLE:	Oxidative Control of Taste and Odour in Drinking Water: The Edmonton Experience
AUTHORS:	Walker, G.; Lee, F.
YEAR:	1983
SOURCE:	35th Annual Convention, Western Canada Water and Sewage Conference
ABSTRACT?:	Y
FULL ARTICLE?:	N
COMMENTS:	This paper can be obtained through AWWA.
ID:	**135**
DISINFECTANT:	Chlorine Dioxide
KEY WORDS:	Potable water, disinfectants, halogen organic compounds, activated carbon, filtration
TITLE:	Alternative Disinfectants for Drinking Water Treatment
AUTHORS:	Lykins, B.; Koffskey, W.; Patterson, K.
YEAR:	1994
SOURCE:	Journal of Environmental Engineering: 120 (4): 754-758
ABSTRACT?:	Y
FULL ARTICLE?:	N
COMMENTS:	This paper discusses a one-year study in Louisiana of the chemical, microbiological, and mutagenic effects of using alternative drinking water disinfectants.
ID:	**136**
DISINFECTANT:	Chlorine Dioxide
KEY WORDS:	
TITLE:	DBP Formation From Chlorine Dioxide/Chlorine/Chloramine Disinfection
AUTHORS:	Krasner, S.; Yates, R.; Liang, S.
YEAR:	2003
SOURCE:	2003 Water Quality Technology Conference Proceedings: AWWA

ABSTRACT?:	N
FULL ARTICLE?:	Y
COMMENTS:	The purpose of this paper was to study the formation of DBPs from pre-disinfection with chlorine dioxide, intermediate chlorination, and post-chloramination.
ID:	**137**
DISINFECTANT:	Chlorine Dioxide
KEY WORDS:	
TITLE:	Advantages of Integrated Application Scenarios for Chlorine, Dioxide, Ozone, and UV
AUTHORS:	Long, B.; Hulsey, R.; Neemann, J.
YEAR:	2003
SOURCE:	2003 Annual Conference Proceedings: AWWA
ABSTRACT?:	N
FULL ARTICLE?:	Y
COMMENTS:	This paper describes the benefits of integrated disinfection.
ID:	**138**
DISINFECTANT:	Chlorine Dioxide
KEY WORDS:	Water, purification, DBPS, Chlorine dioxide, chlorites, chlorates
TITLE:	Sources, occurrence, and control of chlorine dioxide by-product residuals in drinking water
AUTHORS:	Gallagher, D.
YEAR:	1994
SOURCE:	AWWARF
ABSTRACT?:	N
FULL ARTICLE?:	N
COMMENTS:	
ID:	**138a**
DISINFECTANT:	Chlorine Dioxide
KEY WORDS:	
TITLE:	Using Chlorite Ion to Control Nitrification
AUTHORS:	McGuire, M.; Lieu, N.; Pearthree, M.
YEAR:	1999
SOURCE:	JAWWA: 91(10): 52-61
ABSTRACT?:	Y
FULL ARTICLE?:	Y
COMMENTS:	This article reviews problems associated with nitrification and presents laboratory and field evidence for using the chlorite ion to control nitrification in distribution systems.
ID:	**139**
DISINFECTANT:	Booster Chlorination
KEY WORDS:	Chemicals, disinfection, chlorine, trihalomethanes, Connecticut OH, kinetics, residual chlorine
TITLE:	Booster Chlorination: Residual Maintenance and Trihalomethane Formation
AUTHORS:	Boccelli, D.; Tryby, M.; Koechling, M.; Summers, R.; Uber, J.; Rossman, L.;
YEAR:	1998
SOURCE:	1998 Symposium Proceedings; Protecting Water Quality in the Distribution System: What is the Role of disinfection Residuals?
ABSTRACT?:	Y
FULL ARTICLE?:	N
COMMENTS:	This study evaluates the bulk chemical advantages that may be achieved using booster disinfection.
ID:	**140**
DISINFECTANT:	Booster Chlorination
KEY WORDS:	Disinfection, distribution systems, disinfectant residuals, disinfection residuals, chlorination, chlorine, organic carbon, trihalomethanes, modeling, mathematics
TITLE:	Booster Chlorination: Bulk Decay Kinetics, TTHM Formation, and Model Development

AUTHORS:	Boccelli, D.; Tryby, M.; Koechling, M.; Summers, R.; Uber, J.; Rossman, L.;
YEAR:	1998
SOURCE:	1998 Annual Conference Proceedings: AWWA
ABSTRACT?:	Y
FULL ARTICLE?:	Y
COMMENTS:	This study evaluates the bulk chemical advantages that may be achieved using booster disinfection.
ID:	**141**
DISINFECTANT:	Booster Chlorination
KEY WORDS:	Trihalomethanes, chlorination, biofilm, bacteria, HPC, distribution systems, organic carbon, residual chlorine
TITLE:	Microbiological and Trihalomethane Responses to Booster Chlorination
AUTHORS:	Gibbs, R.; Scutt, J.; Croll, B.
YEAR:	1990
SOURCE:	Journal of the Institution of Water and Environmental Management: 4(2): 131-139: Institution of Water and Environmental Management
ABSTRACT?:	Y
FULL ARTICLE?:	N
COMMENTS:	Two zones of a water distribution supply area receiving treated surface water were used to study the effect of booster chlorination on bacterial regrowth.
ID:	**142**
DISINFECTANT:	Booster Chlorination
KEY WORDS:	Chloramines, reservoirs, nitrification, chloramination, distribution systems, Portland OR, sodium hypochlorite, disinfection byproducts, pumping stations, disinfection residuals, chlorine, retention time
TITLE:	Portland's Experience with Chloramine Residual Management
AUTHORS:	Ireland, C.; Knudson, M.
YEAR:	1998
SOURCE:	1998 Symposium Proceedings; Protecting Water Quality in the Distribution System: What is the Role of disinfection Residuals?
ABSTRACT?:	Y
FULL ARTICLE?:	N
COMMENTS:	This paper was selected because it includes a study of booster chlorination and a utility that has utilized chloramines for a significant period of time.
ID:	**143**
DISINFECTANT:	Booster Chlorination
KEY WORDS:	Water distribution, optimization, chlorine, system analysis, chlorination
TITLE:	Water Distribution Systems Analysis: Optimization of Location and Chlorine Dosage of the Booster Chlorination in Water Distribution Network
AUTHORS:	Nace, A.; Harmant, P.; Villon, P.; Suez-Lyonnaise, C.
YEAR:	2001
SOURCE:	Bridging the Gap: Meeting the World's Water and Environmental Resources Challenges: Proceedings of the 2001World Water and Environment Resources Congress
ABSTRACT?:	Y
FULL ARTICLE?:	N
COMMENTS:	This study provides information on optimizing locations of booster disinfection.
ID:	**144**
DISINFECTANT:	Booster Chlorination
KEY WORDS:	Storage tanks, water quality, residual chlorine, chlorination
TITLE:	Maintaining Distribution System Storage Tank Water Quality with a Booster Chlorination and Recirculation System
AUTHORS:	Skov, K.; Gorman, E.; Flynn, J.
YEAR:	1994
SOURCE:	1994 Annual Conference proceedings: AWWA

ABSTRACT?:	Y
FULL ARTICLE?:	Y
COMMENTS:	The results of this study indicate that if the Grassy Hill tanks was not equipped with a booster chlorination and recirculation system, the chlorine residual in the tank would be severely degraded due to long residence time associated with the storage tank and low residual.
ID:	**145**
DISINFECTANT:	Booster Chlorination
KEY WORDS:	Distribution systems, water quality, residual chlorine, chloramines, nitrification, sampling, flushing, ammonia
TITLE:	Excessive Loss of Chloramine Residual: A Case Study
AUTHORS:	Smith, C.; Smith, J.; Milosky, B.
YEAR:	1996
SOURCE:	1996 Annual Conference Proceedings: AWWA: Water Quality
ABSTRACT?:	Y
FULL ARTICLE?:	N
COMMENTS:	This report describes the evaluation of water quality and provides recommendation. The utility operates a booster chlorination facility and uses chloramines for disinfection.
ID:	**146**
DISINFECTANT:	Booster Chlorination
KEY WORDS:	Chlorination, disinfectant residuals, treatment plants, distribution systems, chlorine, regrowth, biofilm, DBPs, T & O, dosage, monitoring, equipment, modeling, kintetics
TITLE:	Booster Chlorination for Managing Disinfectant Residuals
AUTHORS:	Tryby, M.; Boccelli, D.; Kowchling, M.; Uber, J.; Summers, R.; Rossman, L.
YEAR:	1999
SOURCE:	JAWWA: 91 (1): 95-108
ABSTRACT?:	Y
FULL ARTICLE?:	Y
COMMENTS:	The role of booster chlorination is discussed as part of coordinated treatment efforts meant to manage the risks associated with biological regrowth and DBPs.
ID:	**147**
DISINFECTANT:	Booster Chlorination
KEY WORDS:	water distribution, disinfection, optimization, water quality, mathematical models, location
TITLE:	Facility Location Model for Booster Disinfection of Water Supply Networks
AUTHORS:	Tryby, M.; Boccelli, D.; Kowchling, M.; Uber, J.; Summers, R.; Rossman, L.
YEAR:	2002
SOURCE:	Journal of Water Resources Planning and Management: 128 (5): 322 - 323:
ABSTRACT?:	Y
FULL ARTICLE?:	N
COMMENTS:	This paper describes a model that can be used to locate booster stations to minimize dosage and may discuss aspects associated with DBPs.
ID:	**148**
DISINFECTANT:	Booster Chlorination
KEY WORDS:	Water, purification, chlorination, distribution
TITLE:	Feedback Control of Booster Chlorination Systems
AUTHORS:	Uber, J.
YEAR:	2003
SOURCE:	AwwaRF
ABSTRACT?:	N
FULL ARTICLE?:	N
COMMENTS:	
ID:	**149**
DISINFECTANT:	Booster Chlorination
KEY WORDS:	

TITLE:	Strategies for Combining Free Ammonia and Boosting Chloramines in Distribution Systems - Survey of Utility Practice
AUTHORS:	Wilczak, A.; Smith, C.; Cohen, Y.; Martin, P.
YEAR:	2003
SOURCE:	2003 Water Quality Technology Conference Proceedings: AWWA
ABSTRACT?:	N
FULL ARTICLE?:	Y
COMMENTS:	This survey was conducted to obtain data from actual full-scale chloramine boosting applications, review process control and evaluate the success of these applications.
ID:	**150**
DISINFECTANT:	Ozonation
KEY WORDS:	Ozone, dosage, organic carbon, HPC, distribution systems, residual chlorine, monitoring
TITLE:	Effects of Ozone on Biodegradable Dissolved Organic Carbon and Heterotrophic Plate Counts in the Distribution System
AUTHORS:	Albidress, L.; Horwedel, J.; Hill, G.; Borchardt, J.; Price, D.
YEAR:	1995
SOURCE:	Ozone: Science & Engineering: 17 (3): 283-295: Lewis Publishers
ABSTRACT?:	N
FULL ARTICLE?:	N
COMMENTS:	The impacts of ozone dose in the formation of BDOC were studied at the North Bay Regional Water Treatment Plant.
ID:	**151**
DISINFECTANT:	Ozone
KEY WORDS:	
TITLE:	Ozonation of Humic Substances: Effects on Molegular Weight Distributions of Organic Carbon and Trihalomethane Formation Potential
AUTHORS:	Amy, G.; Kuo, C.; Sierka, R.
YEAR:	1988
SOURCE:	Ozone Science and Engineering vol. 10: 39-54
ABSTRACT?:	N
FULL ARTICLE?:	N
COMMENTS:	
ID:	**152**
DISINFECTANT:	Ozone
KEY WORDS:	DBP, GAC, Microbial, DBP
TITLE:	Identification and Occurrence of Ozonation By-Products in Drinking Water
AUTHORS:	AwwaRF
YEAR:	1993
SOURCE:	AwwaRF
ABSTRACT?:	Y
FULL ARTICLE?:	N
COMMENTS:	This project identifies suspected and unknown ozonation by-products.
ID:	**153**
DISINFECTANT:	Ozone
KEY WORDS:	DBP, GAC, Microbial, DBP
TITLE:	Formation and Control of Brominated Ozone By-Products
AUTHORS:	AwwaRF
YEAR:	1997
SOURCE:	AwwaRF
ABSTRACT?:	Y
FULL ARTICLE?:	N
COMMENTS:	This project studies the impacts of water quality and treatment (using ozone) on variables in the formation of bromonated organic DBPs.
ID:	**154**

DISINFECTANT:	Ozone
KEY WORDS:	Contaminant Candidate List, Microbial, DBPs, Microbiology, disinfection, treatment
TITLE:	Inactivation of Waterborne Emerging Pathogens by Selected Disinfectants
AUTHORS:	AwwaRF
YEAR:	2002
SOURCE:	AwwaRF
ABSTRACT?:	Y
FULL ARTICLE?:	N
COMMENTS:	This project determines the efficiencies of conventional and newly developing disinfectants for emerging waterborne pathogens.
ID:	**155**
DISINFECTANT:	Ozone
KEY WORDS:	UV, ozone, treatment
TITLE:	Evaluation of Ozone and Ultraviolet Light
AUTHORS:	AwwaRF
YEAR:	to be published
SOURCE:	AwwaRF
ABSTRACT?:	Y
FULL ARTICLE?:	N
COMMENTS:	This project will compare the relative treatment performance of ozone followed by UV and UV only. This will cover finished and distribution system water quality, including red water and biostability control. Will be finished in June 2004.
ID:	**156**
DISINFECTANT:	Ozone
KEY WORDS:	Cryptosporidium, microbial, DBPs, ozone
TITLE:	Synergistic Effects of Multiple Disinfectants
AUTHORS:	AwwaRF
YEAR:	2000
SOURCE:	AwwaRF
ABSTRACT?:	Y
FULL ARTICLE?:	N
COMMENTS:	Examines and quantifies the presence and significance of synergistic effects through the use of disinfectant strategies.
ID:	**157**
DISINFECTANT:	Ozonation
KEY WORDS:	Filtration, distribution systems, chlorine, ozonation, reactors, biomass, biological filters, residual chlorine, chloramines
TITLE:	Effects of Biological and Conventional Treatment on Biological Stability of Distribution Systems
AUTHORS:	Baribeau, H.; Hacker, P.; Mofidi, A.; Coffeey, B.; De Leon, R.
YEAR:	1998
SOURCE:	1998 Symposium Proceedings; Protecting Water Quality in the Distribution System: What is the Role of disinfection Residuals?
ABSTRACT?:	Y
FULL ARTICLE?:	N
COMMENTS:	This study compares the biological stability of water in simulated distribution systems following three treatment scenarios: ozonation followed by biofiltration; ozonation followed by non-biological filtration, and now ozone followed by non-biological filtration. Chloramination was used as a primary disinfectant.
ID:	**158**
DISINFECTANT:	Ozonation
KEY WORDS:	
TITLE:	A New Awareness of Copper Pipe Failures in Water Distribution Systems
AUTHORS:	Cantor, A.; Bushman, J.; Glodoski, M.

YEAR:	2003
SOURCE:	2003 Water Quality Technology Conference Proceedings: AWWA
ABSTRACT?:	N
FULL ARTICLE?:	Y
COMMENTS:	This paper looks into the causes of copper pipe failures and determines that microbiologically influenced corrosion is the likely cause. The study points out that treatment was switched to ozonation followed by chloramination in 1998.
ID:	**159**
DISINFECTANT:	Ozonation
KEY WORDS:	Oxidants, utilities, oxidation, manganese oxide, surface water, kinetics, ozone, potassium permanganate, chlorine dioxide
TITLE:	Effect of Soluble Mn Concentration on Oxidation Kinetics
AUTHORS:	Dean, G.; Carlson, K.
YEAR:	2003
SOURCE:	JAWWA: (95) 1: 98-108
ABSTRACT?:	Y
FULL ARTICLE?:	N
COMMENTS:	The overall objective of this study was to evaluate chlorine dioxide, potassium permanganate, and ozone for the oxidation of low initial Mn2+ concentrations to Mn2+ of more than 10 micrograms/Liter.
ID:	**160**
DISINFECTANT:	Ozonation
KEY WORDS:	Manganese, oxidation, chlorine dioxide, potassium permanganate, ozone, dosage, organic carbon
TITLE:	Oxidation of Dissolved Manganese in Natural Waters
AUTHORS:	Dean, G.; Carlson, K.
YEAR:	1996
SOURCE:	1996 Annual Conference Proceedings: AWWA
ABSTRACT?:	Y
FULL ARTICLE?:	N
COMMENTS:	The overall objective of this study was to evaluate chlorine dioxide, potassium permanganate, and ozone for the oxidation of low initial Mn2+ concentrations to Mn2+ of more than 10 micrograms/Liter.
ID:	**161**
DISINFECTANT:	Ozonation
KEY WORDS:	DBPs, compliance, regulations, color, distribution systems, softening, lime, filtration, ozonation, trihalomethanes, halogenated organics, sampling
TITLE:	Compliance of a Large South Florida Utility With the Upcoming DBP Rule
AUTHORS:	Dunkelberger, G.; Duran, F.; Carey, E.
YEAR:	1993
SOURCE:	68th Annual Florida Water Resources Conference: Florida Section American Water Works Association
ABSTRACT?:	N
FULL ARTICLE?:	N
COMMENTS:	This paper reviews how the Stage 1 D/DBP will impact a large Florida utility. Distribution system sampling is used to compare DBP levels with and without ozonation.
ID:	**162**
DISINFECTANT:	Ozonation
KEY WORDS:	
TITLE:	Synergistic Impacts of Ozone and UV
AUTHORS:	Edwards, J.; Hulsey, B.; Tran, T.; Sung, W.; Muri, J.; Malley, J.; Spencer, C.; Carthcart, A.;
YEAR:	2003
SOURCE:	2003 Annual Conference Proceedings: AWWA

ABSTRACT?:	N
FULL ARTICLE?:	Y
COMMENTS:	This paper presents early results of the research which will determine the effectiveness of ozone and UV inactivation for viruses, the synergistic impacts of UV on DBP formation, biostability, corrosion control and red water, and recommend optimum primary disinfection schemes.
ID:	**163**
DISINFECTANT:	Ozone
KEY WORDS:	
TITLE:	Effects of Ozonation on Coagulation of NOM Using Polymer Alone and Polymer/Metal Salt Mixtures
AUTHORS:	Edwards, M.; Benjamin, M.; Tobiason, J.
YEAR:	1994
SOURCE:	JAWWA: 89(1): 105-116
ABSTRACT?:	Y
FULL ARTICLE?:	N
COMMENTS:	This study investigated the effects of ozonation on coagulation with cationic polymer and with polymer-metal salt mixtures.
ID:	**164**
DISINFECTANT:	Ozonation
KEY WORDS:	Ozonation, distribution systems, organic carbon, sampling, organics, water quality, corrosion, ozone, bacteria, pipes
TITLE:	Case Study: Ozonation and Distribution System Biostability
AUTHORS:	Escobar, I.; Randall, A.
YEAR:	2001
SOURCE:	JAWWA: 93 (10): 77-89: AWWA
ABSTRACT?:	Y
FULL ARTICLE?:	N
COMMENTS:	The main objective of this project was to collect and analyze full-scale system data on short-term and long-term responses to the implementation of ozonation with respect to its impact on bacterial regrowth potential as quantified by AOC and BDOC.
ID:	**165**
DISINFECTANT:	Ozonation
KEY WORDS:	water quality, chlorine, distribution systems, organic carbon, ozonation, Vancouver BC, turbidity, treatment plants, dissolved oxygen, chlorine demand, DBPs, THMs, halogenated organics, resiudal chlorine, temperature, coliforms, HPC, bacteria, disinfection GIS
TITLE:	Largest Unfiltered Ozonation Plant: Water Quality Impacts
AUTHORS:	Ferguson, A.; Gibson, K.; Smith, J.; MacQuarrie, D.
YEAR:	2001
SOURCE:	2001 Water Quality Technology Conference Proceedings: AWWA
ABSTRACT?:	Y
FULL ARTICLE?:	Y
COMMENTS:	This paper describes the water quality impacts of GVRD's switch to ozonation.
ID:	**166**
DISINFECTANT:	Ozonation
KEY WORDS:	Water quality, distribution systems, Vancouver BC, residual chlorine, management, DBPs, ozone, interagency agreements
TITLE:	Managing Water Quality in Consecutive Systems - The City of Coquitlam/GVRD Experience
AUTHORS:	Gehlen, I.; Wright, K.; Carver, M.; Neden, D.; MacQuarrie, D.; Smith, J.
YEAR:	2002
SOURCE:	2002 Annual Conference Proceedings: AWWA
ABSTRACT?:	Y
FULL ARTICLE?:	Y

COMMENTS:	This paper describes impacts of changes to GVRD's treatment practices, including the implementation of ozonation
ID:	**167**
DISINFECTANT:	Ozonation
KEY WORDS:	Microorganisms, DBPs, distribution systems, biofilm, chlorine demand, pH, temperature, cold weater operations, optimization GAC, BAC, alternative disinfectants, chlorine dioxide, potassium permanganate, ozone
TITLE:	Microbiological Changes in Source Water Treatment: Reflections in Distribution Water Quality
AUTHORS:	Geldreich, E.;
YEAR:	1991
SOURCE:	1991 Annual Conference Proceedings: AWWA
ABSTRACT?:	Y
FULL ARTICLE?:	Y
COMMENTS:	The author uses case studies to discuss distribution system issues associated with treatment changes.
ID:	**168**
DISINFECTANT:	Ozonation
KEY WORDS:	Water treatment, DBPs, research, chlorination, GAC, BAC, alternative disinfectants, biofilm
TITLE:	Microbial Aspects of Water Treatment Processes: A Progress Report
AUTHORS:	Geldreich, E.; Lykins, B.; Clark, R.
YEAR:	1990
SOURCE:	1990 Proceedings of AWWA Seminar on Current Research Activities in Support of USEPA's Regulatory Agenda
ABSTRACT?:	Y
FULL ARTICLE?:	N
COMMENTS:	The paper focuses on the microbial quality of treated and distribution water. The investigation looks at studies that have been performed on alternative disinfectants.
ID:	**169**
DISINFECTANT:	Ozonation
KEY WORDS:	T&O, Oxidants, Ozone
TITLE:	Evaluating Oxidants for the Removal of Model Taste and Odor Compounds from a Municipal Water Supply
AUTHORS:	Glaze, W.; Schep, R.; Chauncey, W.; Rugh, E.; Zarnoch, J.; Aieta, E.; Tate, C.; McGuire, M.
YEAR:	1990
SOURCE:	JAWWA: (82) 5: 79-84
ABSTRACT?:	Y
FULL ARTICLE?:	Y
COMMENTS:	Nine oxidizing agents were evaluated for removing six T&O compounds spiked into Colorado River water.
ID:	**170**
DISINFECTANT:	Ozonation
KEY WORDS:	T & O, aeration, chlorination, chlorine dioxide, filtration, coagulation, activated carbon, ozonation, sedimentation, disinfection
TITLE:	Advances in Taste-and-Odor Treatment and Control: Treatment Trains
AUTHORS:	Hoehn, R.; Mallevialle, J.
YEAR:	1995
SOURCE:	AwwaRF
ABSTRACT?:	Y
FULL ARTICLE?:	N
COMMENTS:	This Chapter discusses the sequences of water treatment processes that can be used to combat taste and odor problems.

ID:	**171**
DISINFECTANT:	Ozonation
KEY WORDS:	Chlorine, Natural Organic Matter, Trihalomethanes, Disinfection Byproducts, chloramines, halogenated organics, ozonation
TITLE:	DBP Yields of Polar NOM Fractions From Low Humic Waters
AUTHORS:	Hwang, Cordelia; Sclimenti, Michael; Bruchet, Auguste; Croue, Jean-Philippe; Amy, Gary
YEAR:	2001
SOURCE:	2001 Water Quality Technology Conference Proceedings: AWWA
ABSTRACT?:	Y
FULL ARTICLE?:	Y
COMMENTS:	In order to better understand the source of common halogenated DBPs, the NOM from four low humic water was evaluated for HAAs and THMs after conventional treatment and ozonation/no ozonation and chloramination. Gives treatment applications.
ID:	**172**
DISINFECTANT:	Ozonation
KEY WORDS:	Treatment plants, DBPs, ozonation, biological filters, coagulation, iron removal, manganese removal, turbidity, organic carbon, optimization, chlorine, pre-treatment, potassium permanganate
TITLE:	Optimizaing Drinking Water Treatment for Conflicting Objectives
AUTHORS:	Ibrahim, E.; Crossland, E.; Dixon, K.
YEAR:	1997
SOURCE:	1997 Water Quality Technology Conference Proceedings: AWWA
ABSTRACT?:	N
FULL ARTICLE?:	Y
COMMENTS:	The main objective of this study was to optimize treatment conditions for iron and manganese removal without a detrimental effect on DBP precursor removal and biological filtration.
ID:	**173**
DISINFECTANT:	Ozone
KEY WORDS:	
TITLE:	The Benefits of Ozone Treatment Prior to Flocculation Processes
AUTHORS:	Jekel, M.
YEAR:	1983
SOURCE:	Ozone Science and Engienering 5 (21)
ABSTRACT?:	N
FULL ARTICLE?:	N
COMMENTS:	
ID:	**174**
DISINFECTANT:	Ozonation
KEY WORDS:	Chlorination, water treatment, trihalomethanes, alternative disinfectants, Netherlands
TITLE:	Chlorination By-Products: Production and Control
AUTHORS:	KIWA
YEAR:	1986
SOURCE:	1986: KIWA (available through AWWA)
ABSTRACT?:	Y
FULL ARTICLE?:	N
COMMENTS:	Describes an analytical investigation of the presence of THMSs in the drinking water of 14 treatment plants in the Netherlands. Discusses use of alternative disinfectants.
ID:	**175**
DISINFECTANT:	Ozonation
KEY WORDS:	Organic carbon, molecular weight, oxidants, iron, iron removal, chlorine, alum, coagulation, chlorine dioxide, ozone, hydrogen peroxide, potassium permanganate, research

TITLE:	Impacts of Dissolved Organic Carbon on Iron Removal
AUTHORS:	Knocke, W.; Shorney, H.; Bellamy, J.
YEAR:	1993
SOURCE:	AwwaRF
ABSTRACT?:	Y
FULL ARTICLE?:	N
COMMENTS:	This research study addressed the need for a better understanding of the interactions between soluble ferrous iron, dissolved organic carbon, and alternative oxidants.
ID:	**176**
DISINFECTANT:	Ozonation
KEY WORDS:	Oxidants, iron removal, manganese removal, chlorine dioxide, potassium permanganate, ozone, chlorine, hydrogen peroxide, humic substances, fulvic acid, oxidation, pH, temperature
TITLE:	Alternative Oxidants for the Removal of Soluble Iron and Manganese
AUTHORS:	Knocke, W.; Van Benschoten, J.; Kearney, M.; Soborski, A.; Reckhow, D.
YEAR:	1990
SOURCE:	AWWARF
ABSTRACT?:	Y
FULL ARTICLE?:	N
COMMENTS:	This study examines the efficiency of various alternative oxidants, for use in removal of soluble iron and manganese.
ID:	**177**
DISINFECTANT:	Ozonation
KEY WORDS:	Ozone, filter media, ozonation, manganese removal, ozidation, Philadelphia PA, distribution systems
TITLE:	Limitations of Ozone for Removal of Manganese to Very Low Levels; A Comparative Study
AUTHORS:	Kohl, P.; Kim, A.; Charlton, N.
YEAR:	2002
SOURCE:	2002 Water Quality Technology Conference Proceedings: AWWA
ABSTRACT?:	Y
FULL ARTICLE?:	Y
COMMENTS:	Pilot studies at two of PWD's water treatment plants demonstrated that ozonation cannot consistently reduce manganese to PWD's internal goal.
ID:	**178**
DISINFECTANT:	Ozonation
KEY WORDS:	ozonation, DBPs, bromide, THMs, halogenated organics, biological filters, CT values, organic carbon, coagulation
TITLE:	The Effects of Ozoneation, Biofiltration, and Secondary Disinfection on DBP Formation
AUTHORS:	Krasner, S.
YEAR:	1996
SOURCE:	1996 Water Quality Technology Conference: AWWA
ABSTRACT?:	Y
FULL ARTICLE?:	N
COMMENTS:	The objective of this paper was to examine the effects of ozonation, biofiltration, and secondary disinfection on the formation of THMs HAAs, and chloral hydrate.
ID:	**179**
DISINFECTANT:	Ozonation
KEY WORDS:	Disinfection byproducts, distribution systems, treatment plants, chlorine, chloramines, ozone, trihalomethanes, weather, chlorinated organics
TITLE:	Monthly Variation of Disinfection By-Products in Water From Treatment Plants Using Three Different Disinfection Processes
AUTHORS:	Level, G.; Benoit, F.; Williams, D.
YEAR:	1994

SOURCE:	Planning for Tomorrow: Proceedings of the Sixth National Conference on Drinking Water: AWWA
ABSTRACT?:	Y
FULL ARTICLE?:	N
COMMENTS:	This study assesses the temporal and spatial variation in DBP levels over a one-year period for three facilities each of which utilizes chlorine- chloramine, chlorine-chlorine, ozone-chlorine.
ID:	**180**
DISINFECTANT:	Ozonation
KEY WORDS:	Stabilization, DBPs, aldehydes, biodegradation, ozone
TITLE:	Stability of DBPs in Drinking Water Systems
AUTHORS:	Levi, Y.; Jammes, C.; Kiene, L; Hochereau, C.; Dumoutier, N.
YEAR:	1993
SOURCE:	1993 Water Quality Technology Conference Proceedings: AWWA
ABSTRACT?:	Y
FULL ARTICLE?:	Y
COMMENTS:	The authors demonstrate the aldehydes and ketones disappear or are being converted when being in contact with a distribution system.
ID:	**181**
DISINFECTANT:	Ozonation
KEY WORDS:	
TITLE:	Advantages of Integrated Application Scenarios for Chlorine, Dioxide, Ozone, and UV
AUTHORS:	Long, B.; Hulsey, R.; Neemann, J.
YEAR:	2003
SOURCE:	2003 Annual Conference Proceedings: AWWA
ABSTRACT?:	N
FULL ARTICLE?:	Y
COMMENTS:	This paper describes the benefits of integrated disinfection.
ID:	**182**
DISINFECTANT:	Ozonation
KEY WORDS:	
TITLE:	Comparison of Treatments Efficiency for Legionella Control in Domestic Water Systems
AUTHORS:	Loret, J.; Robert, S.; Thomas, V.; Levi, Y.; Cooper, A.; Myers, E.
YEAR:	2003
SOURCE:	2003 Water Quality Technology Conference Proceedings
ABSTRACT?:	N
FULL ARTICLE?:	Y
COMMENTS:	In this study, CIRSEE developed a domestic water supply simulation unit to study the Legionella inactivation efficacy of different disinfectants.
ID:	**183**
DISINFECTANT:	Ozonation
KEY WORDS:	Potable water, disinfectants, halogen organic compounds, activated carbon, filtration
TITLE:	Alternative Disinfectants for Drinking Water Treatment
AUTHORS:	Lykins, B.; Koffskey, W.; Patterson, K.
YEAR:	1994
SOURCE:	Journal of Environmental Engineering: 120 (4): 754-758
ABSTRACT?:	Y
FULL ARTICLE?:	N
COMMENTS:	This paper discusses a one-year study in Louisiana of the chemical, microbiological, and mutagenic effects of using alternative drinking water disinfectants.
ID:	**184**
DISINFECTANT:	Ozonation
KEY WORDS:	Ozone, biological filters, granular activated carbon, trihalomethanes, organics, organic carbon, contactors, pilot pltans, chloramines, chlorine

TITLE:	Removal of THM and HAA Precursors Using Ozone and Biologically Active GAC
AUTHORS:	McClellan, J.; Reckhow, D.; Tobiason, J.
YEAR:	1996
SOURCE:	1996 Annual Conference Proceedings: AWWA
ABSTRACT?:	Y
FULL ARTICLE?:	N
COMMENTS:	This study includes results from a long-term pilot study of a post-filtration ozone/biologically active GAC process. Compares DBP formation chloramines disinfection with free chlorine only.
ID:	**185**
DISINFECTANT:	Ozonation
KEY WORDS:	Distribution systems, chlorination, chloramination, ozonation, water quality, public information, trihalomethanes, color, bacteria, taste and odor, treatment plants
TITLE:	Observations of Reconditioning a Chlorinated Water Distribution System to Chloramines
AUTHORS:	McMorran, Carl; Moriarty, Kathy
YEAR:	1997
SOURCE:	1997 Annual Conference Proceedings: AWWA
ABSTRACT?:	Y
FULL ARTICLE?:	N
COMMENTS:	Discusses the conversion of Bangor ME Water District supply from an unfiltered, chlorinated supply to an unfiltered, ozonated, chloraminated system.
ID:	**186**
DISINFECTANT:	Ozonation
KEY WORDS:	Distribution systems, groundwater, surface water, wells, nutrients, bacteria, monitoring, water quality, disinfectant residuals, temperature, pH, turbidity, DBPs, sampling
TITLE:	Impact of Treatment Changes on Dsitribution System Stability
AUTHORS:	Norton, C.; LeChevallier, M.; Volk, C.
YEAR:	1998
SOURCE:	1998 Water Quality Technology Conference Proceedings: AWWA
ABSTRACT?:	N
FULL ARTICLE?:	Y
COMMENTS:	This paper discusses a study to evaluate changes in biostability levels throughout a full-scale distribution system following the conversion from groundwater to biologically active surface water.
ID:	**187**
DISINFECTANT:	Ozonation
KEY WORDS:	Ozone, ozonation, biological filters, filtration, DBPs, bacteria, organics, organic carbon, chlorine, biodegradation
TITLE:	Ozone and Biological Treatment for DBP Control and Biological Stability
AUTHORS:	Price, M.
YEAR:	1994
SOURCE:	AwwaRF
ABSTRACT?:	N
FULL ARTICLE?:	N
COMMENTS:	This project provides an in-depth look at the combined processes of ozone and biological filtration for drinking water treatment.
ID:	**188**
DISINFECTANT:	Ozonation
KEY WORDS:	Color removal, corrosion control, distribution systems, surface water, red water, iron, testing, Tucson AZ
TITLE:	Corrosion Control Strategies for Changing Water Supplies in Tucson AZ
AUTHORS:	Price, S.; Jefferson, T.
YEAR:	1997
SOURCE:	Journal of New England Water Works Asociation: 111 (3): 285-293: New England

	Water Works Association
ABSTRACT?:	N
FULL ARTICLE?:	N
COMMENTS:	A study was conducted to develop a strategy to minimize corrosion and prevent the reoccurrence of colored water problems. Ozone was one of the control variables.
ID:	**189**
DISINFECTANT:	Ozonation
KEY WORDS:	Disinfection, oxidation, iron, manganese, volatile organic chemicals, disinfection byproducts, taste and odor, color removal, ozone, chlorine, chloramines, chlorine dioxide
TITLE:	Disinfection Alternatives
AUTHORS:	Reckhow, D.
YEAR:	1996
SOURCE:	1996 Annual Conference Proceedings: AWWA
ABSTRACT?:	Y
FULL ARTICLE?:	N
COMMENTS:	The purpose of this paper is to highlight some of the important characteristics of the available oxidants and disinfectants.
ID:	**190**
DISINFECTANT:	Ozone
KEY WORDS:	
TITLE:	The Use of Ozone and High Rate of Biological Filtration to Meet Simultaneous Treatment Goals
AUTHORS:	Schneider, O.; Schafer, J.; Nickols, A.; Kurtz, W.
YEAR:	1996
SOURCE:	1996 Water Quality Technology Conference Proceedings: AWWA
ABSTRACT?:	N
FULL ARTICLE?:	N
COMMENTS:	
ID:	**191**
DISINFECTANT:	Ozonation
KEY WORDS:	Ozone, disinfection, Denver, CO, distribution systems, water main rehabilitation
TITLE:	Evaluating Ozone for Disinfecting New or Rehabilitated Water Mains
AUTHORS:	Schulz, C.; Pier, D.; Lohman, S.; Dahm, B.; Dee, S.;
YEAR:	2002
SOURCE:	2002 Water Quality Technology Conference Proceedings: AWWA
ABSTRACT?:	Y
FULL ARTICLE?:	Y
COMMENTS:	This paper presents interim results of an ongoing research project to develop a new method for disinfecting newly constructed or rehabilitated water mains using ozone.
ID:	**192**
DISINFECTANT:	Ozonation
KEY WORDS:	Bromate, ozonation, DBPs, GAC, UV radiation, Bromide
TITLE:	Removal of Bromate After Ozonation During Drinking Water Treatment
AUTHORS:	Siddiqui, M.; Amy, G.; Zhai, W.; Ozekin, K.; Westerhoff, P.; Chandrakanth, M.
YEAR:	1994
SOURCE:	1994 Annual Conference Proceedings: AWWA
ABSTRACT?:	Y
FULL ARTICLE?:	Y
COMMENTS:	This paper evaluates ways to remove the bromate ion (BrO3-) after ozonation.
ID:	**193**
DISINFECTANT:	Ozonation
KEY WORDS:	Ozone, surface water, organics, color, THMs, chloramines, chlorine, disinfection, residual chlorine, turbidity
TITLE:	Ozonation at Belle Glade, Florida: A Case History

AUTHORS:	Singer, P.; Robinson, K.; Elefritz, R.
YEAR:	1990
SOURCE:	Ozone: Science & Engineering: 12 (2): 199-215: Lewis Publishers
ABSTRACT?:	N
FULL ARTICLE?:	N
COMMENTS:	The authors describe experiences of implementing ozonation and chloramination. Implementation occurred in 1984 and 1987, respectively. This appears to provide some long-term water quality information.
ID:	**194**
DISINFECTANT:	Ozonation
KEY WORDS:	Disinfection Byproducts, history, chemistry, modeling, natural organic matter, epidemiology, toxicity, risks, regulations, chloramines, ozone, chlorine dioxide, ultraviolet disinfection, coagulation, softening, adsoprtion, oxidation, membranes, organic carbon, costs, Europe
TITLE:	Formation and Control of Disinfection By-Products in Drinking Water
AUTHORS:	Singer, Philip
YEAR:	1999
SOURCE:	American Water Works Association
ABSTRACT?:	Y
FULL ARTICLE?:	N
COMMENTS:	This book includes discussion of alternative disinfectants on DBP formation.
ID:	**195**
DISINFECTANT:	Ozonation
KEY WORDS:	Ozonation, coagulation, halogens, DBPs, chlorination, chloramines
TITLE:	The Impact of Ozonation and Coagulation on Disinfection By-Product Formation
AUTHORS:	Smith, M.; Cowman, G.; Singer, P.
YEAR:	1993
SOURCE:	1993 Annual Conference Proceedings: AWWA
ABSTRACT?:	Y
FULL ARTICLE?:	Y
COMMENTS:	The objective of this research was to evaluate the impact of ozonation on the formation of halogenated DBPs produced by subsequent chlorination and chloramination with and without coagulation.
ID:	**196**
DISINFECTANT:	Ozonation
KEY WORDS:	Iron, manganese removal, standards, chemistry, methods, training, filtration, hydraulics, microbiology, particle counting, adsorption, pretreatment, ozone, chlorine dioxide, hyrdogen peroxide, testing, equipment, filter media, softening, flotation, cleaning
TITLE:	Iron and Manganese Removal Handbook
AUTHORS:	Sommerfeld, E.
YEAR:	1999
SOURCE:	AWWA
ABSTRACT?:	Y
FULL ARTICLE?:	N
COMMENTS:	A reference covering chlorination, ozonation and manganese and iron.
ID:	**197**
DISINFECTANT:	Ozonation
KEY WORDS:	
TITLE:	Ultraviolet Disinfection Affects on Distribution System Water Quality
AUTHORS:	Spencer, C.; Cathcart, A.; Muri, j.; Hulsey, R.; Malley, J.; Capuzzi, A.
YEAR:	2003
SOURCE:	2003 Water Quality Technology Conference: AWWA
ABSTRACT?:	N
FULL ARTICLE?:	Y

COMMENTS:	The research described in this paper investigates disinfection (ozone followed by UV) effectiveness of Giardia, Crypto, and enteric viruses and synergistic impacts on DBP formation, biostability, corrosion control, and red water.
ID:	**198**
DISINFECTANT:	Ozonation
KEY WORDS:	T & O, oxidation, chlorine dioxide, chloramines, PAC, Alberta CA, North America
TITLE:	Oxidative Control of Taste and Odour in Drinking Water: The Edmonton Experience
AUTHORS:	Walker, G.; Lee, F.
YEAR:	1983
SOURCE:	35th Annual Convention, Western Canada Water and Sewage Conference
ABSTRACT?:	Y
FULL ARTICLE?:	N
COMMENTS:	This paper can be obtained through AWWA.
ID:	**199**
DISINFECTANT:	Ozonation
KEY WORDS:	Chlorine, choramination, nutrients, reactors, biofilm, disinfection, ozone, biofilters, statistical analysis
TITLE:	The Effects of Ozone and Chloramination on Distribution System Regrowth
AUTHORS:	West, T.; Daniel, P.; Labonte, J.; DeGraca, A.; Boozarpour, M.; Chemlick, D.; Hou, S.
YEAR:	1998
SOURCE:	1998 Annual Conference Proceedings
ABSTRACT?:	Y
FULL ARTICLE?:	Y
COMMENTS:	This paper discusses a study in which the primary objectives were to answer two key questions: how will a change from free chlorine to chloramine influence bacterial regrowth; and will chloramine control increase regrowth potential associated with an increase in nutrient levels due to ozonation of an unfiltered water?
ID:	**200**
DISINFECTANT:	Ozonation
KEY WORDS:	Oxidation, organic carbon, biological filters, chloramines, disinfection byproducts, decomposition, testing, ozone, hydrogen peroxide, chlorine dioxide ultrafiltration
TITLE:	Investigations of Oxidation, TOC Removal and Biological Filtration Effects on Chloramine Stability and DBP Formation at EBMUD CA
AUTHORS:	Wilczak, A.; Hoover, l.; Song, D.; Lai, H.; Kachur, S.; Long, C.
YEAR:	1999
SOURCE:	1999 Water Quality Technology Conference
ABSTRACT?:	Y
FULL ARTICLE?:	Y
COMMENTS:	This paper discusses EBMUD's first year of chloramines conversion.
ID:	**201**
DISINFECTANT:	Ozonation
KEY WORDS:	DBPs, chlorination, chloramines, ozone, weather, treatment plants, distribution systems, THMs, surveys, chlorinated organics
TITLE:	A Survey of Disinfection By-Products in Canadian Drinking Water
AUTHORS:	Williams, D.; Lebel, G.; Benoit, F.
YEAR:	1994
SOURCE:	Planning for Tomorrow: Proceedings of the Sixth National Conference on Drinking Water: AWWA
ABSTRACT?:	Y
FULL ARTICLE?:	N
COMMENTS:	This study was designed to determine the concentrations of halogenated DBPs in Canadian drinking water supplies where chlorine was used at some stage in the treatment process.
ID:	**202**

DISINFECTANT:	Ozonation
KEY WORDS:	Ozonation, utilities, costs, biodegradation, THMs, sludge, turbidity, organic carbon, DBPs
TITLE:	Use of Ozonation and FBT to Control THM Precursors
AUTHORS:	Yavich, A.; Masten, S.
YEAR:	2003
SOURCE:	JAWWA 95 (4): 159-171
ABSTRACT?:	Y
FULL ARTICLE?:	N
COMMENTS:	This study was conducted using a continuously operated ozonation-FBT system with no treatment.
ID:	**203**
DISINFECTANT:	UV disinfection
KEY WORDS:	UV disinfection, ozone, hydrogen peroxide, water quality, T & O, organic carbon, trihalomethanes, HPC, aldehydes
TITLE:	UV Oxidation for Drinking Water -- Feasibility Studies for Addressing Specific Water Quality Issues
AUTHORS:	Andrews, S.; Huck, P.; Chute, A.; Bolton, J.; Anderson, W.
YEAR:	1995
SOURCE:	1995 Water Quality Technology Conference Proceedings: AWWA
ABSTRACT?:	Y
FULL ARTICLE?:	Y
COMMENTS:	In this study, the feasibility of using UV irradiation, alone and in combination with hydrogen peroxide or ozone, to effect changes in selected drinking water quality parameters was investigated at bench scale.
ID:	**204**
DISINFECTANT:	UV disinfection
KEY WORDS:	Contaminant Candidate List, Microbial, DBPs, Microbiology, disinfection, treatment
TITLE:	Inactivation of Waterborne Emerging Pathogens by Selected Disinfectants
AUTHORS:	AwwaRF
YEAR:	2002
SOURCE:	AwwaRF
ABSTRACT?:	Y
FULL ARTICLE?:	N
COMMENTS:	This project determines the efficiencies of conventional and newly developing disinfectants for emerging waterborne pathogens.
ID:	**205**
DISINFECTANT:	UV disinfection
KEY WORDS:	DBPs, UV disinfection, treatment
TITLE:	UV Disinfection and Disinfection By-Product Characteristics of Unfiltered Water
AUTHORS:	AwwaRF
YEAR:	to be published
SOURCE:	AwwaRF
ABSTRACT?:	Y
FULL ARTICLE?:	N
COMMENTS:	This project will investigate the impact of variable water quality on the effectiveness of UV disinfection and address the impacts of turbidity, algae and total organic carbon. Was completed May 2003.
ID:	**206**
DISINFECTANT:	UV disinfection
KEY WORDS:	UV, ozone, treatment
TITLE:	Evaluation of Ozone and Ultraviolet Light
AUTHORS:	AwwaRF
YEAR:	to be published
SOURCE:	AwwaRF

ABSTRACT?:	Y
FULL ARTICLE?:	N
COMMENTS:	This project will compare the relative treatment performance of ozone followed by UV and UV only. This will cover finished and distribution system water quality, including red water and biostability control. Will be finished in June 2004.
ID:	**207**
DISINFECTANT:	UV disinfection
KEY WORDS:	Distribution system, microbiology, UV disinfection
TITLE:	Impact of UV Disinfection on Biological Stability
AUTHORS:	AwwaRF
YEAR:	To be published
SOURCE:	AwwaRF
ABSTRACT?:	Y
FULL ARTICLE?:	N
COMMENTS:	This project will evaluate the impact of UV disinfection on biofilm accumulation on drinking water distribution systems. Was completed in May 2003.
ID:	**208**
DISINFECTANT:	UV disinfection
KEY WORDS:	DBPs, groundwater, microbial, DBPs, UV
TITLE:	Evaluation of the By-products Produced by the Treatment of Groundwater with UV Radiation
AUTHORS:	AwwaRF
YEAR:	1996
SOURCE:	AwwaRF
ABSTRACT?:	Y
FULL ARTICLE?:	N
COMMENTS:	Reports if and what by-products are produced by UV disinfection and if UV influences DBP formation.
ID:	**209**
DISINFECTANT:	UV disinfection
KEY WORDS:	Cryptosporidium, microbial, DBPs, ozone
TITLE:	Synergistic Effects of Multiple Disinfectants
AUTHORS:	AwwaRF
YEAR:	2000
SOURCE:	AwwaRF
ABSTRACT?:	Y
FULL ARTICLE?:	N
COMMENTS:	Examines and quantifies the presence and significance of synergistic effects through the use of disinfectant strategies.
ID:	**210**
DISINFECTANT:	UV disinfection
KEY WORDS:	UV disinfection, biofilm, humic substance, organic carbon, oxidation, biodegradation, oxidants, dosage, NOM
TITLE:	Effects of UV Disinfection on Humic Stubstances and Biofilms
AUTHORS:	Camper, A.; Buls, J.; Goodrum, L.
YEAR:	2001
SOURCE:	2001 Water Quality Technology Conference: AWWA
ABSTRACT?:	Y
FULL ARTICLE?:	Y
COMMENTS:	The goal of this project was to investigate the use of UV light for disinfection of drinking water. The work plan included determining whether water containing NOM that is treated with UV light is more likely to support microbial growth on pipe surfaces and assessing changes to humic substances upon UV exposure.
ID:	**211**

DISINFECTANT:	UV disinfection
KEY WORDS:	Reactors, distribution systems, chemicals, UV disinfection, reduction, chlorine, biofilm, chlorine dioxide, disinfection residuals
TITLE:	Impact of UV Disinfection on Biological Stability in Distribution Systems
AUTHORS:	Dykstra, T.; O'Leary, K.; Chauret, C.; Andrews, R.; Gagnon, G.;
YEAR:	2002
SOURCE:	2002 Water Quality Technology Conference Proceedings
ABSTRACT?:	Y
FULL ARTICLE?:	Y
COMMENTS:	This study evaluates the synergistic effects of UV coupled with secondary disinfectants using a model distribution system.
ID:	**212**
DISINFECTANT:	UV disinfection
KEY WORDS:	
TITLE:	Synergistic Impacts of Ozone and UV
AUTHORS:	Edwards, J.; Hulsey, B.; Tran, T.; Sung, W.; Muri, J.; Malley, J.; Spencer, C.; Carthcart, A.;
YEAR:	2003
SOURCE:	2003 Annual Conference Proceedings: AWWA
ABSTRACT?:	N
FULL ARTICLE?:	Y
COMMENTS:	This paper presents early results of the research which will determine the effectiveness of ozone and UV inactivation for viruses, the synergistic impacts of UV on DBP formation, biostability, corrosion control and red water, and recommend optimum primary disinfection schemes.
ID:	**213**
DISINFECTANT:	UV disinfection
KEY WORDS:	Disinfectant residuals, chloramines, trihalomethanes, Wisconsin, ultraviolet radiation, chlorine, chlorine demand
TITLE:	Effects of UV Irradiation on Disinfectant Residuals and the THM Formation in Chlorinated and Chloraminated Drinking Water
AUTHORS:	Fleming, K.; Harrington, G.
YEAR:	2001
SOURCE:	2001 Microbial/Disinfectio By-Products Health Effects Symposium
ABSTRACT?:	Y
FULL ARTICLE?:	N
COMMENTS:	The effects of UV irradiation on disinfectant residuals and the formation of four THMS in chlorinated and chloraminated drinking waters, produced by three Wisconsin water utilities, were studied.
ID:	**214**
DISINFECTANT:	UV disinfection
KEY WORDS:	Biofilm, bacteria, distribution systems, UV disinfection, modeling
TITLE:	Impact of Ultraviolet Disinfection on Biofilm Growth in a Model Distribution System
AUTHORS:	Giese, N.; Darby, J.
YEAR:	2002
SOURCE:	2002 Water Quality Technology Conference Proceedings
ABSTRACT?:	Y
FULL ARTICLE?:	Y
COMMENTS:	This paper investigates the impact of UV disinfection on water distribution system biofilms using two model distribution systems.
ID:	**215**
DISINFECTANT:	UV disinfection
KEY WORDS:	T&O, Oxidants, Ozone
TITLE:	Evaluating Oxidants for the Removal of Model Taste and Odor Coumpounds from a

	Municipal Water Supply
AUTHORS:	Glaze, W.; Schep, R.; Chauncey, W.; Rugh, E.; Zarnoch, J.; Aieta, E.; Tate, C.; McGuire, M.
YEAR:	1990
SOURCE:	JAWWA: (82) 5: 79-84
ABSTRACT?:	Y
FULL ARTICLE?:	Y
COMMENTS:	Nine oxidizing agents were evaluated for removing six T&O compounds spiked into Colorado River water.
ID:	**216**
DISINFECTANT:	UV disinfection
KEY WORDS:	chlorine, filtration, pressure, trihalomethanes, cryptosporidium, organic halogen, haloacetic acids, GAC, DBPs, aldehydes, UV disinfection
TITLE:	Microbial Inactivation Strategies for the Future: UV, Chlorine, and DBPs
AUTHORS:	Kashinkunti, R.; Metz, D.; Linden, K.; Sobsey, M.; Moran, M.; Samuelson, A.
YEAR:	2002
SOURCE:	2002 Water Quality Technology Conference Proceedings
ABSTRACT?:	Y
FULL ARTICLE?:	Y
COMMENTS:	This paper describes results of Cincinnati Water Works evaluation of technologies for meeting requirements for microbial inactivation/removal in drinking water. Includes bench-scale tests and evaluates aldehydes and AOC concentrations.
ID:	**217**
DISINFECTANT:	UV disinfection
KEY WORDS:	Odor, T & O, oxidation, utilities, disinfection, Lake Michigan, hydrogen peroxide
TITLE:	Direct UV and UV Oxidation Processes for Treatment of Taste and Odor Causing Compounds in Water
AUTHORS:	Linden, K.; Rosenfeldt, E.; Johnson, S.; Melcher, B.
YEAR:	2002
SOURCE:	2002 Water Quality Technology Conference Proceedings
ABSTRACT?:	Y
FULL ARTICLE?:	Y
COMMENTS:	In an effort to control T & O, the application of UV treatment was examined for utilities using Lake Michigan as a drinking water supply source.
ID:	**218**
DISINFECTANT:	UV disinfection
KEY WORDS:	Oxidation, aldehydes, chlorination, NOM, DBPs, UV disinfection, THMs, HAAs
TITLE:	Bench-Scale Investigation into Comparative Evaluation of DBP Formation from Different UV/H2O2 Technologies
AUTHORS:	Liu, W.; Andrews, S.; Sharpless, C.; Stefan, M.; Linden, K.; Bolton, J.
YEAR:	2002
SOURCE:	2002 Water Quality Technology Conference Proceedings
ABSTRACT?:	Y
FULL ARTICLE?:	Y
COMMENTS:	Four classes of DBPs were examined to evaluate the impacts of different UV technologies combined with H2O2 irradiation.
ID:	**219**
DISINFECTANT:	UV disinfection
KEY WORDS:	
TITLE:	Advantages of Integrated Application Scenarios for Chlorine, Dioxide, Ozone, and UV
AUTHORS:	Long, B.; Hulsey, R.; Neemann, J.
YEAR:	2003
SOURCE:	2003 Annual Conference Proceedings: AWWA
ABSTRACT?:	N

FULL ARTICLE?:	Y
COMMENTS:	This paper describes the benefits of integrated disinfection.
ID:	**220**
DISINFECTANT:	UV disinfection
KEY WORDS:	Groundwater, disinfection, disinfection by-products, alternative disinfectants, UV radiation, UV disinfection, regulations, compliance
TITLE:	Evaluation of By-Products Produced by Treatment of Groundwater with Ultraviolet Irradiation
AUTHORS:	Malley, J.; Shaw, J.; Ropp, J.
YEAR:	1995
SOURCE:	AWWARF
ABSTRACT?:	Y
FULL ARTICLE?:	N
COMMENTS:	Performed UV irradiation on 20 groundwater and 10surface water supplies. Looked at DBP formation and compared with chlorine and chloramines used as post disinfectants.
ID:	**221**
DISINFECTANT:	UV disinfection
KEY WORDS:	Disinfection UV disinfection, hypochlorites, water treatment, residual chlorine, distribution systems, England
TITLE:	Letters - UV vs. Chlorine
AUTHORS:	McCarty, D.;
YEAR:	1986
SOURCE:	JAWWA: 78 (6): 6 pages
ABSTRACT?:	Y
FULL ARTICLE?:	Y
COMMENTS:	This letter was written in response to an earlier article. This letter describes the benefits realized by the British water treatment practice of using UV as a disinfectant.
ID:	**222**
DISINFECTANT:	UV disinfection
KEY WORDS:	Bacteria, chloramines, environment, pathogens, water storage, Escheria Coli, reservoirs, UV disinfection
TITLE:	Bacterial Survival after Ultraviolet Light Disinfection: Resistance, Regrowth and Repair
AUTHORS:	Mofdi, A.; Rochelle, P.; Chou, C.; Mehta, H.; Linden, K.; Malley, J.
YEAR:	2002
SOURCE:	2002 Water Quality Technology Conference Proceedings
ABSTRACT?:	Y
FULL ARTICLE?:	Y
COMMENTS:	This paper describes bench-scale experiments conducted with low-and medium-pressure UV lamps to inactivate pathogenic and non-pathogenic strains of E.coli and mixtures of naturally occurring heterotrophic bacteria.
ID:	**223**
DISINFECTANT:	UV disinfection
KEY WORDS:	Disinfection, drinking water, oxidation, T & O, bromate, water quality, hydrogen peroxide, dosage, UV radiation, organics
TITLE:	Using Ultraviolet Light to Achieve Multiple Water Quality Objectives
AUTHORS:	Mofidi, A.; Coffee, B.; Chou, C.; Liang, S.; Green, J.
YEAR:	2000
SOURCE:	2000 Water Quality Technology Conference Proceedings: AWWA
ABSTRACT?:	Y
FULL ARTICLE?:	Y
COMMENTS:	This study was conducted to determine whether UV light will provide multiple benefits, such as removing bromate, NDMA, MTBE, and MIB and geosmin.
ID:	**224**
DISINFECTANT:	UV disinfection

KEY WORDS:	Monitoring, water quality, South Berwick ME, groundwater, chemicals, sampling, bacteriophages, reactors, Indianapolis IN, UV disinfection, databases, O&M
TITLE:	Full-Scale Implementation of Ultraviolet Disinfection in Groundwater Treatment Systems
AUTHORS:	Petri, B.; Hunter, G.; Moran, D.; Nadeau, M.; Leach, J.; Malley, J.
YEAR:	2001
SOURCE:	AWWARF
ABSTRACT?:	Y
FULL ARTICLE?:	N
COMMENTS:	This study provides a long-term water quality and economic monitoring database on UV efficiency with two very different source waters.
ID:	**225**
DISINFECTANT:	UV disinfection
KEY WORDS:	UV radiation, organic carbon, water quality
TITLE:	The Effect of Ultraviolet Irradiation on Organic Matter in Natural Waters
AUTHORS:	Shaw, J.; Malley, J.
YEAR:	1995
SOURCE:	1995 Annual Conference Proceedings: AWWA
ABSTRACT?:	Y
FULL ARTICLE?:	Y
COMMENTS:	Experiments were conducted to evaluate the effects of UV irradiation on biological regrowth potential of two surface waters and two groundwaters.
ID:	**226**
DISINFECTANT:	UV disinfection
KEY WORDS:	Disinfection, chemicals, cultures, cryptosporidium, chlorine, alternative disinfectants, parasites, pathogens, residual chlorine, bacteria, viruses, microorganisms, UV radiation
TITLE:	Sequential Disinfection with UV Irradiation and Chlorine Species on Several Important Pathogens
AUTHORS:	Shin, G.; Ishida, G.; Linden, K.; Sobsey, M.
YEAR:	2002
SOURCE:	2002 Water Quality Technology Conference Proceedings: AWWA
ABSTRACT?:	Y
FULL ARTICLE?:	Y
COMMENTS:	This study determined the kinetics and extent of inactivation of pathogenic microorganisms by various doses of monochromatic low-pressure UV followed by free and combined chlorine.
ID:	**227**
DISINFECTANT:	UV disinfection
KEY WORDS:	Bromate, ozonation, DBPs, GAC, UV radiation, Bromide
TITLE:	Removal of Bromate After Ozonation During Drinking Water Treatment
AUTHORS:	Siddiqui, M.; Amy, G.; Zhai, W.; Ozekin, K.; Westerhoff, P.; Chandrakanth, M.
YEAR:	1994
SOURCE:	1994 Annual Conference Proceedings: AWWA
ABSTRACT?:	Y
FULL ARTICLE?:	Y
COMMENTS:	This paper evaluates ways to remove the bromate ion (BrO_3^-) after ozonation.
ID:	**228**
DISINFECTANT:	UV disinfection
KEY WORDS:	Flow, reservoirs, UV disinfection, Albany NY, water quality, water storage, regulations, compliance
TITLE:	UV Disinfection at a Finished Water Reservoir; the "In's and Out's" of Albany's New UV Facility
AUTHORS:	Simcoe, W.; Daviero, G.; Yonkin, M.
YEAR:	2002
SOURCE:	2002 Water Quality Technology Conference Proceedings

ABSTRACT?:	Y
FULL ARTICLE?:	Y
COMMENTS:	This paper details the complexities and challenges of using UV disinfection taking into account bi-directional flow and future change in operations when covered storage is implemented. This discusses the recent implementation of UV disinfection.
ID:	**229**
DISINFECTANT:	UV disinfection
KEY WORDS:	Disinfection Byproducts, history, chemistry, modeling, natural organic matter, epidemiology, toxicity, risks, regulations, chloramines, ozone, chlorine dioxide, ultraviolet disinfection, coagulation, softening, adsoprtion, oxidation, membranes, organic carbon, costs, Europe
TITLE:	Formation and Control of Disinfection By-Products in Drinking Water
AUTHORS:	Singer, Philip
YEAR:	1999
SOURCE:	American Water Works Association
ABSTRACT?:	Y
FULL ARTICLE?:	N
COMMENTS:	This book includes discussion of alternative disinfectants on DBP formation.
ID:	**230**
DISINFECTANT:	UV disinfection
KEY WORDS:	Ultraviolet Disinfection, dosage, groundwater, water quality, chloramines, chlorine, pilot plants, viruses, copper pipe, PVC pipe, cast iron pipe, performance, fouling, costs, biofilm
TITLE:	Evaluation of Ultraviolet Technology in Drinking Water Treatment
AUTHORS:	Snicer, G.; Malley, J.; Margolin, A.; Hogan, S.
YEAR:	1996
SOURCE:	1996 Water Quality Technology Conference: AWWA
ABSTRACT?:	Y
FULL ARTICLE?:	N
COMMENTS:	The major objectives of the study were to: evaluate the long-term operational parameters and problems associated with UV disinfection; examine the effects of groundwater quality on UV effectiveness; evaluate chlorine and chloramines as possible secondary disinfectants and determine their effectiveness on copper, iron, and PVC pipe.
ID:	**231**
DISINFECTANT:	UV disinfection
KEY WORDS:	
TITLE:	Ultraviolet Disinfection Affects on Distribution System Water Quality
AUTHORS:	Spencer, C.; Cathcart, A.; Muri, j.; Hulsey, R.; Malley, J.; Capuzzi, A.
YEAR:	2003
SOURCE:	2003 Water Quality Technology Conference: AWWA
ABSTRACT?:	N
FULL ARTICLE?:	Y
COMMENTS:	The research described in this paper investigates disinfection (ozone followed by UV) effectiveness of Giardia, Crypto, and enteric viruses and synergistic impacts on DBP formation, biostability, corrosion control, and red water.
ID:	**232**
DISINFECTANT:	UV Disinfection
KEY WORDS:	
TITLE:	UV Disinfection Guidance Manual: Proposal Draft
AUTHORS:	USEPA
YEAR:	2003
SOURCE:	USEPA
ABSTRACT?:	N
FULL ARTICLE?:	N
COMMENTS:	

ID:	**233**
DISINFECTANT:	UV disinfection
KEY WORDS:	UV disinfection, DBPs, water contamination, microorganisms, oxidation, performance
TITLE:	UV Treatment for the Control of Chemical and Microbiological Contaminants
AUTHORS:	Wolfe, R.; Stewart, M.; Hwang, C.; Yates, R.
YEAR:	1995
SOURCE:	1995 Water Quality Technology Conference Proceedings: AWWA
ABSTRACT?:	Y
FULL ARTICLE?:	Y
COMMENTS:	The purpose of this paper is to document the merits of UV treatment for the control of microbial and chemical contaminants in water.
ID:	**234**
DISINFECTANT:	UV disinfection
KEY WORDS:	THMs, HAAs, UV radiation, biological filters, ozonation, DBPs, dosage
TITLE:	Impacts of Medium Pressure UV on THM and HAA Formation in Pre-UV Chlorinated Drinking Water
AUTHORS:	Zheng, M.; Andrews, S.; Bolton, J.
YEAR:	1999
SOURCE:	1999 Water Technology Conference Proceedings: AWWA
ABSTRACT?:	Y
FULL ARTICLE?:	Y
COMMENTS:	This study describes an investigation of the impact of medium-pressure UV irradiation on the removal or formation of THM or HAA precursors. This is included because it discusses DBPs and UV.
ID:	**235**
DISINFECTANT:	Nanofiltration
KEY WORDS:	
TITLE:	Nanofiltration: Long-Term Effects on Water Quality
AUTHORS:	Gatel, D.; Cotte, M.; Pele, B.;
YEAR:	2003
SOURCE:	2003 Annual Conference Proceedings: AWWA
ABSTRACT?:	N
FULL ARTICLE?:	Y
COMMENTS:	This is a study documenting the water quality changes over three years after nanofiltration implementation. Data are compared with water quality conditions for two years previous to nanofiltration.
ID:	**236**
DISINFECTANT:	General information
KEY WORDS:	
TITLE:	Excerpt from the Community Water System Survey 2000
AUTHORS:	USEPA
YEAR:	2000
SOURCE:	USEPA
ABSTRACT?:	N
FULL ARTICLE?:	N
COMMENTS:	A series of 8 tables indicating how many utilities use selected treatment processes, including pre and post disinfection.
ID:	**237**
DISINFECTANT:	Taste
KEY WORDS:	
TITLE:	A Comparison of French and American Consumer Sensitivities to Chlorinous Flavors
AUTHORS:	Mackey, E.; Piriou, P.; Bruchet, A.
YEAR:	2003
SOURCE:	2003 Water Quality Technology Conference Proceedings: AWWA

ABSTRACT?:	N
FULL ARTICLE?:	Y
COMMENTS:	
ID:	**238**
DISINFECTANT:	Taste
KEY WORDS:	
TITLE:	Comparison of Customer Sensitivity and Comparison with Tap Water Quality
AUTHORS:	Mackey, E.; Baribeau, H.; Suffet, I.
YEAR:	2002
SOURCE:	2002 Water Quality Technology Conference Proceedings: AWWA
ABSTRACT?:	Y
FULL ARTICLE?:	Y
COMMENTS:	This study looks at the question: to what extent are chlorinous flavors a problem with public perception of tap water aesthetics and safety?

APPENDIX B
ISSUE PAPER NO. 1
CHANGES IN MICROBIAL QUALITY

AwwaRF PROJECT NO. 2940
LONG-TERM EFFECTS OF DISINFECTION CHANGES
ON WATER QUALITY

Issue Paper
on
Changes in Microbial Quality
by
Anne Camper

CHLORAMINES

Key Factors Affecting Water Quality Changes

When utilities switch from free chlorine to monochloramine, there are several water quality changes that may impact the microbiology of the system. Monochloramine has much less "disinfectant power" than free chlorine at the same concentration; this is both an advantage by allowing for a higher residual at greater water ages and a disadvantage because of the CT required for disinfecting any contaminants introduced into the distribution system. For example, LeChevallier et. al, (1988) showed that viable bacterial counts of drinking water bacteria were reduced by 99% in one minute with 0.08 mg/L free chlorine, but the same level required 94 mg/L of monochloramne. However, monochloramine may be more effective at controlling heterotroph and coliform regrowth than free chlorine (LeChevallier et. al, 1988). This is likely due to the slower reaction rate of monochloramine, which permits greater penetration into the biofilm than free chlorine. This hypothesis is substantiated by observations in the field where particle-associated coliforms were more protected from chlorination if the particles were greater than 7 micrometers, while the performance of monochloramine was unaffected (Berman et. al, 1988). More detailed laboratory experiments with defined populations of bacteria have also demonstrated that there is a reaction-diffusion limitation for free chlorine that may not exist for monochloramine in biofilms (Xu et. al, 1996; Chen and Stewart 1996; Stewart and Raquepas 1995).

The decay of monochloramine to produce free ammonia or the over-addition of ammonia during monochloramine formation can lead to nitrification problems. Nitrification can be difficult to control and commonly leads to an increase in the number of heterotrophic bacteria found in the water. Recent research suggests that the monochloramine decay is pH dependant and will occur as water age increases (Valentine et. al, 1998; Harrington and Noguera 2003). Therefore, attention must be paid to these and other issues of good distribution system maintenance and operation (flushing and cleaning, reducing dead ends, etc.) to ensure that microbiological quality is maintained when chloramination is practiced.

Positive Impacts on Water Quality

In theory, and based upon some highly publicized early investigations, one would expect that monochloramine would be a better disinfectant in the control of distribution system biofilms. In a demonstration field scale study, chloramines were more effective at reducing the number of biofilm total coliforms and HPC than chlorine and the water also had a more stable residual (Neden et. al, 1992). In another study where statistical evaluation of the influence of chloramine concentration on attached microbial populations in a distribution system was made, an inverse relationship was established (Donlan and Pipes 1988). In a field survey of 31 utilities, those that used chloramines had 0.51% of 35,159 samples with coliforms, while chlorinated systems had 0.97% of 33,196 samples. More importantly, the average density of coliforms in the water of the chlorinated systems was 35 times higher than that in the chloraminated systems (LeChevallier et. al, 1996a). In this case, however, only the bulk water and not the biofilms were analyzed. In a case study where a distribution system that was in violation of the Total Coliform Rule switched to monochloramine and corrosion control, their coliform problem was eliminated (Schreppel et. al, 1997). These studies suggest that the use of monochloramine as a secondary disinfectant is at least as effective as free chlorine and in general tends to provide better microbiological control. This is not always true, however; in a study where a system switched from monochloramine back to free chlorine, no overall deterioration in microbial water quality (HPC or coliforms) was detected (Muylwyk et. al, 2001).

The advantage of monochloramine over free chlorine was also shown in a pilot plant study where biofilms were grown on corroding mild steel pipes (Camper 1996). The presence of a low (0.2 mg/L) chlorine residual produced a consistent pattern of elevated coliform counts in the biofilm over control systems without the disinfectant, while the monochloraminated systems (0.4 mg/L) had decreased coliform counts when compared to the control. A similar trend was seen with the heterotrophs in the biofilms; compared to the control, the counts increased in the presence of low residual concentrations of free chlorine and decreased when the disinfectant was monochloramine. In other pilot experiments using the same mild steel system, similar trends were seen. In three of six experiments where monochloramine was used, coliform counts decreased and in the other three instances the counts were unaffected. This is compared to the experiments where chlorine was used, where in all cases the presence of low concentrations of free chlorine increased coliform and heterotroph biofilm numbers (Camper and Jones, 2000). These results may be partially explained by the observation that free chlorine enhanced the presence of γ-subclass of the *Proteobacteria* (Batte et. al, 2000), the branch in which the coliforms reside, over that of monochloraminated reactors.

There are limited published reports on the relative efficacy of monochoramine vs free chlorine on pathogenic bacteria. In the one report found in the peer reviewed literature, a case study of the occurrence of *Legionella* in hospitals, the authors claim that there may be up to a 90% reduction in infections caused by this organism if the disinfectant residual is monochloramine instead of free chlorine (Kool et. al, 1999). The authors do caution, however, that these studies should be confirmed by other research. A follow-up study by the Centers for Disease Control has quoted that the percentage of buildings positive for this organism in their drinking water fell from 20% to 6% within one month after implementation of chloramination (Barry Fields, Centers for Disease Control). Another laboratory study demonstrated the presence of *Legionellae* in drinking water biofilms in the presence of either chlorine or monochloramine (Williams and Braun-Howland 2003), but the concentrations of the disinfectants were vastly different (0.2 mg/L and up to 4 mg/L, respectively). Also related to public health is the presence of endotoxins in water, which are a component (lipopolysaccharides) of the gram-negative cell wall. Monochloramine was found to be less effective than free chlorine at inactivating endotoxin (0.7 vs 1.4 units/ml hr), but the rates for both disinfectants are quite slow (Anderson et. al, 2003).

In other cases, the distinction between chlorine and monochloramine on distribution system biofilms has been less clear. LeChevallier et. al, (1990) utilized a model pipe system to determine the influence of chlorine and monochloramine disinfectant efficacy on biofilms which had accumulated on several pipe materials. Biofilms on galvanized, copper, or PVC surfaces were readily disinfected by free chlorine or monochloramine (1 mg/L). Iron pipe surface-associated bacteria were more susceptible to monochloramine than free chlorine, but higher monochloramine concentrations (4 mg/L) were required than on the other surfaces. In other experiments where monochloramine and free chlorine were used to disinfect biofilms grown on polycarbonate (plastic), the reduction of biofilm culturable cell counts by either disinfectant was approximately 99.9%. However, this reduction required substantially more time with the monochloraminated system (46 days) than with the chlorinated system (approximately 14 days) (Morin and Camper 1997). To assess the interactions between pipe materials, organic carbon levels and disinfectants, studies using annular reactors with ductile iron, polyvinyl chloride (PVC), epoxy, and cement-lined coupons were carried out (Camper et. al, 2003). Laboratory annular reactors were fed biologically treated water with and without 0.2 mg/L residual free chlorine or monochloramine, in the presence or absence of 0.5 or 2.0 mg/L humic substances. In this study, the type of disinfectant did not lead to significant differences in effluent and biofilm counts. In comparison, work done by Gagnon et. al, (submitted) showed that chorine was more effective at disinfecting bulk and biofilm organisms than monochloramine in reactors with either polycarbonate or cast iron coupons. Similarly, work done at the U.S. EPA (Clark and Sivaganesan 1999) showed that chlorine was consistently more effective than monochloramine in reducing bulk and biofilm total direct counts on cement, PVC and polyethylene coupon surfaces. In this study, the free chorine residual was 0.2 - 0.5 mg/L while the monochloramine residual was 1 mg/L. Another laboratory study showed that 0.5 mg/L free chlorine adequately controlled bacterial numbers on polycarbonate and ductile iron, but monochloramine levels at 2.0 mg/L were required for the same level of control on the ductile iron (Ollos et. al, 1998).

One of the difficulties in making comparisons between chlorine and monochloramine is the presence/absence of corroding pipe surfaces. Since chlorine is so reactive, it is possible that its activity against organisms may be limited not only by reaction with biofilm material, but also with the metal surface itself. Chen et. al, (1993) showed that *Pseudomonas aeruginosa* biofilms treated with 4 mg/L monochloramine were most resistant to disinfectant when grown on mild steel vs. stainless steel. They also determined that corrosion products deterred the disinfection of homogenized biofilm bacteria. To see if corrosion control may improve disinfectant efficacy, work has been done with reactors exposed to free chlorine or monochloramine with and without the presence of corrosion control chemicals (orthophosphate, polyphosphate, pH adjustment). In ductile iron annular reactors without the presence of corrosion control, there was a substantial demand for both disinfectants. Influent concentrations of 3.25 mg/L in both cases resulted in barely measurable residuals. When compared to the control reactor (no disinfectant), the chlorinated reactor had higher numbers of culturable cells in the biofilm. The monochloraminated counts were lower than that of the parallel control system, and lower than the numbers obtained from the previous experiment where chlorine was used. Overall, monochloramine appeared to be a slightly better disinfectant under these conditions and the addition of a corrosion control chemical improved biofilm control, especially in the chloraminated reactor, with orthophosphate providing the best result (Abernathy and Camper 1998). These results support field observations that monochloramine may be more effective at controlling biofilms grown on corroding ferrous metal distribution system materials, and the combination of monochloramine and corrosion control for improving microbial water quality has been demonstrated (Schreppel et. al, 1997). Since corroding surfaces are often considered to be related to problems in full-scale distribution systems, the use of monochloramine as a secondary disinfectant may be indicated when the primary material in the network is unlined cast or ductile iron pipes. In addition to the use of chloramines, provision of adequate corrosion control may also improve biofilm control. However, even this observation must be extrapolated with care. As shown by Batte et. al, (2000), phosphate added to a reactor system containing non-corroding material (polycarbonate) seemed to protect the biofilms from both chlorine and monochloramine (both at an applied dose of 0.5 mg/L) when compared to systems with the disinfectants but lacking the phosphate.

Another potential advantage of monochloramine over free chlorine relates to its relative lack of reaction with organic matter. When the concern is production of biodegradable organic matter from recalcitrant natural organics in water that may then support bacterial growth, chloramines have been shown to produce the lowest levels at realistic drinking water doses of four technologies (ozone, UV, chlorine, monochloramine) when soil derived humic substances were the NOM source (Buls 2000). This is in spite of the observation that a significant mechanism for monochloramine decay is the oxidation of natural organic matter (Vikesland et. al, 1998).

With all of these investigations, it is crucial that the reader understand the exact conditions under which the data were collected. Water quality, including temperature, organic matter, pH, etc. will impact the results. In full scale systems, the point of addition of chlorine and ammonia will impact suspended bacterial numbers. In a pilot plant study, it was shown procedures other than pre-ammioniation resulted in lower numbers (Means et. al, 1986). The synergistic effect of chlorine and monochloramine has been shown with *E. coli*; better disinfection of suspended cells occurred when free chlorine existed in the presence of monochloramine (Kouame and Haas 1991). Similarly, Momba and Binda (2002) showed that primary chlorination followed by chloramination retarded biofilm formation by HPCs longer than chlorine alone. Ward et. al, (1984) also showed that pH, order of application of chlorine and ammonia and chlorine to ammonia ratios had an impact on the disinfection of suspended bacteria.

The method of determining bacterial numbers is also very important. It has been shown that monochloraminated bacteria retain significant respiratory activity even though the plate counts decreased dramatically; under conditions where an established two species biofilm was exposed to 4 mg/L monochloramine, culturable counts decreased by 99.9% and respiring cells by only 10-15% (Stewart et. al, 1994). And as shown above, the pipe material will also have a notable affect.

Potential Adverse Impacts on Water Quality

There are several potential adverse impacts of chloramination on microbial water quality. The first of these is the distribution of water with a lower capacity for inactivating suspended organisms. Consequently, utilities generally trend to have a higher monochloramine residual in their distribution systems than a free residual. If the distribution system is intact, then this concern is minimal. The difference in disinfection efficacy is more serious if there are either inadvertent or deliberate breaches of the integrity of the system. In these cases, there may be inadequate CT between the point of contamination and the next consumer. This problem may also occur in distribution systems with inadequate free chlorine or chlorine dioxide residuals and is therefore not unique to monochloramine.

A potential adverse impact on water quality that is limited to chloraminated systems is the potential for nitrification. Although it is beyond the scope of this document to expand on this phenomenon, it is clear that the potential for nitrification is of crucial concern for the industry. During nitrification events, bacteria that can utilize ammonia as an energy source produce nitrite, which is then converted by other organisms to nitrate. An outcome is further depletion of the disinfectant residual, leading to increased numbers of heterotrophic bacteria. In addition, it may be possible for the heterotrophs to proliferate by utilizing the organic carbon that is fixed by the autotrophic nitrifying bacteria. Control of nitrification may be gained by intermittent conversion to free chlorine, provided that undesirable water quality issues associated with free chlorine (disinfection by products, etc.) can be avoided. Although this was once a reasonably common approach, utilities are learning that nitrification can be managed by controlling pH, water age, and the initial chlorine to ammonia ratio.

Unknown Issues/Information Gaps

- Need for study where chlorine and chloramine levels are varied to determine at what concentrations equivalent results for biofilm and suspended cell control are seen (if they are seen). Study should include corroding surface, inert surface, and corrosion control.
- Better data/more complete studies on the survival of opportunistic pathogens in the presence of both disinfectants.
- More information on the impact of these disinfectants on microbial water quality in health care facilities, other buildings.

OZONE

Key Factors Affecting Water Quality Changes

Ozone is a powerful oxidizing agent that may be used as a primary disinfectant in water treatment because of its ability to inactivate cysts and oocysts. Because it is so reactive, it is highly unlikely that there are true "ozone resistant" organisms in the environment, ensuring that this will be a viable disinfectant over time. However, care must be taken to ensure that ozone is applied at sufficiently high doses and for adequate time to facilitate inactivation. Ozone may also reduce taste and odor compounds, remove color, or reduce disinfection demand/disinfection by products by oxidizing some of the organic matter. The oxidizing capacity of this disinfectant has two key impacts on microbial water quality; (1) it inactivates organisms in the bulk water that are difficult to disinfect using other technologies and (2) it may enhance the possibility of downstream microbial growth due to the conversion of recalcitrant organic matter to a more biologically utilizable form. The second possible impact may be mediated by additional water treatment steps downstream from ozonation (biological filters) or by careful optimization of secondary disinfection strategies. It is also possible that the NOM in certain waters is low enough or of a form that is not modified by ozone and subsequent microbial activity is not impacted.

Positive Impacts on Water Quality

The key positive impact of ozone on microbial water quality has been good disinfection of the bulk water prior to entering the distribution system. Good inactivation of source water organisms that may be associated with disease is always desirable. However, it is known (as described below) that ozonation may increase the biodegradabilty of the natural organic matter present in water, and this may lead to biofilm proliferation in the distribution system. If the water is further treated after ozonation, or if an adequate secondary residual is maintained even in the absence of subsequent treatment, it is possible that microbial growth in the distribution system can be maintained or lowered. For example, Sharp et. al, (2001), in a pilot study, showed that the biological stability of ozonated water could be greatly enhanced if a chlorine residual was retained. As expected with this particular water, if the chlorine was not present, there was more biofilm accumulation after ozonation than in non-ozonated water. Similarly, in a survey of full scale utilities that reduced biodegradable organic content of their water, bulk water counts were more sensitive to chlorine residual than the content of degradable organic matter (Najm et. al, 2000). This point illustrates that one cannot consider the effect of ozone alone; the downstream processes will also ultimately influence microbial water quality. It is also possible that ozonation will not adversely influence the biodegradable fraction of the organic matter, and therefore would not have any particular impact on biofilm growth.

Potential Adverse Impacts on Water Quality

Water that has been ozonated often has elevated levels of lower molecular weight organic compounds; these compounds have been associated with increased biofilm development downstream (van der Kooij et. al, 1989; Price 1994; LeChevallier et. al, 1996b). Goel et. al, (1995) reported that the fraction of recalcitrant natural organic matter in water made available for microbial growth was increased after ozonation, but the numerical value varied from site to site. This has also been substantiated by van der Kooij et. al, (1982), Werner and Hambsch (1986), Servais et. al, (1987), and Speitel et. al, (1993). Buls (2000) also showed that ozone applied to soil-derived humic substances increased the bio-available fraction. In pilot studies, it was shown that ozonated water supported biofilm growth, but if the same non-ozonated source water was chlorinated at a low level (0.05 mg/L) no biofilm was formed on plastic pipes (Lund and Ormerod 1995).

If biodegradable organic matter is not reduced, or if secondary disinfection is not adequate, there may be the chance for regrowth of opportunistic pathogens in biofilms. In a field study of eight distribution systems, it was shown that mycobacteria in distribution system water samples and biofilms were positively correlated with elevated assimilable organic carbon and biodegradable organic carbon levels (Falkinham et. al, 2001). Also, these organisms are resistant to disinfection by chlorine and monochloramine (Taylor et. al, 2000). If the general biofilm growth is enhanced by increased levels of degradable organic matter, it is possible that other organisms of health concern that reside in that biofilm may exhibit enhanced persistence or protection from disinfection or predation.

Unknown Issues/Information Gaps

- Better methods for assessing the biodegradable fraction of NOM produced by ozonation or other disinfectants; improved ability to predict microbial growth based on measurable water quality parameters.
- An understanding of how utilities may use corrosion control to mitigate potential increases in microbial growth due to ozonated NOM.

CHLORINE DIOXIDE

Key Factors Affecting Water Quality Changes

Chlorine dioxide is an oxidant with high disinfection capability against *Cryptosporidium* (Chauret et. al, 2001) and lower chlorinated DBP formation potential than free chlorine. When chlorine dioxide decays, it produces chlorite, which has been shown to reduce the activity of ammonia oxidizing bacteria and subsequent nitrification (McGuire et. al, 1999). Disinfection efficacy of chlorine dioxide is less pH dependent than free chlorine, posing a considerable benefit for waters at elevated pH. It has also been shown that chlorine dioxide may have less impact on the corrosion of cast iron pipe than free chlorine or monochloramine, resulting in lowered concentrations of iron in the water (Eisnor et. al, 2001). Since corrosion has been shown to be a key factor in regrowth (see discussion on monochloramine above), there may be an added benefit from chlorine dioxide over other secondary disinfectants in controlling biofilms in distribution systems.

Positive Impacts on Water Quality

Because of its oxidizing capability, one would expect that chlorine dioxide would have the ability to effectively disinfect the bulk water, and in most cases this is seen. For example, strains of *Mycobacterium avium* were most susceptible to ozone and chlorine dioxide at CT levels orders of magnitude lower than that for chlorine (Taylor et. al, 2000). Botzenhart et al (1993) demonstrated that *Legionella pneumophila* was more susceptible to chlorine dioxide than *Escherichia coli*. In a hospital setting, a shock does of chlorine dioxide (50-80 mg/L for 1 hr) followed by a constant dose of 3-5 mg/L was used to disinfect a water system. The authors report no *Legionella* in the water and reduced biofilm numbers (Walker et. al, 1995). Limoni and Telsch (1985) reported that water containing a 0.2 mg/L chlorine dioxide residual had low suspended heterotrophic counts. Chlorite, a degradation product of chlorine dioxide, has also been seen to have bacteriostatic properties with a positive influence on microbial counts in drinking water (Masschelein 1991). McGuire et. al, (1999) have shown that chlorite can reduce nitrification and biofilm growth in distribution systems at concentrations below the current MCL. Recently completed work by Gagnon et. al, (submitted) showed that chlorine dioxide had the greatest disinfection efficiency in a model reactor system, followed by free chlorine and monochloramine. Per unit of disinfectant, chlorine dioxide was 3-5 fold more effective on suspended cells than chlorine and 6-9 fold more effective than monochloramine. For biofilms, the results generally were a 2-4 and 5-6 fold increase in efficacy. However, chlorine dioxide appeared to have the same effect on biofilm HPC on iron surfaces as chlorine, with a slight control of biofilm numbers.

As with all other disinfectants described, there are circumstances where experience deviates from the expected result. Berg et. al, (1988) showed that there was a resistant fraction of *Legionella pneumophila* and *Escherichia coli* at constant chlorine dioxide doses that could not be attributed to aggregation of the cells.

Potential Adverse Impacts on Water Quality

There are relatively few published full-scale or experimental studies where chlorine dioxide has been compared to other disinfectants. The few that have been published tout the advantages of chlorine dioxide and no significant negative impacts are noted. There are some potential adverse impacts, however. The ability of this disinfectant to provide selective pressure for undesirable bacteria in the distribution system may exist. It is also possible that this disinfectant may alter the biodegradability of organic matter in the distribution system. These potential impacts are speculative at this time, and research work is needed to confirm or deny them.

Unknown Issues/Information Gaps

- Changes in microbial ecology created by chlorine dioxide; is there a selection for opportunistic pathogens in distribution systems and health care facilities.
- Further exploration of the possible synergistic interactions between ozone and UV with chlorine dioxide.
- Better insight on behavior of chlorine dioxide in a wider variety of water qualities and seasonal effects.
- Interaction with adsorbed natural organic matter at the pipe wall and impact on corrosion is unknown; both would have an affect on the growth of biofilms on pipe surfaces.
- Better understanding of the effects of a change from free chlorine or monochloramine to chlorine dioxide on biofilms and microbial populations in controlled settings.

UV

Key Factors Affecting Water Quality Changes

UV light as a primary disinfectant is attractive because it does not add any chemicals to the water, and halogenated by-products are not formed. As with ozone, UV light, when applied properly, has greater efficacy against cysts and oocysts than free chlorine. There have been some concerns about the ability for UV light to influence the biodegradability of natural organic matter and the potential for decreasing the biological stability of the water. As discussed below, this concern appears to be relatively unfounded. In one case, (Lund and Omerod 1995) reported less biofilm accumulation in a distribution system receiving UV-treated water compared to ozonated water, which the investigators attributed to inhibitory hydroxyl radicals produced during UV treatment. However, this phenomenon was not universal; in systems with well established biofilms, UV treatment increased biofilm growth (Lund and Hongve 1994). These points illustrate the contradictory evidence that may arise when comparing disinfectants, and demonstrates the need to understand the circumstances under which the data were collected.

Positive Impacts on Water Quality

When properly applied, UV light can be effective at reducing bacterial numbers and inactivating pathogenic organisms (Harris et. al, 1987; Parrota and Bekdash 1998; Wolfe 1990). In a recently published report (Lehtola et. al, 2003), 90% of bacteria in drinking water were inactivated at doses below 50 mWs/cm2. Practical experience has shown that this may not always be the case. In a study where household UV units were tested, water samples downstream from the UV unit contained levels of bacteria only slightly lower than those before treatment (Burr et. al, submitted). Some of these samples were also positive for coliforms. A field study also showed that in one facility, counts were higher after UV treatment (Lehtola et. al, 2003). In both cases, the lamps may have been fouled or biofilm organisms may have detached downstream from the UV lamps.

One expectation of UV treatment would be that the water entering the distribution system would carry a much lower microbial load and therefore biofilm growth may be decreased. When

UV treated tap water was fed to reactors, the rate and amount of biofilm accumulated was the same as that in non-UV treated systems (Burr et. al, submitted; Dykstra et. al, 2002). This is in contrast to the findings of Lund and Omerod (1995) who reported less biofilm accumulation in a distribution system receiving UV-treated water.

In the cases listed above, there was no secondary disinfection, and this may have a pronounced effect on the ability to maintain microbial water quality in the distribution system. There have been two published reports on the interactions between UV and chorine, chloramines, and/or chlorine dioxide. When UV treated water was subsequently treated with chorine or chloramines, there was no difference between the performance of the two secondary disinfectants, but biofilm growth was reduced compared to the UV-treated only water (Snicer et. al, 1996). In another study where water was subsequently treated with chlorine, chlorine dioxide or chloramine, synergistic effects between UV and the secondary disinfectant were noted (Dykstra et. al, 2002). In reactors, chlorine dioxide and chlorine were most effective at reducing suspended counts followed by chloramines. In all cases, the counts were lower than if the secondary disinfectants were used alone. For biofilms, a combination of UV and chlorine dioxide or monochloramine gave better results than those obtained without UV treatment. For chlorine, there was no difference. As a control, it was noticed that UV alone did not result in elevated levels of suspended or biofilm heterotrophs. These results and all others listed in this section show the need to evaluate UV light along with other relevant water quality conditions, including the need for secondary disinfection.

Potential Adverse Impacts on Water Quality

It has been hypothesized that UV treatment may increase the bioavailability of natural organic matter in water. For example, Kieber et. al, (1990) demonstrated that natural UV irradiation of humic substances resulted in the production of low molecular weight carbonyl groups. If this occurred, then one might expect to see the same potential for regrowth that often accompanies ozonation. One investigation with well established biofilms that were subjected to UV treated water showed increased biofilm growth (Lund and Hongve 1994). In another case, a one log increase in biofilm numbers was seen post-UV treatment (Snicer et. al, 1996). However, this phenomenon is not universal. In a study specifically designed to look at biofilm growth in a model distribution system (Shaw et. al, 2000), four source waters were treated with UV light with a low pressure mercury vapor source at 254 nm and at a dose of 130 mW-s/cm^2. There was no statistical evidence for increased assimilable organic carbon or biodegradable organic carbon after UV treatment, although there were slight upward trends in these characteristics in a few samples. These authors suggest that utilities should test their sources before adopting UV disinfection. In recently submitted work (Burr et. al, submitted) it was shown that reasonable doses of UV did not increase the bioavailability of humic substances, and that biofilm growth was not enhanced by UV treatment of the organic carbon. In a batch study, Lehtola et. al, (2003) showed that UV treatment actually decreased AOC and small molecular weight organic molecules, and that the release of microbially available phosphorus was not enhanced at reasonable UV doses.

Unknown Issues/Information Gaps

- More information/controlled studies on the synergistic effects of UV treatment followed by chlorination/chloramination/chlorine dioxide application.
- Greater understanding of the impact of UV light on the degradable fraction of NOM.
- For small systems and point of use/point of entry UV units, a better understanding of the conditions that lead to decreased efficacy over time against suspended cells.
- Controlled studies comparing the efficacy of realistic UV wavelengths, including 254 nm, for inactivation of organisms.

BOOSTER CHLORINATION

Key Factors Affecting Water Quality Changes

Booster chlorination should have the same impact on distributed water quality as the ability to maintain adequate chlorine residual throughout the distribution system. There are no published accounts of the impact of booster chlorination on microbial water quality, but one would assume that the results would be equivalent to those seen when a measurable residual reaches parts of a distribution system where inadequate disinfection occurred. In such cases, there may be the reduction of suspended bacteria, but there may be temporary increases due to destabilization of surface deposits and entrainment of biofilm bacteria.

It is critical to keep in mind that the interaction between the organisms and the disinfectant is also influenced by water quality, the organic matter, and the pipe surface. Even with free chlorine, the most well understood disinfectant, the case is complicated. For example, in one field scale study, no correlation was found between free chlorine residuals and the number of HPC organisms per unit surface area (Hudson et. al, 1983). Many reports exist that demonstrate the relative lack of sensitivity of biofilm cells to disinfectants. This effect is even more pronounced if the biofilms are grown on reactive iron surfaces (Kerneis et. al, 1994; LeChevallier et. al, 1990; Chen et. al, 1993). It has been noted that increased corrosion rates decrease the efficacy of free chlorine against biofilm organisms, (LeChevallier et. al,1993), pointing to the importance of pipe materials.

Potential impacts are discussed in the section on chloramines, where the primary comparison has been with free chlorine.

Unknown Issues/Information Gaps

- Studies illustrating the response of model or actual distribution systems with established biofilms to the sudden or gradual increase in free chlorine residuals.
- Better insight on the interaction between free chlorine, corrosion, and microbial water quality.

APPENDIX C
ISSUE PAPER NO. 2
CHANGES IN CHEMICAL QUALITY - DBPS

AwwaRF PROJECT NO. 2940
LONG-TERM EFFECTS OF DISINFECTION CHANGES ON WATER QUALITY

Issue Paper
on
Changes in Chemical Quality - DBPs
by
Philip C. Singer

CHLORAMINES

If systems using free chlorine as a primary and secondary disinfectant convert to combined chlorine (monochloramine) for secondary disinfection, they should be able to effectively control the continuing formation of trihalomethanes (THMs) and haloacetic acids (HAAs) in their distribution systems. If applied properly, monochloramine will not produce additional THMs or trihaloacetic acids, but it will react with residual organic carbon in the water to some degree to produce dihaloacetic acids, such as dichloroacetic acid (Cowman and Singer, 1996). The extent of dihaloacetic acid formation, however, is relatively small and much less than that produced from free chlorine. There will also be some continuing formation of other organic halides, measurable as total organic halides (TOX), that are as yet unidentified, but the extent of TOX formation from combined chlorine will be appreciably less than TOX production from free chlorine (Johnson and Jensen 1986). Some organic halides will increase in concentration when free chlorine or other oxidants and primary disinfectants, such as ozone, are coupled with combined chlorine for secondary disinfection (e.g. cyanogen chloride, CNCl; Krasner et. al, 1991; Singer et. al, 1999; iodomethanes and dichloroacetaldehyde, Weinberg et. al, 2002), but most of the more common and more widely-identified disinfection by-products (DBPs) will not change in concentration after switching from free chlorine to combined chlorine as a secondary disinfectant. One recently identified DBP of concern associated with conversion to combined chlorine is nitrosodimethyl-amine (NDMA) and the possible formation of other nitrosamines can be anticipated.

Key Factors Affecting Water Quality Changes

There are a number of key water quality and operational factors that influence the effectiveness of conversion to combined chlorine as a secondary disinfectant for controlling the formation of DBPs. Most significant among these are:

- The ammonia dosage relative to the residual chlorine concentration (ammonia N to Cl_2 ratio)
- pH
- Temperature
- The location at which the ammonia is applied relative to the point of Cl_2 addition and the removal of precursors (organic material)
- Mixing at the point of ammonia addition

Conversion of free chlorine to monochloramine conforms to the following stoichiometric reaction:

$$HOCl + NH_3 = NH_2Cl + H^+ + Cl^- \quad (1)$$

Under typical drinking water conditions, monochloramine (NH_2Cl) is the dominant chloramine species. According to this reaction, one mole of ammonia must be added for each mole of free chlorine. This produces one mole of monochloramine. This translates, on a weight basis, to 1 mg/L of ammonia as N to 5 mg/L of free chlorine as Cl_2 to produce 1 mg/L of monochloramine as Cl_2 (a 1:5 N:Cl_2 ratio or a 5:1 Cl_2:N ratio). If less ammonia than the 1:5 ratio is applied, then a portion of the monchloramine is converted to dichloramine which is unstable and decomposes primarily to nitrogen gas and chloride, with a corresponding loss in chlorine residual. As a result, it is common-place to add a slight excess of ammonia above the 1:5 ratio to assure that there is complete conversion of the free chlorine or monochoramine (combined chlorine) with no loss in chlorine residual. Most often, ammonia to chlorine ratios of 1:4.5 or 1:4 ($N:Cl_2$) are employed. This results in a residual ammonia concentration in the treated water. Recommended practice is to keep this residual ammonia concentration below 0.2 mg/L in order to avoid nitrification problems. In some cases, because of the presence of other reducing agents in the water, such as sulfide, ferrous iron, manganous manganese, or organic nitrogen, it may be necessary to keep the ammonia nitrogen to chlorine ratio less than 1:5 to account for the chlorine demand of these other impurities. However, this is usually not the case if free chlorine is used as a primary disinfectant within the treatment plant where it also oxidizes these reducing agents prior to the addition of ammonia. In general, TOX formation from chloramination decreases as the chlorine-to-ammonia ratio decreases (Speitel 1999).

The pH is also an important consideration. The optimal pH for the formation of monochloramine from free chlorine and ammonia is about 8-8.3. At this pH, the kinetics of monochloramine formation are fastest, and monochloramine is the only active chlorine product in accordance with Reaction 1. At lower pH values, dichloramine ($NHCl_2$) and nitrogen trichloride (NCl_3) may be formed, and these species are associated with undesirable taste and odor.

Likewise, temperature impacts the kinetics of monochloramine formation from free chlorine and ammonia. The reaction is slowest in colder water, so monochloramine is formed more slowly in the winter months.

The location at which ammonia is added to the water is an important consideration with respect to the ability of monochloramine to halt subsequent formation of THMs and HAAs. If ammonia is added to the water while it still contains highly reactive DBP precursors, appreciable levels of THMs and HAAs may be formed (Jacangelo et. al, 1989, Singer et. al, 1998). This would be the case if ammonia was added to the water prior to the point of free chlorine addition, concurrently with free chlorine, or immediately downstream of the point of chlorine addition. If ammonia is added at a downstream location such that many of the more reactive DBP precursors have had an opportunity to react with free chlorine, e.g. addition of ammonia at the high service pumps at the point of entry to the distribution system, subsequent formation of THMs and HAAs is negligible. Additionally, for ammonia to effectively stop subsequent formation of THMs and HAAs, it must be rapidly dispersed and mixed into the water. If there are appreciable gradients in the concentration of ammonia or free chlorine due to incomplete mixing, the stoichiometric formation of monocloramine may not occur throughout the solution and undesirable byproducts (e.g. dichloramine) may form and/or a portion of the desired chlorine residual may be lost. If ammonia is added at the point of entry to the distribution system and effective mixing is achieved, little additional formation of THMs and HAAs can be expected in the distribution system, and the THM and HAA concentrations at all locations in the system should be similar to the concentration at the point of entry to the system (Boulos et. al, 2003).

Bromamines and bromine-containing DBPs may also be formed if ammonia is added to a chlorinated water containing appreciable levels of bromide. This is especially true if ammonia is added to the bromide-containing water prior to the point of free chlorine addition, concurrently with free chlorine, or immediately downstream of the point of chlorine addition. Bromamines form rapidly, relative to chloramines, and dissipate rapidly in the environment (Speitel 1999), so they are not expected to be significant in water distribution systems.

NDMA

NDMA is deserving of special attention as it is a relatively new DBP that is not currently regulated. It is a known carcinogen and is associated to a greater degree with the use of combined chlorine than with the use of free chlorine. NDMA formation appears to be primarily attributable to the reaction of dimethylamine and other secondary amines with combined or free chorine (Choi and Valentine 2002, Mitch and Sedlak 2002); greater levels appear to be generated from combined chlorine. NDMA concentrations increase with increasing monochloramine concentration (Choi and Valentine 2002, Najm and Trussell 2001). The source of dimethylamine is believed to be certain types of cationic polymers (Wilczak et. al, 2002).

Little information is available about the fate of NDMA in full-scale distribution systems. Studies suggest that NDMA concentrations may increase in distribution systems with increasing water age and may be associated with transient nitrite formation occurring during the course of nitrification.

Positive Impacts on Water Quality

Theoretical

It is expected that ammonia would convert residual free chlorine to monochloramine and stop subsequent formation of THMs and HAAs. In comparison to free chlorine, monochloramine is a weaker oxidant and has a lesser tendency to participate in chlorine substitution reactions. Monochloramine is expected to form a more stable and persistent residual in the distribution system than free chlorine. Like free chlorine, the persistence of the combined chlorine residual is impacted by residual dissolved organic carbon (DOC) in the water and on the pipe walls and residence time in the distribution system, including storage and operation of storage tanks.

The hydrolysis of monochloramine (the reverse of Reaction 1) can generate some hypochlorous acid (HOCl), which may be involved in the formation of some halogenated organic compounds (TOX), including THMs and HAAs (Speitel 1999), but such formation is expected to be minor.

Actual experience

In most cases, little additional formation of THMs and HAAs is observed in the distribution system if combined chlorine is used as a secondary disinfectant, so that the THM and HAA concentrations at all locations in the system are relatively similar to the concentration at the point of entry to the system (Boulos et. al, 2003). Utilities that have converted to combined chlorine as a secondary disinfectant find a relatively small coefficient of variation among their four quarterly compliance samples (Obolensky et. al, 1999). Analysis of ICR data (Obolensky and Frey 2002) indicated that THM and HAA concentration ranges were narrower and more comparable in systems using combined chlorine than in systems using free chlorine.

In some cases, additional THM and HAA formation is observed after ammonia is added (Jacangelo et. al, 1989, Singer et. al, 1998). This may be due to poor mixing of the ammonia at the point of addition, slow kinetics of monochloramine formation due to low temperature or improper pH control, or the continuing presence of reactive organic precursors. This is especially true if ammonia is added earlier in the treatment process, e.g. before or immediately after filtration, before the most reactive organics are oxidized or removed.

Potential Adverse Impacts on Water Quality

Theoretical

The presence of excess ammonia associated with the use of a $N:Cl_2$ ratio greater than 1:5 can lead to nitrification in the distribution system and subsequent loss of combined residual, increase in heterotrophic plate counts (HPCs), and formation of biofilms. Nitrification may also be associated with NDMA formation. Chloraminated systems are expected to have higher levels of NDMA levels and more frequent occurrence than systems with free chlorine.

Loss of combined chlorine residual, or a reduction in the combined chlorine residual to less than about 0.5 mg/L, can lead to decreases in HAA concentrations due to biodegradation (Williams et. al,1995, Baribeau et. al, 2001, Speight and Singer 2003). This same phenomenon occurs when the free chlorine is depleted to levels below 0.2 mg/L. While this is not necessarily an "adverse" effect, it does lead to erroneous implications concerning HAA occurrence and a misrepresentation of the average systemwide HAA concentration associated with regulatory compliance monitoring.

Actual Experience

Nitrification problems have been reported in chloraminated water distribution systems. The extent of nitrification is exacerbated by high temperatures, high free ammonia levels (associated with high $N:Cl_2$ ratios), and reduced combined chlorine residuals.

Greater levels of NDMA have been found in chloraminated systems compared to systems with a free chlorine residual.

Lower concentrations of HAAs have been found in remote parts of water distribution systems (Wiliams et. al, 1995, Baribeau et. al, 2001a, Speight and Singer 2003) where the chlorine residuals have been reduced to low levels, e.g. less than about 0.5 mg/L for combined chlorine, less than 0.2 mg/L for free chlorine. This is reported to be a microbiological phenomenon and is found primarily in waters with warmer temperatures where microbial activity occurs. In some cases, HAAs have been completed eliminated due to microbial activity.

Weinberg et. al, (2002), in a limited survey intended to assess the occurrence and stability of other halogenated DBPs of potential toxicological concern that have recently been identified, found that some of these newly identified DBPs were present to a significant degree in chloraminated distribution systems. These included iodo-trihalomethanes and dichloroacetaldehyde. They suggested, not unsurprisingly, that the formation and control of the four regulated THMs is not necessarily an indicator of the formation and control of other halogenated DBPs.

Unknown Issues/Information Gaps

The following items are deserving of increased research:

- What is the nature of the TOX produced by combined chlorine?
- What is the linkage, if any, between chloramination, nitrification, and NDMA production?
- Are there other nitrosamines that may be formed in distribution systems in addition to NDMA and are they associated with chloramination practice?
- How widespread is the occurrence of the newly identified halogenated DBPs of potential toxicological concern? Many were not studied in the Information Collection Rule. The study by Weinberg et. al, (2002) was quite limited in scope and was intended to provide a snapshot of their occurrence rather than a comprehensive occurrence survey.
- From an analytical perspective, how do you quench samples containing monochloramine without altering the by-products present? Most analytical schemes for measuring DBPs involve quenching the residual chlorine with ammonium sulfate. This works reasonably well for free chlorine, but the quenched sample still contains monochloramine which allows for some continuing formation of halogenated organic material (TOX). Likewise, quenching a sample containing combined chlorine with ammonium sulfate does not quench the combined chlorine and allows for some continuing formation of halogenated organics.

Data/Information Needed to Fill Gaps

The following types of information are needed to fill these information gaps:

- Measurement and characterization of TOX in chloraminated systems.
- Analysis of NDMA in chloraminated systems of treatment plants that use cationic polymers or anion exchangers. The same should be done for treatment plants that do not use cationic polymers or anion exchangers. Measurements need to be made for various water ages. Corresponding measurements of ammonia, nitrite, and nitrate need to be made.
- Based on the findings of Weinberg et. al, (2002), a more comprehensive survey should be conducted to assess the occurrence of the newly identified halogenated DBPs of potential toxicological concern.

OZONE

Ozone is a very effective oxidant and disinfectant. It is capable of rapidly oxidixing bromide to hypobromous acid which can react with NOM in the water to produce brominated DBPs, such as bromoform, dibromoacetic acid, and tribromoacetic acid. The hypopromous acid can also be subsequently oxidized by ozone to bromate, a potential carcinogen. Ozone is also capable of oxidizing organic material in the water to produce lower molecular weight and more polar organic by-products, such as aldehydes, acids, and aldo- and ketoacids, that tend to be more biodegradable than the original organic material. Accordingly, for systems using ozone as a primary disinfectant in place of free chlorine, the following DBPs are of concern:

- bromate
- brominated THM and HAAs
- biodegradable organc material

Although ozone is thermodynamically capable of oxidizing chloride to chlorine, the kinetics are too slow for this to occur under the conditions of water treatment practice, so no chlorinated by-products are formed as a result of ozonation.

Key Factors Affecting Water Quality Changes

The key water quality and operational factors that influence the effectiveness of utilizing ozone as a primary disinfectant and its impact on DBP formation are:

- pH
- the amount and nature of the organic carbon in the water
- bromide ion concentration
- bicarbonate/carbonate concentration (alkalinity)
- the point of ozone addition (pre-, intermediate, or post-ozonation)
- temperature

Ozone is unstable in water and decomposes to produce hydroxyl radicals and other radical species en route to the ultimate formation of oxygen (Staehelin and Hoigne, 1982). These radicals catalyze ozone decomposition so that the decomposition process is autocatalytic, and the rate increase with time. pH markedly affects the rate of ozone decomposition, with the decomposition rate increasing as pH increases. pH also affects the speciation of hypobromous acid (HOBr) that is formed when the bromide ion is oxidized. The acidity constant of HOBr is $10^{-8.8}$, indicating that HOBr dominates at pH values below 8.8 whereas OBr^- (hypobromite) dominates at pH values greater than 8.8. Ozone decomposition kinetics (and therefore pH) are critical because the formation of bromate proceeds through a number of pathways involving the hydroxyl radical which is generated via ozone decomposition (von Gunten and Oliveras, 1998). Bromate formation increases with increasing pH, whereas the formation of brominated organic DBPs, such as bromoform, increases with decreasing pH. Under most practical ozonation scenarios, the formation of bromine-containing organic DBPs is relatively small (Song et. al, 1997).

Ozone reacts with NOM and other organic impurities in water through direct and indirect reaction pathways (Hoigne and Bader 1976, Singer and Reckhow 1999). The former involves molecular ozone which is a relatively selective oxidant in terms of the types of organic substrates with which it reacts. The latter involves the radical species arising from ozone decomposition which tend to be relatively non-selective oxidants in that they react rapidly with most organic substrates. Accordingly, NOM exerts an ozone demand on the water, resulting in increased requisite ozone dosages with increasing concentrations of NOM or TOC. The hydrophobic (humic) fraction of NOM is more reactive toward ozone than the hydrophilic fraction, and ozone tends to be very reactive toward ultraviolet (UV)-absorbing organics, reflective of unsaturated carbon-carbon double bonds (Westerhoff et. al, 1999). Because hydrophobic organic carbon and UV-absorbing organics are the primary precursors of halogenated DBPs, such as THMs and HAAs (Reckhow et. al, 1990a, Reckhow and Singer 1990, Liang and Singer 2003), oxidation of these materials by ozone is expected to lower the THM and HAA formation potential of the water. The by-products of these oxidation reactions tend to be lower in molecular weight than the parent compounds, and tend to be more polar in nature. Depending upon the ozone to organic carbon ratio employed, the by-products consist primarily of low molecular weight acids such as oxalic acid, aldehydes such as formaldehyde and glyoxal, and aldo- and ketoacids such as pyruvic acid and ketomalonic acid (Reckhow, 1999). Many of these by-products are biodegradable and can lead to biofilm problems in the distribution system if they are not removed by a biofiltration or biostabilization process prior to entry into the distribution system. The production of biodegradable organic material, sometimes characterized as BDOC (biodegradable organic carbon) or AOC (assimilable organic carbon), increases with increasing O_3/C ratios (O_3/C ratios of 0.5 to 1.0 by weight are common in practice).

The formation of bromate during ozonation is impacted greatly by the bromide concentration of the raw water. Bromate formation increases with increasing bromide ion concentration, as well as increasing pH (see above) and increasing temperature (see below). Bromate formation occurs through a variety of pathways, involving both molecular ozone and the hydroxyl radical, and it is widely acknowledged that the hydroxyl radical plays a key role in bromate formation (von Gunten and Oliveras 1998). This is one of the reasons why pH is so important; as pH increases, ozone decomposes more rapidly, producing the hydroxyl radicals which play a dominant role in bromate formation.

Bicarbonate and carbonate act as hydroxyl radical scavengers (Staehelin and Hoigne 1982). In so doing, they decrease the rate of ozone decomposition, thereby increasing the persistence of molecular ozone and driving the reactivity of ozone toward the direct reaction pathway (Reckhow et. al, 1986). Because they scavenge hydroxyl radicals, they alter the kinetics and pathway of bromate formation and the degree to which ozone oxidizes and alters DBP precursors. However, the carbonate and bicarbonate radicals produced from the scavenging of hydroxyl radicals serve as secondary oxidants that are also capable of contributing to bromate formation (von Gunten and Oliveras 1998).

The effectiveness of ozone as a disinfectant, oxidant, and DBP control strategy depends upon the point of ozone application. Because of the ozone demand exerted by NOM, much higher doses are required when ozone is used in a pre-ozonation mode (applied to raw water) rather than in an intermediate ozonation mode (applied to settled water, before filtration) or in a post-ozonation mode (applied to filtered water). Accordingly, higher levels of biodegradable organic material are produced when ozone is used in a pre-ozonation mode compared to the other modes of application. When ozone is used in a post-ozonation mode (after filtration), the biodegradable organic material produced becomes a substrate for microorganisms in the distribution system, which can lead to undesirable biofilm formation. The most judicious use of ozone is in an intermediate ozonation mode because the ozone demand of the water is less than that of the raw water, and the biodegradable organic material produced by the process can be stabilized in a biologically active filter.

Temperature impacts the kinetics of ozone mass transfer and reaction, and the solubility of ozone in water. Reaction kinetics are slowest in colder water, but ozone solubility increases with decreasing temperature.

Other significant factors that might impact water quality include the efficiency of ozone generation (e.g. the ozone concentration in the gas phase), the mode of ozone application (i.e. multi-stage countercurrent bubble contactors vs. in-line injection of ozone), staging of ozone application (i.e. applying the entire dose in the first stage of a multi-stage contactor vs. distribution of the ozone dose among several stages), the method of quenching residual ozone (i.e. use of peroxide, bisulfite, thiosulfate, and the relative kinetics of each), and contact time.

Positive Impacts on Water Quality

Theoretical

- Ozonation does not oxidize chloride to free (active) chlorine so ozonation should not produce chlorine-containing DBPs.
- A switch to ozonation for primary disinfection should result in the partial oxidation of DBP precursors and a corresponding reduction in the formation of THMs and HAAs upon subsequent chlorination. In fact, if ozone is used for primary disinfection, combined chlorine can be used as a secondary disinfectant and there is no need to use free chlorine at all. If this is the case, then THMs and HAAs can be reduced to very low levels.
- If free chlorine continues to be used after ozonation, there should be a reduction in the chlorine demand of the water so that less chlorine should be required to meet a desired free chlorine residual.
- More effective inactivation of pathogenic microorganisms should be achieved, especially for *Cryptosporidium* where free chlorine is ineffective.
- Ozone should be more effective when it is coupled with biological filtration. Some of the DBP precursors that remain after ozonation are biodegradable so that the DBP formation potential should be lowered further after biofiltration.
- Color should be reduced (and eliminated) as a result of ozonation. Ozone is very effective in oxidizing the chemical moieties in NOM that are responsible for the yellowish hue it imparts to water.
- Ozone should be effective in the oxidation of reduced iron (Fe(II)) and manganese (Mn(II)). The kinetics of Fe(II) and Mn(II) by ozone are fast.

Actual Experience

- Ozonation does not produce chlorine-containing DBPs.
- Ozonation results in only a minor reduction (on the order of 10-15%) in the THM and HAA formation potential of most waters when applied at a dose of 0.5-1.0 mg ozone per mg DOC (Chang and Singer 1991, Schechter and Singer 1995, Singer et. al, 1999, Chaiket et. al, 2002). Greater reductions are achieved when ozonation is coupled with biofiltration.
- Some chlorinated DBPs (e.g. halonitromethanes, such as chloropicrin (trichloronitromethane) and bromopicrin (tribromonitromethane), 1,3 dichloropropanone) have been found to increase as a result of pre-treatment with ozone prior to chlorination (Hoigne and Bader 1988, Singer et al 1999, Weinberg et. al, 2002). Some others (e.g. cyanogen chloride, dichloracetaldehyde) have been

found to increase as a result of pre-treatment with ozone prior to chloramination (Krasner et. al, 1991, McKnight and Reckhow 1992, Singer et. al, 1999, Weinberg et. al, 2002).

- There is only a minor reduction in the subsequent dosage of chlorine that must be applied to meet a target chlorine residual following ozonation (Singer et. al, 1999). In some cases, an apparent increase in the requisite chlorine dose has been reported. Part of this may be due to the presence of residual ozone in the water which consumes chlorine, or to the presence of residual reducing agents (e.g. peroxide, bisulfite) that were added to the water in excess to consume any residual ozone. Subsequent chlorine doses are reduced noticeably when ozonation is followed by biofiltration prior to the application of chlorine.
- Ozone oxidizes the humic substances responsible for the yellowish hue in naturally colored waters (Singer and Johnson 2003).
- Ozone oxidizes reduced iron(II) and manganese(II) to particulate ferric hydroxide and particulate manganese dioxide, respectively. The ferric hydroxide tends to be removed easily, but the manganese dioxide tends to be colloidal in nature and has been found to be difficult to remove by filtration (Wilczak et al. 1993)

Potential Adverse Impacts on Water Quality

Theoretical

- Ozonation of bromide-containing waters should result in the production of bromate. Bromate production is expected to increase with increasing bromide ion concentration, increasing pH, and increasing ozone exposure (ozone residual times contact time; Driedger et. al, 2001). As ozone exposure increases to meet the CT required for inactivation of *Cryptosporidium*, bromate levels are expected to increase, even for waters with low to moderate bromide ion concentrations. The MCL for bromate is currently 10 μg/L, which is also the practical quantitation level (PQL) for bromate. Analytical detection limits for bromate have been improved, so it might be anticipated that the PQL will be decreased to 5 or even 2 μg/L. If this happens, the MCL for bromate can be expected to be lowered to 5 μg/L (the 10^{-4} cancer risk level) in the near future.
- Ozonation of bromide-containing waters are also expected to result in the production of brominated organic DBPs, such as bromoform and the brominated acetic acids. Production of these bromine-containing organic DBPs is expected to increase with increasing bromide concentration and decreasing pH.
- Ozonation should result in the production of low molecular weight oxidation by-products (e.g. aldehydes, acids, aldo- and ketoacids), many of which are biodegradable. If not biologically eliminated, and if the water is not biologically stablilized, these biodegradable by-products can enhance biofilm formation in distribution systems. Conversely, if the ozonated water is biologically filtered to remove these biodegradable oxidation by-products, this can be considered a positive effect as overall organic carbon concentrations should be lowered.
- It has been suggested that ozonation can lead to the formation of epoxides and organic peroxides that may have adverse human health impacts (Glaze 1986).

- Some of the ozonation by-products can lead to an enhanced production of some chlorinated DBPs upon subsequent chlorination, e.g. chloral hydrate, chloropicrin, and upon subsequent chloramination, e.g. cyanogen chloride.
- Ozone is thermodynamically capable of oxidizing reduced manganous manganese (Mn(II)) to permanganate (Mn(VII)), resulting in "pink" water. This would require that adequate contact time be provided to allow the permanganate to be chemically reduced to particulate manganese dioxide (Mn(IV)), and that a suitable solid-liquid separation step (sedimentation or filtration) be provide for removal of the manganese dioxide.

Actual Experience

- Bromate is produced when bromide-containing waters are ozonated, but bromate formation can be controlled by lowering the pH of the water or by adding ammonia or chlorine and ammonia to tie up the bromide as bromamines. To date, the most widely used approach for bromate control has been to reduce the pH to 6-6.5. Recent work at the Southern Nevada Water Authority and by Buffle et. al, (2003) demonstrates that bromate formation can be effectively controlled, even in waters with very high bromide concentrations, by adding chlorine to the water to rapidly oxidize the bromide to hypobromous acid and then adding ammonia to convert the hypobromous acid to monobromamine prior to the application of ozone. Using this approach, ozone is not contacted with bromide and essentially no bromate is formed.
- Minor amounts of bromine-containing organic DBPs have been found in ozonated waters (Song et. al, 1997). More recent work by Weinberg et. al, (2002) found the presence of a number of brominated acids, such as brominated propanoic, propenoic, butanoic, butenoic, oxopentanoic, heptanoic, nonanoic, butenedioic acids, in finished water fromaa treatment plant using ozone and treating water with a high level of bromide.
- Minor increases in the concentration of chloral hydrate and chloropicrin have been observed following chlorination of ozonated waters compared to parallel experiments in chlorinated waters that were not preozonated (Logsdon et. al, 1992, McKnight and Reckhow 1992). Similarly, minor increases in the concentration of cyanogen chloride have been observed following chloramination of ozonated waters compared to parallel experiments in chloraminated waters that were not preozonated (Krasner et. al, 1991, Singer et. al, 1999).
- Ozonation results in the formation of a variety of oxidation by-products. These include aldehydes (primarily formaldehyde, acetaldehyde, glyoxal and methyl glyoxal), acids (oxalic acid and formic acid), aldo- and ketoacids (pyruvic acid, ketomalonic acid). On a mass balance basis, less than 5% of the DOC measured in ozonated waters is accounted for by the known oxidation by-products identified to date (Krasner et. al, 1996). The majority of these identified by-products are biodegradable and have been shown to be removed by biological filtration using either anthracite or granular carbon. GAC works best in colder waters.
- No evidence has been found that epoxides or organic peroxides produced by ozonation are stable in ozonated water.

- Manganese control is difficult when intermediate ozone is used. Wilczak et. al, (1993) has shown that oxidation of manganese(II) by ozone results in the production of colloidal manganese dioxide which is not easily removed by filtration. They recommend the addition of free chlorine after ozonation in order to facilitate the retention of manganese in the filter bed. This practice works effectively for dual media and anthracite filters, but cannot be used when GAC filters are used because the chlorine will be chemically reduced by the GAC and will, in time, destroy the GAC.

Unknown Issues/Information Gaps

The following items are deserving of increased research:

- What, if any, are the potential health risks associated with the oxidation by-products of ozone?
- How widespread is the occurrence of the newly identified halogenated DBPs of potential toxicological concern in plants using ozonation? As noted above, the study by Weinberg et. al, (2002) was quite limited in scope and was intended to provide a snapshot of their occurrence rather than a comprehensive occurrence survey.
- Can we improve on the carbon balance of identified oxidation by-products in ozonated water? Can we improve our identification of oxidation by-products so that we can account for more than 5% of the measured DOC in ozonated waters?
- Can the bromate produced by ozonation of bromide-containing waters be effectively controlled to levels as low as 5 or even 2 μg/L?

Data/Information Needed to Fill Gaps

The following types of information are needed to fill these information gaps:

- The nature and characteristics of the oxidation by-products of ozonation need to be identified and quantified. The majority of these by-products are polar materials, but new separation and analytical methods have been developed that will facilitate this endeavor.
- New bromate control technologies, especially the application of chlorine and ammonia developed by the Southern Nevada Water Authority and expanded upon by Buffle et. al, (2003) need to be demonstrated on a plant-scale basis at a number of different water utilities.
- Based on the findings of Weinberg et. al, (2002), a more comprehensive survey should be conducted to assess the occurrence of the newly identified halogenated DBPs of potential toxicological concern.

CHLORINE DIOXIDE

Chlorine dioxide is a strong oxidant as well as a good disinfectant. It can be used for both primary and secondary disinfection, although it is used infrequently for secondary disinfection as it is difficult to maintain a persistent chlorine dioxide residual in the distribution system because of its strong oxidizing properties. When it is applied to water for either oxidation or disinfection purposes, about 50-70% of the applied chlorine dioxide is reduced to chlorite (Werdehoff and Singer 1987) which has an MCL of 1.0 mg/L. Chlorate is also a common oxyhalide species encountered when chlorine dioxide is used. There is currently no MCL for chlorate. It usually occurs as a by-product of chlorine dioxide generation, but it is also produced when free chlorine is applied to waters that have been pre-treated with chlorine dioxide; chlorine oxidizes the residual chlorite, resulting from the reduction of chlorine dioxide, to chlorate. Because chlorine dioxide is a strong oxidant, it reacts with NOM in the water to produce oxidation by-products similar to those produced by ozonation. Chlorine dioxide does not tend to participate in substitution reactions like free chlorine so that halogenated by-products are not formed to any appreciable extent; chlorine dioxide does not react with NOM to produce THMs and HAAs. Additionaly, chlorine dioxide also reacts only very slowly with bromide (Hoigne and Bader 1994) so that brominated by-products, either organic or inorganic (e.g. bromate) are not a concern following treatment with chlorine dioxide. Accordingly, for systems using chlorine dioxide as a primary or secondary disinfectant in place of free chlorine, the following DBPs are of concern:

- Chlorite (ClO_2^-) and chlorate (ClO_3^-)
- Biodegradable organic material

Key Factors Affecting Water Quality Changes

The key water quality and operational factors that influence the effectiveness of utilizing chlorine dioxide as a primary and secondary disinfectant and its impact on DBP formation are:

- Dose of ClO_2
- Chlorine dioxide generation
- Presence and addition of other oxidants
- Amount of TOC at the point of ClO_2 addition
- Point of ClO_2 addition
- Residual TOC in the distribution system
- Temperature
- pH

The dose of chlorine dioxide has a marked effect on the concentration of chlorite produced. Because chlorine dioxide participates in oxidation reactions primarily through a one-electron transfer,

$$ClO_2 + e^- = ClO_2^- \quad (2)$$

The majority of the applied chlorine dioxide is reduced to chlorite. Hence, chlorite concentrations increase with increasing doses of chlorine dioxide. Werdehoff and Singer (1987) found that chlorite comprises about 50-70% of the chlorine dioxide consumed; this finding has been reproduced by many other investigators. Because the MCL for chlorite is 1.0 mg/L, chlorine dioxide doses need to be limited to about 1.5 mg/L to stay below the MCL for chlorite, unless some type of chlorite removal process is instituted. Iatrou and Knocke (1992) have shown that the addition of ferrous iron is capable of reducing the chlorite ion, but this can only be done prior to clarification or filtration to provide for removal of the particulate ferric hydroxide produced by the reaction. Reduced sulfur compounds and GAC are also capable of reducing chlorite.

Another potential source of chlorite is via the generation of chlorine dioxide. Chlorine dioxide is typically generated on-site from chlorite and either gaseous chlorine or hypochlorous acid, under mildly acidic conditions. If the stoichiometry associated with the feed streams to the chlorine dioxide generator is not carefully controlled, unreacted chlorite may be found in the chlorine dioxide product stream. Likewise, if the pH in the chlorine dioxide generator is not carefully controlled, chlorate may be formed. Again, although chlorate is not currently regulated, it may be a concern in the future. Chlorite and chlorate can both be formed from the disproportionation of chlorine dioxide at elevated pH values.

Other oxidants used in conjunction with chlorine dioxide can lead to a variety of oxyhalide by-products. For example, if chlorine dioxide is applied to pre-chlorinated waters containing a free chlorine residual, chlorate will be formed. If chlorine is applied to waters that have been pre-treated with chlorine dioxide, chlorine will oxidize the residual chlorite to re-form chlorine dioxide; chlorate is also a by-product of the reaction between chlorite and free chlorine (Singer and O'Neil 1987). Such reactions can be anticipated when chlorine dioxide is used for oxidative pre-treatment and/or primary disinfection and free chlorine is used for secondary disinfection. Likewise, if ozone is applied to waters that have been pre-treated with chlorine dioxide, ozone will oxidize the residual chlorite to chlorate.

While chlorine dioxide does not participate in substitution reactions with NOM to any significant degree to form THMs, HAAs, and most other commonly observed halogenated DBPs that result from chlorination, Richardson et. al, (1994) and others have identified a number of by-products arising from chlorine dioxide treatment. The majority of these are the same kinds of oxidation by-products seen with ozone, e.g. aldehydes, carboxylic acids. The extent to which these oxidation by-products are formed depends upon the TOC concentration of the water at the point of chlorine dioxide addition and the dose of chlorine dioxide applied. Because settled water tends to have a lower TOC concentration than raw water and therefore a lower oxidant demand, lower chlorine dioxide doses are required to achieve primary disinfection objectives in settled water than in raw water. Less chlorite will be produced because less chlorine dioxide is applied, and lower concentrations of oxidation by-products are formed because the TOC concentration is lower.

The residual TOC concentration in the treated water also has a major impact on the ability to maintain a disinfectant residual in distribution systems in which chlorine dioxide is used as a terminal (secondary) disinfectant. Because chlorine dioxide is a strong oxidant, it readily reacts with TOC in the water and on the walls of the pipes and can be easily depleted in systems with high organic carbon concentrations. Unlike chlorine, the dose of chlorine dioxide cannot be increased to higher levels to account for the dissipation of the chlorine dioxide residual because of the chlorite MCL which limits the chlorine dioxide dose to about 1.5 mg/L. Maintenance of a persistent chlorine dioxide residual is easier to achieve in colder waters because of slower reaction kinetics.

As noted earlier, pH is an important factor because it influences: the efficiency of chlorine dioxide generation and the potential formation of chlorate, the stability of the chlorine dioxide relative to disproportionation, and the nature of the oxyhalide species produced when chlorine is added as a secondary disinfectant following pre-treatment with chlorine dioxide.

Positive Impacts on Water Quality

Theoretical

- When converting to chlorine dioxide as a primary and/or secondary disinfectant, there should be minimal formation of THMs, HAAs, and very little formation of other halogenated organics (TOX). The only source of formation of any of these halogenated DBPs would be from excess chlorine added during the generation of chlorine dioxide to assure the complete utilization of chlorite.
- There should be a reduction in the formation of halogenated DBPs upon subsequent chlorination because, like ozone, chlorine dioxide should partially oxidize and alter DBP precursors. If chlorine dioxide is used as a primary disinfectant and chloramines are used for secondary disinfection, there should be minimal formation of THMs and HAAs.
- There should be a corresponding reduction in the subsequent chlorine demand following partial oxidation of the TOC in the water.
- Chlorine dioxide should be an effective oxidant of select contaminants of concern, e.g. Fe(II), Mn(II), taste and odor-causing organics, and color.
- Chlorine dioxide should produce a persistent residual when it is used as a secondary disinfectant if the finished water is relatively low in reactive organic carbon and has a relatively low temperature.

Actual Experience

- Most utilities using chlorine dioxide as an oxidant or disinfectant experience essentially no production of THMs or HAAs, and little formation of TOX as a result of chlorine dioxide addition.
- Most utilities using chlorine dioxide as an oxidant or disinfectant experience a reduction in the formation of halogenated DBPs upon subsequent chlorination. The extent of this decrease is dependent on the dose of chlorine dioxide applied. It has been hypothesized (Werdehoff and Singer 1987) that this is due in part to re-formation of chlorine dioxide as a result of the oxidation of chlorite by free chlorine; the re-formed chlorine dioxide has another opportunity to oxidize DBP precursors before they can react with chlorine to produce additional DBPs.
- A reduction in the chlorine demand of the water has been seen following pre-treatment with chlorine dioxide (Werdehoff and Singer 1987). The extent of the reduction is dependent on the dose of chlorine dioxide applied.
- It has been found to be difficult to maintain a persistent chlorine dioxide residual in most distribution systems due to residual organic carbon in the treated water and to the wall demand of the pipes. In some cases, utilities erroneously believed that they were measuring a true chlorine dioxide residual, but this turned out to be an artifact of the analytical method they were using and positive coliform counts were observed (Singer 1988). Some Canadian systems (Laval) appear to effectively employ chlorine dioxide as a terminal disinfectant, but this may be attributable to the low temperature of the water.

- Although not a DBP issue per se, residual chlorite has been shown to control nitrification problems in distribution systems.
- Chlorine dioxide is widely used as an effective oxidant for purposes of iron and manganes oxidation, taste and odor destruction, and oxidation of color.

Potential Adverse Impacts on Water Quality

Theoretical

- Based on the stoichiometry of the one-electron transfer reactions of chlorine dioxide with various reducing agents (Singer and Reckhow 1999), 1 mole of chlorite should be produced from the reduction of 1 mole of chlorine dioxide. This is also equivalent to a 1:1 ratio by weight.
- During the on-site generation of chlorine dioxide, addition of excess chlorite beyond the stoichiometric requirement leads to the presence of chlorite as a contaminant in the chlorine dioxide product stream. Addition of excess chlorine beyond the stoichiometric requirement in order to assure the complete utilization of chlorite leads to the presence of free chlorine as a contaminant in the chlorine dioxide product stream. This chlorine can produce some halogenated DBPs, but because the chlorine concentration is typically small, DBP production is limited.
- Chlorite produced from the chemical reduction of chlorine dioxide or as a contaminant in the chlorine dioxide feed stream is thermodynamically capable of being oxidized to chlorate by the application of free chlorine for secondary disinfection. Chlorate might also be found in the chlorine dioxide feed stream if the pH of the chlorine dioxide generator is not controlled properly.
- Oxidation of TOC by chlorine dioxide should lead to production of biodegradable organic material (Richardson et. al,1994) that must be controlled to avoid biofilm problems in distribution systems.

Actual Experience

- Most utilities using chlorine dioxide report the production of 0.5-0.7 mg ClO_2^- per mg ClO_2 consumed. This causes them to limit their chlorine dioxide doses to about 1.5 mg/L. Some utilities (Evansville, IN) have added ferrous salts to their rapid mix basins after applying chlorine dioxide as a raw water oxidant at the intake. The ferrous iron is oxidized to ferric hydroxide which then assists in the coagulation process and is removed by sedimentation and filtration. pH is an important operational control parameter to assure that the ferrous iron is oxidized by chlorite and not by dissolved oxygen. This means that the most effective pH is between 6 and 7 (Iatrou and Knocke 1992).
- Well-operated modern chlorine dioxide generators are capable of producing chlorine dioxide with little chlorite, chlorate, or free chlorine contamination.
- Application of chlorine to water pre-treated with chlorine dioxide results in production of some new chlorine dioxide which can have a positive influence as a disinfectant, and some chlorate which may be a negative outcome (Singer and O'Neil 1987).

- Application of monochloramine as a secondary disinfectant to water pre-treated with chlorine dioxide does not appear to produce any chlorine dioxide or chlorate, and does not appear to effect the residual chlorite concentration. The residual chlorite, in turn, has been shown to control nitrification.
- Few utilities encounter biofilm problems as a result of production of biodegradable organic material from chlorine dioxide treatment. This may be a result of the relatively low doses of chlorine dioxide typically used.

Unknown Issues/Information Gaps

- How much bodegradable organic material is produced from chlorine dioxide?
- Are there other by-products from chlorine dioxide that may prove to be problematic (Richardson et. al, 1994).

Data/Information Needed to Fill Gaps

- A survey of water treatment plants using chlorine dioxide should be conducted to determine levels of biodegradable organic material (BDOC, AOC), and other chlorine dioxide by-products of the type suggested by Richardson et. al, (1994).

UV

UV irradiation is an emerging primary disinfectant. Like ozone, it does not produce a persistent disinfectant residual and must be followed by free or combined chlorine for secondary disinfection. Unlike the other disinfectants discussed in this issue paper, it is a physical process as opposed to a chemical process. Accordingly, there are no anticipated DBPs of concern, unless UV irradiation is used in conjunction with a chemical disinfectant/oxidant as part of an advanced oxidation process. In the latter case, biodegradable oxidation by-products can be expected, and bromate formation is possible in bromide-containing waters.

Key Factors Affecting Water Quality Changes

The key water quality and operational factors that influence the effectiveness of UV irradiation for primary disinfection and its impact on subsequent DBP formation are:

- UV dose, type of UV source and reactor design
- UV absorbance of the water
- Presence of particulate material
- The location in which UV is positioned in the treatment train
- Temperature
- Foulants that may deposit on the UV lamps

Different types of UV lamps and processes are available for primary disinfection. These include low pressure monochromatic lamps, medium pressure broadband polychromatic lamps, and pulsed UV systems. The effectiveness of UV irradiation for disinfection depends upon the intensity (I) of the radiation emitted by the lamp over the wavelengths which match the absorbance spectrum of the organism's DNA and the contact time (T) with which the organism is in contact with the radiation. Most UV reactors for drinking water treatment have mean residence times of less than 1 sec.

The disinfection effectiveness of the UV radiation is impacted by the concentration of UV-absorbing dissolved organic and inorganic substances in the water which can prevent it from penetrating very far into the solution, and by particulate material which is capable of scattering the radiation and preventing it from reaching the target organisms. Accordingly, to be most effective, UV reactors are most often positioned downstream of processes such as coagulation ands filtration that are intended to remove UV-absorbing TOC and particulate material. The UV lamps are also capable of being fouled by various other constituents of water, such as iron and manganese oxides and calcium carbonate that can impact their effective emission of radiation.

Like all processes, temperature impacts the kinetics of the disinfection reactions.

Positive Impacts on Water Quality

Theoretical

- UV irradiation should not produce THMs, HAAs, or any other halogenated organic DBPs. Likewise, it should not produce bromate or chlorite like some of the other alternative primary disinfectants. It is supposed to effectively inactivate *Giardia* and *Cryptosporidium* at relatively low doses and, in so doing, should eliminate the need for free chlorine as a primary disinfectant, thereby eliminating the production of halogenated DBPs during primary disinfection.
- UV irradiation should not increase the DBP formation potential of a water.

Actual Experience

- There is limited plant-scale experience to demonstrate the fact that UV irradiation does not produce THMs, HAAs, other halogenated organic DBPs, chlorite, or bromate. Numerous laboratory studies and pilot plat investigations show that UV irradiation does not produce these DBPs (Malley 1999).
- Laboratory studies (Malley 1999) have demonstrated that UV irradiation does not increase the DBP formation potential of water from various sources.

Potential Adverse Impacts on Water Quality

Theoretical

- At the doses typically used for disinfection, UV irradiation should not exhibit oxidizing properties. Therefore it should be ineffective for the destruction of taste and odor-causing organics, color removal, and iron and manganese oxidation.
- At the doses typically used for disinfection, UV irradiation should be incapable of oxidizing DBP precursors. As a result, DBP precursors remain in the water and are subject to subsequent reaction with secondary disinfectants added to the water later in the treatment train.

Actual Experience

- There is limited plant-scale experience to demonstrate the fact that UV irradiation does not exhibit oxidizing properties at the doses typically used in practice.
- There is limited experience to demonstrate how effectively UV irradiation controls DBP formation when it is integrated with other disinfection practices.
- UV irradiation has been shown to produce nitrite when nitrate-containing waters are irradiated (Sharpless and Linden 2001). The significance of this for finished drinking water is not clear.

Unknown Issues/Information Gaps

- Plant-scale experience is needed to see how effectively UV irradiation performs on a long-term basis with respect to fouling, how it can be integrated with oxidative pre-treatment strategies for waters that are in need of an oxidant for iron and manganese oxidation and taste and odor destruction, how effectively it controls DBP formation when it is integrated with various secondary disinfection practices, and how effectively it does indeed inactivate pathogens of concern.
- Comparison of different types of UV designs (e.g. monochromatic, polychromatic, pulsed UV) need to be evaluated in a variety of waters with different chemical characteristics.
- A methodology is needed to guarantee that a given piece of UV irradiation equipment (lamps and reactor) is performing up to its specifications and continues to perform effectively over the lifetime of the design.

Data/Information Needed to Fill Gaps

Plant-scale demonstrations of UV irradiation on different types of waters should be conducted. These demonstrations should be integrated with different oxidative pre-treatment and secondary disinfection processes. Methodologies to guarantee that a given piece of UV irradiation equipment is performing up to specifications should be developed and tested.

BOOSTER CHLORINATION

Booster chlorination can be used as an approach to more evenly distribute the chlorine residual throughout the distribution system rather than relying solely on chlorination at the point of entry to the system to achieve this objective. In the latter case, there are pronounced gradients in the residual chlorine concentration from the point of entry to regions of the system with high water ages. Because the formation of THMs, HAAs, and most other halogenated DBPs is driven by the concentration of free chlorine, chlorination at the point of entry leads to relatively high rates of halogenated DBP formation in parts of the system with low water ages and correspondingly high free chlorine concentrations, and relatively low rates of DBP formation in parts of the system with high water ages and correspondingly low free chlorine concentrations. By distributing the chlorine residual more evenly in the system, booster chlorination would tend to dampen this variation in rates of DBP formation. Booster chlorination can also be used in conjunction with combined chlorine. In this case, chlorine and ammonia may need to be added at the booster chlorination stations in order to achieve the desired 1:5 ratio of ammonia N to free chlorine by weight that is required for monochloramine formation, and pH control may also be necessary. In the case of booster chlorination, the DBPs of concern are the THMs, HAAs, and other halogenated organics represented by TOX.

Key Factors Affecting Water Quality Changes

The key water quality and operational factors that influence the impact of booster chlorination for secondary disinfection on DBP formation are:

- The typical chlorine residual profile in the system.
- Residual TOC concentration and concentration of UV-absorbing substances in the finished water.
- Temperature
- Distribution system architecture as it affects the residence time distribution of water in the system .
- The layout of the booster stations in the system, the relative amounts of chlorine added at each booster station, and the algorithm for operating the booster stations.
- In the case of booster chloramination, the presence of residual ammonia and chlorine at the inlet to the booster station.

The need to use booster chlorination depends upon the pattern of residual chlorine concentrations in the distribution system. Chlorine dissipates in the system due to reactions with residual organic materials in the water, reactions with organic materials on the walls of the pipe, and reactions with the pipe walls themselves. The extent of these chlorine-depleting reactions depends upon the concentration and characteristic of the residual TOC in the water, the extent of biofilm development on the pipe walls, the pipe materials themselves, and temperature which affects the kinetics of these reactions and therefore the rate of chlorine consumption. The residual chlorine concentration at any point in the system at any time is determined by the chlorine concentration at the point of entry to the system, the dissipation reactions noted above, water demand patterns in the system, and architecture of the system (pipe sizes, materials and layout; volume, location, and operation of storage facilities).

THM and HAA concentrations, and the concentrations of most other halogenated DBPs, tend to increase with increasing residence time. However the rate of this increase is dependent upon the free chlorine concentration, as well as the residual concentration of precursors. Hence, continuing THM and HAA formation in the system can be expected to track the chlorine residual and, correspondingly, the residence time or age of the water in the system. One can therefore expect higher rates of THM and HAA formation at locations with relatively high chlorine residuals and relatively low water ages, and lower rates of THM and HAA formation at locations with relatively low chlorine residuals and relatively high water ages. In some portions of the system, where the residual is essentially depleted, one can expect a loss in HAA concentrations due to biodegradation (Williams et. al, 1995, Baribeau et. al, 2001a, Speight and Singer 2003). Additionally, some halogenated DBPs, e.g. the haloacetonitriles and haloketones, are intermediate halogenated DBP species and, after forming to some level, their concentrations decrease with increasing water age and contact with free chlorine (Reckhow and Singer 1985, Krasner et. al, 1989b) as a result of hydrolysis and continuing reactions with free chlorine to form other DBPs. In any case, the spatiotemporal distribution of free chlorine in the system is complex, as is the water age distribution, leading to complex spatiotemporal patterns of DBP concentrations. Lowering the free chlorine concentration at the point of entry will tend to lower the rate of DBP formation in portions of the system with low water ages, but boosting the chlorine concentration at other locations in the system with higher water ages will increase the rate of formation of most DBPs. Accordingly, THM and HAA concentrations can be expected to be lowered in portions of the system with low water ages, and increased in portions of the system with high water ages.

In the case of chloramination, booster station operations are also complex. The stoichiometry of formation of monochloramine in accordance with Reaction 1, i.e.1 mg N/5 mg Cl_2 must be maintained. Accordingly, the dosage of chorine needed at the booster station to meet the target residual depends on the chlorine residual measured at the booster station inlet and the ammonia concentration measured at the inlet. Most chloraminating systems apply excess ammonia, and there is usually some additional ammonia released as the chloramine residual dissipates in the water. This is usually sufficient to meet the 1:5 N:Cl_2 ratio. If the ammonia concentration is insufficient, then more ammonia needs to be applied at the booster station.

Positive Impacts on Water Quality

Theoretical

- Cl_2 residuals can be expected to be lower at locations close to the point of entry to the system. This can be expected to cause fewer consumer complaints about the chlorinous "taste" of the water.
- Cl_2 residuals can be expected to be higher at locations with higher water ages. This can be expected to provide better protection against microbial contamination and biofilm formation at these locations.
- THM and HAA concentrations, and the concentrations of most halogenated DBPs, can be expected to be reduced in portions of the system with low water ages and increased in portions of the system with high water ages. The extent of these changes depends upon placement and operation of the booster stations.

Actual Experience

While there is extensive information and a long history of use of booster stations to increase the concentration of free chlorine in distribution systems, there is limited information concerning the impact of booster chlorination on DBP concentrations. A laboratory study intended to simulate booster chlorination (Boccelli et. al, 1998) showed no reduction in overall DBP production when chlorine was added in increments over the same time period compared to adding all of the chlorine at time zero to simulate point of entry chlorination. Speight and Singer (2003) showed distribution system data from Vancouver, BC indicating that, by adopting a booster chlorination program and increasing the penetration of free chlorine to more remote portions of their service area, HAA concentrations increased because the utility was able to overcome the depletion of their residual that led to HAA biodegradation.

Potential Adverse Impacts on Water Quality

Theoretical

- Use of booster chlorination stations requires careful operation and control to assure the continued maintenance of an adequate disinfectant residual in the system. If not operated properly, microbial protection of the water could be compromised. A reliable data acquisition and management system is needed at key points in the system to monitor residual chlorine and control the dose of chlorine at the booster stations. A reliable water quality model of the distribution system would be of great assistance in operating the booster station(s) properly.
- There needs to be a reliable source of chlorine available at each of the booster stations, requiring careful operational oversight by the utility.
- DBP concentrations in portions of the distribution system with high water ages will increase as more free chlorine gets to these locations. This will impact the regulatory compliance posture for many utilities.

Actual Experience

- There is limited experience in the waterworks industry and few, if any, publications specifically involving the impact of booster chlorination stations on DBP levels, and how the use of such stations has impacted the regulatory compliance picture for these utilities, both for the Stage 1 and Stage 2 DBP Rules.

Unknown Issues/Information Gaps

- Information needs to be developed to determine how different booster system scenarios impact the concentrations of regulated DBPs in a variety of water distribution systems, and how these scenarios impact regulatory compliance for different regulatory paradigms. Tryby et. al, (1999, 2002) have demonstrated such an approach.

Data/Information Needed to Fill Gaps

- Distribution system models should be used to predict how different booster system scenarios might impact the concentrations of THMs and HAAs in a variety of water distribution systems, and how these scenarios impact regulatory compliance for different regulatory paradigms such as the Stage 1 and Stage 2 DBP Rules. These models should then be verified with real data from utilities employing booster chlorination stations in their distribution systems.

APPENDIX D
ISSUE PAPER NO. 3
CHANGES IN CHEMICAL QUALITY - CORROSION, METAL RELEASE, COLOR-PRODUCING COMPOUNDS AND PARTICLE CHARACTERISTICS

AwwaRF PROJECT NO. 2940
LONG-TERM EFFECTS OF DISINFECTION CHANGE
ON WATER QUALITY

Issue Paper
on
Changes in Chemical Quality - Corrosion and Particle Characteristics
by
Jonathan Clement

INTRODUCTION

A major element in understanding the impacts of disinfection on corrosion in the distribution system is differentiating between electrochemical or classical corrosion and solubility. Together, both electrochemical corrosion and solubility both relate to metal release. The first part of this, electrochemical corrosion, involves the transfer of electrons through oxidation. For example, when a fresh metal corrodes as in the following equation it forms lead (Pb) (II) from Pb (0).

$$Pb(s) <-> Pb^{+2} + 2e^{-}$$

The Pb+2 is then free to react with other dissolved constituents including carbonate, hydroxide, or orthophosphate. This action produces a film over the underlying metal that is in contact with drinking water. An example is the formation of lead carbonate $PbCO_3$. The release of lead is then governed by the water chemistry and to a lesser degree by the oxidant, except in certain cases.

While this overview is simplified, it is very important to make the distinction between the two processes, as it has caused significant confusion in the industry. For example, new pipe with fresh metal will be significantly affected by changes in disinfectant, whereas old pipe with a layer of oxidized, relatively insoluble material may be affected by other parameters in solution chemistry. How the two processes relate to each other is complex and not well understood. It is possible that the pipe surface may have a very low metal release rate due to loss of solubility, but it may have a high electrochemical corrosion rate.

CHLORAMINES

Key Factors Affecting Water Quality Changes

There are two primary water quality effects from chloramines - the impact of increased oxidizing conditions in the distribution system and the impact that nitrification may have in the system with respect to changes in pH, alkalinity, and loss of free chlorine residual and dissolved oxygen.

- Treatment using chloramines primarily consists of providing a residual of monochloramine in the distribution system in the range of 1.5 – 4 mg/L. The primary impact, relative to free chlorine, is that the monochloramine residual persists into the distribution system providing an oxidized environment over a greater portion of the service area. Monochloramine has been shown to be less reactive than free chlorine with iron (Valentine et. al, 2000), a major contributor to chlorine residual loss in systems that have significant amounts of iron in the infrastructure. As a result, redox potential could be higher than with free chlorine over larger areas of the distribution system, creating different chemical conditions at the pipe surface.

- In the processes of chloramine decay and nitrification, monochloramine reacts to form hypochlorite and ammonia. The free ammonia, in turn, is converted via microbiologically mediated oxidation to nitrite and then to nitrate. The process consumes oxygen and releases hydrogen ions, lowering the pH and alkalinity of the affected water. While there is very little in the peer-reviewed research to support this, there are field and model distribution studies that demonstrate this (Black & Veatch, 2000, nitrification study for MWRA).

Positive Impacts on Water Quality

Theoretical

With the introduction of chloramine in the system the redox potential at the pipe surface may theoretically be higher than with free chlorine due to the more selective reactivity of monochloramine compared to hypochlorous species. Data have shown that chloramines persist into biofilms more effectively than free chlorine (LeChevallier 1996a).

Iron. Recent research has shown that higher levels of oxygen (greater than 2 mg/L) or free chlorine (greater than 0.2 mg/L) may slow iron release and the formation of red water (Clement et. al, 2002, Snoeyink 2003). This is based on the iron release model that involves the release of iron from the pipe surface that is in the soluble form (AwwaRF and DVGW, 1996). Most of the iron in unlined cast iron pipe scales is in the form of Fe^{+2}. At the surface, this Fe^{+2} can be oxidized to Fe^{+3} and form insoluble crystals that are adherent to the pipe wall. In this process, oxygen and disinfectants are consumed. With the lowering of the redox potential, Fe^{+3} can be converted to soluble and mobile Fe^{+2}. The Fe^{+2} then can migrate into the bulk water and be oxidized to form a particle causing red water. Free chlorine has been found to be significantly more reactive with iron scales than monochloramine (Valentine et. al, 2000).

Copper. Several reports of microbial mediated corrosion in copper pipe have been published (cited in AwwaRF 1996 and DVGW). Chloramines have demonstrated the ability to control biofilms more successfully than free chlorine, thus the benefit of mitigation of microbial mediated copper corrosion would be expected with the change to chloramines from free chlorine.

Lead. There has been little research on the beneficial or detrimental effects of disinfectants on lead solubility. However, disinfectants such as hypochlorite and monochloramine can act as primary oxidants of lead much as dissolved oxygen does. The only potential benefit to a conversion to chloramine from free chlorine would be if biofilm were reduced on the lead surface. As biofilms are controlled, lead corrosion control that relies on the formation of insoluble films may also demonstrate improved solid stability. The actions of this could be related to areas with different coverage of biofilm with different levels of redox potential. Different levels of oxygen concentration from one point to another could induce galvanic corrosion (Walker and Oliphant 1982).

Particle Changes. As chloramine has a much slower reaction rate than chlorine for several compounds, it is likely that certain metals may be oxidized more slowly with chloramine. This may be an issue with manganese which has a slow oxidation rate compared to iron. The oxidation of iron at most pH (7-9) conditions encountered by utilities will take seconds whereas the oxidation of manganese may require hours and even days with free chlorine (Knocke et al. 1990). Further reductions in the manganese oxidation rate may be experienced with chloramine. As a result, the use of chloramine may slow the manganese oxidation rate sufficiently for manganese to pass through the distribution system un-oxidized and thus prevent the formation of discoloration.

Actual Experience

From the literature and unpublished utility studies, there is a growing body of information on the effects experienced with a conversion to chloramines.

Iron. No water systems were found in the reported literature or within the body of anecdotal information, which reported on any positive benefits with regard to red water formation or iron release resulting from conversion to chloramines. This may be due to the fact that few water systems collect enough objective discoloration data (iron, apparent color, and turbidity) to detect an impact. However, the lack of any observed benefit may be associated more with the fact that monochloramine reactions with iron are slower than the reactions of free chlorine with iron.

Copper. In the last decade, a rather surprising finding was made by Reiber (1996). Blue water complaints associated with elevated copper levels were experienced at a number of water systems. After investigation, the high levels of copper were found to be the result of biological growth on copper pipe. The biological growths would occasionally slough off carrying high levels of copper. The sites that experienced high copper were found to have very low free chlorine residuals. In that same study, the introduction of higher disinfectant residuals reduced and in most cases eliminated the problem. It was concluded that higher residuals would help to prevent the problem. Higher residuals would prevent bacterial regrowth preventing the sloughing problem.

Lead. Recently there has been renewed interest in the effect of distribution system disinfectants due to an increase in lead levels in Washington D.C. that has been potentially linked to the conversion of free chlorine to monochloramine. This is discussed in the next two sections.

Particle Changes. In a study (Black & Veatch 1996) for the Portland Water Bureau, the use of chloramines for disinfection of the groundwater supply was recommended as a possible solution to manganese discoloration. Chloramine has a slower reaction rate than free chlorine and would slow the oxidation of manganese sufficiently to prevent the creation of significant amounts of oxidized manganese species. Whether chloramine would be viable in offsetting the discoloration impact with manganese would largely depend on the pH, temperature and travel time in the distribution system.

Potential Adverse Impacts on Water Quality

Theoretical

There are some basic driving elements where chloramines would induce adverse effects on corrosion and water quality:

- Decreases in pH, alkalinity, and dissolved oxygen and increases in ammonia, nitrite, and nitrate due to nitrification reactions associated with the breakdown of chloramines.
- An increase in the redox potential at the pipe surface favoring the formation of more oxidized particles.
- The potential for formation of metal complexes with ammonia or amines introduced through chloramination.
- The sloughing-off of biofilms containing metal solids from the pipe surface.

Iron. Unlined iron pipe can support considerable amounts of biofilms with potentially great diversity, partly by providing suitable sites for the initiation and growth of attached bacteria and partly by adsorbing natural organic matter from the bulk water (Camper 1996, Clement et. al, 2003). Iron pipe has been demonstrated to adsorb organic material from the bulk water (Gu et al, 1994), supporting the growth of higher numbers of heterotrophs than other pipe materials. Nitrification reactions taking place at the pipe surface result in the loss of chloramine residual and dissolved oxygen and a localized decrease in pH. Iron release is favored under such conditions.

The introduction of chloramine has the potential to significantly reduce biofilms (LeChevallier, 1996a). It is possible that the presence of these biofilms may act to inhibit the release of iron by a stable film (Videla 1996). This may be through the formation of a stable biofilm that traps iron, or through the formation of a protective mineral coating. Loss of the stable biofilm may contribute to instability of the iron pipe surface and result in iron release.

Copper. Copper release from copper pipe has been modeled and is relatively well understood (Schock et. al,1995). Copper levels in water are primarily a function of pH, Dissolved Inorganic Carbon (DIC), and orthophosphate as well as the redox potential of the water. Copper can exist as Cu^{+1} and Cu^{+2}. Cu^{+1} has a much lower solubility than Cu^{+2}. Data clearly show that most copper solids are in the Cu^{+2} state. Very few water systems may have conditions where the redox potential is so low that they would support the formation of Cu^{+1} based solids. However, the potential exists that in remote points in a distribution systems or in long service lines or low flow areas, the residual and dissolved oxygen levels may become sufficiently low to allow Cu^{+1} solids to exist. Therefore, in a situation where chloramine was introduced into a free residual depleted area, the increase in redox potential may cause a shift from Cu^{+1} to Cu^{+2} solids and increase solubility.

Copper is capable of forming a number of complexes with ammonia (AwwaRF and DVGW 1996). Decay of chloramines results in the release of ammonia and could increase the solubility of copper in the system by promoting the formation of copper-ammonia complexes.

Lead. Most of the lead compounds identified on piping systems in drinking water have been derivatives of Pb(II), primarily $PbCO_3$ and $Pb(CO_3)_2(OH)_2$(Shieman and Jackson, 1981). The formation of Pb(IV) forms under only extreme oxidizing conditions. As a result, most researchers have explored the solubility of various Pb(II) compounds, resulting in very little information for Pb(IV) compounds. Theoretically, under conditions with high redox potential, this could lead to the conversion of Pb (II) to Pb(IV)(AwwaRF and DVGW, 1996). This in turn could lead to the formation of PbO_2 (platternite) which may have a lower solubility than some

Pb(II) compounds depending upon specific water chemistry conditions. Therefore, the potential exists for a water system that converted from free chlorine to chloramine to have a reduction in redox potential significant enough to induce a phase change from Pb(IV) to Pb(II).

Particle Transformation. Theory has not yet been formulated to elucidate what may be expected with regard to particle transformation when a utility changes from free chlorine to chloramines for disinfection.

Actual Experience

Iron. The vast number of large utilities that have converted from free chlorine to chloramine have not reported problems with red water. For example, one water system that completed a major conversion from free chlorine to chloramine was East Bay Municipal Utility District in northern California. The target monochloramine residual was 2 mg/L. Despite this dramatic change and the fact that the distribution system was comprised of approximately 50% unlined cast-iron pipe, the system experienced no red water problems related to the conversion (Hunsinger 2000).

Copper. A study (Black & Veatch 2000) conducted for the Irving, Texas water system demonstrated that chloramines could have an adverse effect on copper levels. Data collected over a six-month period linked areas displaying nitrification with low pH values at those locations. The nitrification was very severe, with the nitrite concentration reaching 0.4 mg/L . In the central parts of the distribution system, the pH ranged from 7.6 to 8.2. Locations subject to nitrification had pH values in the range of 6.2 to 6.8. With the lower pH levels, the copper levels increased from < 0.50 mg/L to approximately 1.0 mg/L. The water system has approached the action level for copper of 1.3 mg/L.

In the absence of nitrification it appears that chloramine has no effect on the release of copper relative to free chlorine. Reiber (1993) conducted parallel tests with copper pilot loop systems comparing free chlorine and chloramine. No difference in the leaching of copper was noted between the two systems.

Lead. The above-referenced study in Irving, Texas also looked at lead solubility associated with the lowered pH resulting from nitrification. Lead levels increased with the lowering of pH from <0.05 mg/L to 0.10 – 0.015 mg/L.

Recent increases in lead levels in the water system in Washington, D.C. have generated a series of investigations. The research is on going and is yet to be published. The implications, potentially, demonstrate the most significant link between metal release and disinfection changes. Work at EPA's research center in Cincinnati (Schock and Giani 2004) suggests that the presence of PbO_2 may be part of the explanation. Until recently, the Washington, D.C. system maintained a relatively high level of free chlorine (> 2 mg/L) before it switched its secondary disinfectant to chloramines. The high levels of free chlorine may have been responsible for the formation of PbO_2. The formation of PbO_2 is only possible under highly oxidizing conditions. After the conversion took place, this highly oxidized film potentially became destabilized and increased lead levels. Similarly, PbO_2 has been found in other systems with free chlorine, including Cincinnati (Schock and Giani 2004).

The implications of this would also suggest that more solubility modeling would be needed for Pb(IV) compounds. The current guidance is based on water treatment for Pb(II) compounds.

It is clear that the recent increased lead levels in Washington, D.C. have generated interest in determining whether chloramines were the cause of the increase in lead levels. At present it remains unclear whether chloramines are the cause, but future research will help explore this important issue, as there appears to be a mechanism.

Summary

A summary of the corrosion and particle related issues associated with converting from chlorine to chloramines is shown in Table D.1.

Table D.1
Summary of issues for corrosion and particle impacts related to conversion to chloramines

Parameter	Positive Impact	Negative Impact	New Information Required
Iron	Maintaining oxidized insoluble surface	Temporary removal of biofilm Lowering of DO and pH with nitrification	Need well-controlled study to determine impacts Low pH from nitrification may cause iron release
Copper	Reduction in bio-induced corrosion	Lowering of pH with nitrification Formation of amine complexes	More data is needed on the pH impact from nitrification, and formation of amine complexes
Lead	None expected	Lowering of pH	Same as copper
Particle Changes	Slowing of the manganese oxidation rate	None expected or unknown	Not considered to be a high priority

OZONE

Key Factors Affecting Water Quality Changes

Ozone induces many changes to water chemistry primarily by altering organic matter and oxidation of some organic compounds. The reactions are very complex and depend highly on the nature of the water quality. The key factors associated with the introduction of ozone are:

- Changes in the nature of particles
- The introduction of oxygen through O_3 generation
- The alteration of organic molecules and their potential effect on metal release

Positive Impacts on Water Quality

Theoretical

Iron. Ozone may affect iron release by altering the nature of organic matter. Natural organic matter (NOM) has a high affinity for reaction with iron particles and iron compounds and it is well documented that iron oxides present on cast-iron pipe can adsorb organic matter (Stumm and Morgan 1985, Davis and Hayes 1986).

While it is unclear how these changes could affect iron release and discoloration, it is possible that there could be some effect. It is possible that ozone could render NOM more or less complexable with iron. Some researchers (Winzenbacher et at. 2000) found that ozonation prevents Fe-NOM complexes from forming and stabilizes iron hydroxide particles. It was also noted that ozone increased the negative charge on organics, which could lead to better adsorption to FeOOH (goethite) scales in drinking water. Goethite makes up a large majority of pipe scales with unlined cast iron pipe and is primarily positively charged below a pH of 8. As such, ozonation could increase the adsorption of NOM onto pipe scales, and slow iron release. The accumulation of NOM on the surface could lead to increase in biofilms.

Some ozone systems that use oxygen as the primary make-up gas can introduce supersaturated levels of oxygen to the treated water. In some cases, 18 mg/L of oxygen have been reported. The high oxygen concentration has an effect on iron release at pipe surfaces (Kuch 1984). From a recent model developed by Snoeyink, (Snoeyink 2003) higher O_2 levels at the pipe surface could slow the iron release rate significantly for unlined cast iron that has been in place for a long time. The scales of such pipe consist of a broad mixture of Fe^{+2} and Fe^{+3} compounds and crystals. Most of the outer protective scale consists of relatively stable insoluble Fe^{+3} compounds. As oxygen is depleted near the pipe surface during periods of stagnation, the Fe^{+3} compounds are reduced to soluble Fe^{+2} compounds and then move from the scale surface to the bulk water. Higher levels of oxygen could theoretically lead to lower iron levels in solution. This was demonstrated in lab testing (Clement et. al, 2002).

Copper. There are no expected positive impacts from ozone. The only remote possibility is a change in the composition of organics that may alter the copper release.

Lead. As with copper, there are no expected positive impacts from ozonation.

Particle Changes. While the impacts of ozone as a process to aid filtration have been studied, little work has been done on how particles have been altered and how this may affect distribution system water quality. Some of the past work (Jekel 1983, Schick, 2002) has shown that ozone breaks up the surface groups on organic molecules rendering them more absorbable in some cases. In other cases, ozone stabilizes particles. It depends largely on the nature of the organics, ozone dose and level of calcium. These reactions are outside the scope of this document as it pertains to distribution system water quality

Actual Experience

Iron. (See discussion below under adverse impacts.)

Lead and Copper. While data have been collected as part of the Lead and Copper Rule (LCR) with water systems that have converted to ozone, very little information has been published. Three major systems that have added ozone as a last treatment step or as a stand alone process include Bangor, ME; Portland, ME; and Milwaukee, WI. None of these systems reported significant changes in lead or copper concentrations that they could attribute to initiation of ozone addition. It should be noted that these systems do not collect lead and copper data frequently enough to determine subtle changes.

Potential Adverse Impacts on Water Quality

Theoretical

Iron. The greatest risk thought to be caused by ozone is the introduction of supersaturated levels of oxygen from oxygen-based ozone feed systems.

Lead and Copper. Ozone could induce changes in organic molecules as stated in the above section. Organics have been noted to affect metal uptake in some cases, but no data are available on the impacts related to ozone.

Actual Experience

Iron. The impact of ozone on red water has been studied only recently (Clement et. al, 2002). A study conducted at the Massachusetts Water Resources Authority in Boston, MA (MWRA) looked at the impact of ozonation on red water and iron release over a two-year period. The combination of high pH (9.0) or orthophosphate (1.5 to 3.0 mg/L) was tested with and without ozonation. The study used actual excavated unlined cast-iron pipe from the distribution system to form seven 100 ft long flow-through pipe racks. The experiment compared iron release and red water (measured by turbidity and color). For both the orthophosphate and high pH loops, the introduction of ozone did not change iron release and red water. The ozone dose was 1.5 mg/L and the oxygen level of the water increased from approximately 8 mg/L to 22 mg/L. The result that iron levels did not increase may be due to the fact the water was initially saturated with dissolved oxygen and additional increases would not have any effect.

Summary

A summary of the corrosion and particle related issues associated with converting from chlorine to ozone is shown in Table D.2.

Table D.2
Summary of issues for corrosion and particle impacts for ozone

Parameter	Positive Impact	Negative Impact	New Information Required
Red Water/ Iron Release	Added oxygen may help stabilize iron scales	Potential interference with high pH corrosion control treatment	Several studies have been completed
Copper	None expected	None expected	Insufficient concern to continue data collection
Lead	None expected	Non Expected	Same as copper
Particle Changes	Highly complex and variable	None expected or unknown	Has been well studied with NOM. Studies should be conducted to evaluate changes to inorganics

CHLORINE DIOXIDE

Key Factors Affecting Water Quality Changes

Information and data with regard to the impact of chlorine dioxide on corrosion and particle transformations are very limited. Like ozone, chlorine dioxide is a primary disinfectant and is unlikely to impact corrosion in the distribution system. The residual chlorite could act as an oxidant, potentially oxidizing some metals.

Work by some researchers has attempted to compare chlorine dioxide to other disinfectants with regard to iron release (Eisnor et. al, 2001). In the study, the researchers found that chlorine dioxide resulted in the lowest corrosion rate and iron release compared to chlorine and monochloramine. As the study used new pipe, it would be expected that the highest residual of oxidant present would result in the highest corrosion rate. This may not be the case with an actual aged corroded pipe in a distribution system as the release mechanism is driven by solubility

UV

Key Factors Affecting Water Quality Changes

Of all the possible disinfectants used, UV disinfection is considered least likely to have an impact on distribution system water quality as it leaves no residual. Ultraviolet disinfection at the dosages used for disinfection (25 – 120 mJ/cm^{-2}) does not greatly alter the organic nature of the water (Kruithof, 2000). One area that has had little study is the ability of ultraviolet illumination to act as a reductant for some constituents. Nitrate has been shown to reduce to nitrite when exposed to certain UV doses. Other constituents may be reduced which may affect their fate in the distribution system.

One other area where UV disinfection may have an impact is on groundwater supplies where UV illumination would replace chlorine as the disinfectant. A system that changes to UV in place of free chlorine will not see the same redox potential and may experience a change in metal behavior.

BOOSTER CHLORINATION

Key Factors Affecting Water Quality Changes

In order to maintain a residual at remote points, many water systems maintain high chlorine residuals at the point of entry as the residual is depleted over time through the system. The alternative of booster chlorination attempts to increase residuals at remote locations in the distribution system by rechlorinating the water at booster pump stations or storage tanks. Booster chlorination assists in keeping residuals more uniform spatially.

The effect of booster chlorination would be expected to be locally limited to the area served by the pump station or storage tank. During the first months after installation there may be higher oxidant levels at points served by the booster station than before booster chlorination was practiced. This may cause a temporary change in the surface chemistry of pipes resulting in some metal release. There are few publications addressing secondary effects of booster chlorination, thus impacts on corrosion and particle release can only be speculated.

Summary

Chloramines have probably the most significant impact of any major disinfectant change. This is largely due to the fact that monochloramine is more selective in its reaction than free chlorine and it may be present at high concentrations near the pipe surface resulting in a change in the redox condition there. Also nitrification can lower the pH, alkalinity, and disinfectant residual promoting low pH and low redox conditions leading to higher metal release rates.

A summary of the ranking of the overall issues with disinfectants and their research priorities are listed below in Table D.3.

Table D.3
Summary of impacts of disinfectants on corrosion and particle changes

Disinfectant Change	Importance of Issue Relative to Corrosion/Particles	Future Data Collection/ Research	Comments
Chloramine	4	5	Chloramines have greatest potential impact on corrosion, likely through nitrification or changes in redox potential
Ozone	3	5* 3**	Data suggests little or no impact from ozone on corrosion. Greatest impact is on particle changes.
UV	1	1	UV has no known direct connection with corrosion. Potential reduction in some constituents (Nitrate to Nitrite)
Chlorine Dioxide	2	2	Little evidence to suggest this is an issue. Impact on particle change relative to ozone is unknown
Booster Chlorinatio n	2	3	Little evidence to suggest that this is an issue

Scale 1 to 5 ; 5 having the highest impact or importance; 1 the lowest
*Greatest importance on particle changes.
** Less potential for corrosion

APPENDIX E
ISSUE PAPER NO. 4
CHANGES IN AESTHETIC QUALITY – TASTE & ODOR

AwwaRF PROJECT NO. 2940
LONG-TERM EFFECTS OF DISINFECTION CHANGES
ON WATER QUALITY

Issue Paper
on
Change in Aesthetic Quality - Tastes & Odor
by
Bob Hoehn

CHLORAMINES

Key Factors Affecting Water Quality Changes

Key factors affecting changes in aesthetic water quality properties following substitution of chloramines for chlorine in the distribution system can be grouped in five categories: These are:

1. The chloramine-formation procedure at the WTP, including the ratio of chlorine to ammonia, pH and the sequence in which the reactants are added;
2. The chemical characteristics of the finished water, including pH, dissolved oxygen concentration, corrosion inhibitors, natural organic matter (NOM) concentration and composition, inorganic ions such as bromide and iodide, and phenolic compounds;
3. The chloramine dose at the treatment plant, which influences the extent to which chloramines will persist throughout the distribution system,
4. Physical and chemical factors associated with the distribution system such as pipe materials, corrosion, water velocity, and residence time in pipes and storage tanks; and,
5. Changes in distribution system water quality induced by blending chloraminated water with chlorinated water or by booster chlorination.

Positive Impacts on Water Quality

Theoretical

Taste-and-odor complaints should be reduced. The most immediate, positive impact should be a reduction in customer complaints, especially of chlorinous odors but also of other odors related to the formation of organic and inorganic by-products of chlorination. Proper controls on the chloramine-generation process (e.g. pH and chlorine-to-ammonia ratio) will generate primarily monochloramine and minimize dichloramine, which is much more odorous than either free chlorine or monochloramine.

In addition, taste-and-odor complaints caused by organic and inorganic DBPs that form during reactions with chlorine should be greatly reduced when the switch is made to chloramines because chloramines are not as reactive as chlorine. The chlorine contact time prior to ammoniation and NOM concentrations in the water are particularly important variables influencing the formation of odorous compounds.

Chloramine residuals should be more persistent in the distribution system. Because chloramines are weaker oxidants, they should persist longer and at higher concentrations in the distribution system than chlorine, thus reducing the number of taste-and-odor problems associated with the loss of oxidant residual. Examples include the slight masking of odor-causing compounds afforded by chloramines and inhibition of microorganisms in low-velocity portions of the distribution system that produce odorous metabolites.

Actual Experience

Reduction in chlorinous taste-and-odor complaints. The technical literature contains numerous examples of utilities using chloramines in efforts to reduce DBP formation, but the aesthetic impact of converting from free chlorine to chloramines has not been well documented. Usually, the subject of customer complaints is addressed only in general terms. An exception is the excellent paper presented by Burlingame et. al, (2003) who detailed extensive studies of customer attitudes conducted at the Philadelphia Water Department (PWD) where chloramines are used in the distribution system.

Burlingame et. al, (2003) used a "forced-choice triangle test" to assess consumer sensitivity to chlorinous flavor in separate water samples containing chlorine and chloramines. The "average" consumer sensitivity to free chlorine and chloramines was 0.8 mg/L and 2.5 mg/L, respectively, and only 15 percent of the panelists could detect chloramine at concentrations less than 2.5 mg/L. The authors also reported a significant impact of pH on dichloramine concentrations. Panelists found that the chlorinous (bleachy) odor, which is typical of dichloramine, was more noticeable at finished-water pH 7.4 to 7.6 than it was at pH 8.0 to 8.2. The authors concluded that PWD could maintain customer satisfaction with the drinking water if the dose was maintained at 2.0 mg/L and the pH at approximately 8.0-8.2.

Odor- and flavor threshold concentrations (OTCs and FTCs) of the various chlorine and chloramine species diluted in odor-free water have been determined by trained panels (Krasner and Barrett 1984; Khiari et. al 2002; Shen et. al, 1997), and the data support the presumption that odor complaints should decrease once a utility converts its final disinfectant from chlorine to chloramines. Table E.1 shows FPA data developed by Krasner and Barrett (1984). The absolute values of the published OTCs and FTCs vary somewhat, but the inference one can make from the data is the same, namely that complaints should diminish if the chloramine residual is comprised primarily of monochloramine.

Table E.1
Odor threshold concentrations (OTC) and flavor threshold concentrations (FTC) of chlorine and chloramines in odor-free water (Krasner & Barrett 1984)

Chlorine Species	Temp	pH	OTC, mg/L	FTC, mg/L	Descriptor
Hypochlorous Acid	25 °C	<6	0.28	0.24	Chlorinous
Hypochlorite Ion	25 °C		0.36	0.30	Chlorinous
Monochloramine	25 °C		0.65	0.48	Swimming Pool
Dichloramine	25 °C		0.15	0.13	Swimming Pool
Trichloramine	25 °C		0.02	--	Geranium

The authors of one report (Kirmeyer et. al 1993) stated without citing evidence that monochloramine rarely causes taste-and-odor problems in drinking water unless the concentration exceeds 5 mg/L, which exceeds the existing MRDL and is several times higher than published chloramine OTCs and FTCs.

Customer acceptance of chloraminated drinking water will be influenced not only by the chloramine concentrations but also by site-specific water characteristics (pH, temperature, nature and concentration of organic and inorganic substances), the presence or absence of other highly odorous and unpleasant-tasting substances in the water (e.g. algal metabolites), and the past experience of the population being served (Kirmeyer et. al 1993; Mackey et. al, 2003). Mackey et. al, (2003) found that French panelists were more sensitive to chloramines than American panelists.

Masking of odorous compounds. Chloramines generally do not oxidize odorous compounds, but they can slightly mask those that either are not eliminated during treatment or are produced in the distribution system (Burlingame et. al, 1992; Khiari et. al 2002).

Prevention of chlorophenolic odors. The earliest reports of chloramine use for water treatment were published in the 1920s and described how utilities used the "ammonia chlorine" process to combat highly offensive medicinal odors caused by chlorination of phenolic substances in river water (White 1999). The process usually involved adding ammonia to the raw water followed later by chlorine addition. Chlorophenolic odors were eliminated not only because chloramines are weaker oxidants than chlorine but also because they "have a greater tendency to participate in chlorine substitution reactions rather than oxidation" (Speitel 1999). Suffet et. al, (AwwaRF and Lyonnaise des Eaux 1995), citing others, noted that chloramines can chlorinate phenol, but at much slower rates than chlorine (days vs. hours). Today, chlorophenolic odors pose few problems than in former years because discharges of phenolic compounds into rivers have been considerably curtailed by regulations.

Reduction in the formation of odorous DBPs. Most of the literature dealing with chloramine use for DBP reduction has focused on the relative DBP concentrations in the finished water rather than on changes in organoleptic properties and reductions in the number of customer complaints. Kirmeyer et. al, (1993) provided a comprehensive review of DBP formation at utilities where the chloramines have been used for distribution-system residual, but taste-and-odor reduction was not discussed.

Potential Adverse Impacts on Water Quality

Theoretical

Ammonia- and biofilm-related tastes and odors. Excess ammonia in chloraminated water can be the result of either improper chloramine generation procedures at the treatment plant (i.e. overfeeding ammonia or underfeeding chlorine) or as a result of chloramine decomposition by hydrolysis and acid catalysis (Valentine et. al 1998), which accounts for loss of residual in distribution systems. Taste-and-odor problems related to excessive ammonia in the distribution system would likely be associated with the depletion of dissolved oxygen during nitrification reactions rather than with ammonia itself. Biofilm growth is influenced by numerous conditions, including organic matter concentrations, pipe materials, storage time, flow velocity, corrosion, and corrosion inhibitors (Abernathy and Camper 1997; Smith et. al, 1996; Camper 2003). Abernathy and Camper (1997) stated that biofilms are very resistant to disinfectants and "complete inactivation will not likely occur at chloramine residuals less than 4.0 mg/L."

An AwwaRF report by Kelley et. al, (2003) showed that fungi in the distribution systems are most likely associated with biofilms rather than growing in the flowing water. While the production of taste-and-odor compounds was not an objective of this particular study, the authors did isolate a fungal species that could produce 2,4,6-trichloroanisole, a known earthy-smelling compound, from 2,4, 6-trichlorophenol. Malleret and Bruchet (2003) also suspected fungi as having produced the highly odorous 2,4,6-tribromoanisole (OTC 30 picograms per liter) from 2,4,6-trichlorophenol that leached from a tank lining. In a limited laboratory study, Kelley et. al, (2003) found that ozone and chlorine dioxide were more effective than either monochloramine or free chlorine for killing one type fungus, though fewer fungi were found in pipe biofilms collected from distribution systems where chloramines rather than chlorine were present.

Kelley et. al, (2003) did not attempt to isolate actinomycetes during their study because these organisms are filamentous bacteria, not fungi. Many actinomycete species, particularly those belonging to Genus *Streptomyces*, can produce numerous odorous compounds, including geosmin (earthy odor) and 2-methylisoborneol (MIB, musty odor) (Zaitlin et. al, 2003). Because they are bacteria, actinomycetes, like the true fungi, probably grow in distribution- system biofilms but escape routine microbiological examination of water samples. While actinomycete-related taste-and-odor problems are more likely to occur in surface raw water supplies, their contribution to problems in distribution systems has not been researched, and, in fact, little, if anything, is known about the subject. A study of this topic would involve isolating actinomycetes on special agars, subculturing the isolates, and evaluating their potential for producing taste-and-odor compounds.

Formation of odorous di- and trichloramines during the chloramine-generation process. The chemistry of chloramine generation, as reviewed by Singer and Reckhow (1999), is described by the following equations:

$$NH_3 + HOCl = NH_2Cl + H_20 \quad [1]$$

$$NH_2Cl + HOCl = NHCl_2 + H_2O \quad [2]$$

$$NHCl_2 + HOCl = NCl_3 + H_2O \quad [3]$$

As the authors noted, monochloramine (equation 1) will be the predominant product if the molar ratio of chlorine to ammonia is less than 1:1 (weight ratio of chlorine-to-ammonia-N < 5) and the pH is approximately 8. They also noted that "at higher Cl:N ratios and lower pH, dichloramine becomes important" and "…at mildly acidic pH values in the range of 6-6.5, nitrogen trichloride can be formed." The discussion of threshold tastes and odors resulting from formation of dichloramine was presented earlier.

Incomplete oxidation of odorous compounds. Odorous compounds, such as sulfides and algal metabolites that are not oxidized at the treatment plant because the free-chlorine contact time is too short may penetrate the distribution system once ammonia is added because chloramines are weaker oxidants than chlorine.

Production of Odorous DBPs by NOM Reactions with Chloramines. Despite the fact that chloramines are weaker oxidants than chlorine, they can form a variety of DBPs during reactions with NOM, not all of which are odorous. The types and concentrations of DBPs that form depend on several factors, including the chlorine-contact time before ammonia is added, pH, and contact time. Speitel (1999) presented an excellent overview of the types of DBPs formed during chloramine reactions with NOM and described conditions under which each is likely to form in significant concentration. Suffet et. al, (AwwaRF and Lyonnaise des Eaux, 1995) also discussed DBPs found in chloraminated water and their associated odors. Kirmeyer et. al, (1993), citing Bull and Kopfler (1991), listed THMs, halogenated acids, haloacetonitriles, halogenated aldehydes, chloropicrin, chlorinated phenols, and cyanogen chloride as being among the compounds that have been detected in chloramine-treated water. Burlingame et. al, (2003) reported that the cyanogen chloride concentration ranged from 2 μg/L to 8 μg/L in the Philadelphia Water Department's (PWD) distribution system during an 18-month period, but no mention was made of its relationship to taste-and-odor complaints. Chloramine residuals leaving the PWD treatment plants during this period were 2 mg/L.

Singer and Reckhow (1999) postulated that some of the DBPs isolated from chloraminated water may been formed by during reactions between organic matter and HOCl generated by reverse reactions shown in equations [1], [2}, and [3]. They also noted, however, that "...monochloramine is too weak to oxidize reduced iron and manganese and most taste- and odor-causing compounds, and it is too weak to eliminate natural organic color. While monochloramine will oxidize natural organic material to some degree and produce some halogenated organic material, it does not generally produce trihalomethanes."

Singer and Reckhow (1999) also noted that part of the oxidation potential of monochloramine is derived from the hydrolysis reactions represented by the reactions that form mono-, di-, and trichloramine (equations 1-3), stating that "the free chlorine liberated [by these reactions], although often in small amounts, may be responsible in part for the formation of some halogenated by-products associated with chloramines."

The presence of bromide ion and iodide ion can also influence the extent to which odorous DBPs are formed. Iodide ion is most often a component of groundwater rather than surface water, and iodonated DBPs form slowly. Thus, one would not expect to find iodonated DBPs in chloraminated drinking water (surface water), only in systems where the chlorine contact time prior to ammoniation was quite long (Khiari et. al 2002). On the other hand, bromide ion is common in both groundwater and surface water, especially around costal regions, and odorous brominated DBPs may form under certain conditions determined by the chlorine dose and contact time and the NOM concentration. Speitel (1999) and Valentine et. al, (1998) discussed the special role bromide ion plays in DBP chemistry, and Valentine et. al, (1998) described how it can accelerate chloramine decay rates. Bromide ion concentrations during their model-development studies varied from 0.1 mg/L to 0.5 mg/L.

Chloramines can also form odorous chloro-substituted DBPs during reactions with organic nitrogen compounds, including amino acids and aldehydes (Hrudey et. al, 1988; Speitel 1999 1999; Khiari et. al 2002). Nitriles and cyanogen chloride are among the organic nitrogen compounds found in chloraminated waters (Khiari et. al 2002). With the exception of nitriles, the odor thresholds of most DBPs isolated from water treated with chloramines have not been well documented in the literature. Speitel (1999), citing other researchers, noted that monochloramine can transfer chlorine to organic nitrogenous compounds as it degrades by acid catalysis reactions. Reaction rates were high at pH levels as high as pH 8.5.

Potential Adverse Effects of Blending. Adding chloraminated water to water containing chlorine leads to changes in the total chlorine residual and shifts in the relative proportions of the various chloramine species. Great care must be taken to avoid both a depletion of the total residual and a shift in the chloramine species toward greater concentrations of odorous dichloramines.

Actual Experience

Ammonia. Free-ammonia residuals can approach 0.5 mg/L in some water supplies (AwwaRF and Lyonnaise des Eaux, 1987; Valentine et. al 1998). Burlingame et. al, (2003) reported an average 0.2 mg/L ammonia-nitrogen in water following ammonia addition to form chloramines at PWD's water treatment facilities. Ammonia OTCs as low as 0.037 mg/L (AwwaRF and Lyonnaise des Eaux, 1987) and as high as 0.9 mg/L (Khiari et al. 2002) have been reported by FPA panels. During these evaluations, panelists described the ammonia odor as rubbery, plastic, cardboard, and reedy, while the sensations evoked during tastes tests were described as "tingling" and "bitter."

Problems associated with nitrification. Intuitively, one would expect nitrification-associated taste-and-odor problems to intensify when chloramines are used for distribution-system residual. Increased bacterial plate counts are common, and many utilities have been forced to revert to chlorine for weeks or even months to combat the problems. Kirmeyer et. al, (1993) presented case histories where utilities were forced to make this change. Increased taste-and-odor complaints could be expected following the changeover because customers quickly become accustomed to the absence of chlorinous odors when the utility switches to chloramines, and the sudden change in taste-and-odor following the conversion to chlorine would be starkly evident.

Changes in water quality that forced utilities to revert to free chlorine are usually related to factors other than increased taste-and-odor complaints. Several utilities that formerly converted their systems from chloramines to free chlorine to alleviate biofilm and regrowth problems have found that they could solve the problems by maintaining a higher chloramine residual and, therefore, no longer need to convert the system residual back to free chlorine, which, in itself, is likely to generate additional complaints.

Persistence of taste-and-odor compounds. Glaze et. al, (1990) demonstrated that monochloramine was relatively ineffective for removing geosmin and MIB from Colorado River water and State Project Water during bench-scale and pilot-plant studies at the Metropolitan Water District of Southern California (MWD). Monochloramine oxidized only 15 percent and 27 percent, respectively, of 100 ng/L geosmin and MIB in Colorado River Water at a dosage of 3 mg/L and contact period of two hours. Two odorous aldehydes (1-hexanal and 1-heptanal) not only were not oxidized by monochloramine, their concentrations actually increased during the chloramine-contact period.

The earthy and musty odors caused by geosmin and MIB are frequently masked by free chlorine but not by chloramines. In the presence of chloramines, these compounds seem to contribute to "background" odors that intensify as the chloramine residual diminishes to zero. The phenomenon of increased background odors with decreasing oxidant residuals is also quite evident in systems using free chlorine, especially at distant reaches and deadends in the distribution system.

Burlingame et. al, (1992) described the persistence of the cucumber odor caused by nonadienal when chloramines were formed in the treated water. The problem was eliminated by extending the free-chlorine contact period prior to ammoniation. Krasner et. al, (1989a) found that highly odorous organic sulfides in raw water at one of MWD's treatment facilities plants were not removed by chloramines, requiring the utility to extend the free-chlorine contact time before ammoniation. The most-odorous organic sulfide recovered from the raw water was dimethyltrisulfide, which produced garlic-like odors. They determined it was being produced in the anaerobic hypolimnion of the water-supply reservoir and escaped oxidation because the chlorine-contact time ahead of ammoniation was quite short. In later studies at MWD, Glaze et. al, (1990) found that dimethyltrisulfide concentrations were reduced by more than 90 percent by dosing the water with 3 mg/L chloramine and allowing a two-hour contact period. This result is somewhat surprising based on the studies of Krasner et. al, (1989a), but the higher dosage and contact time Glaze et. al, used may explain the difference.

Aldehydes and nitriles. Hrudey et. al, (1988) reported finding low-molecular-weight amino acids in the water supply at Edmonton, Alberta that produced odorous aldehydes when chloramine was added. Khiari et. al, (2002) presented OTCs of 14 aldehydes evaluated by expert panels; values ranged from 30 ng/L to 47 μg/L.

The most comprehensive odor evaluations of nitriles and halonitriles were those presented by Khiari et. al, (2002). They reported nitrile OTC concentrations ranging from a low of 0.13 μg/L for octanenitrile to a high of 64 μg/L for isobutyronitrile, and the odor descriptors were generally fruity to sweet. Halonitrile OTCs were much higher (50-400 μg/L) than those of the unchlorinated nitriles, and the authors stated that halonitriles in drinking water should pose no particular problems from a taste-and-odor standpoint.

Blending. Blending of chlorinated and chloraminated waters poses special organoleptic problems that are associated with chloramine chemistry and the organoleptic properties of the individual water sources at the time of blending. (Barrett et. al, 1984; Barrett et. al, 1985; Khiari et. al, 2002). Taste-and-odor problems are more pronounced when breakpoint chlorination is attempted prior to blending chloraminated water with chlorinated water. The problems that developed when MWD sold chloraminated water to member agencies that used free chlorine and the approaches that were taken to minimize problems were discussed by Barrett et. al,(1984), who developed a model that determined the appropriate blend ratios that should be observed to avoid depletion of residual and formation of dichloramines.

Unknown Issues/Information Gaps.

- Not much is known about the impacts of biofilms on taste-and-odors and other aesthetic properties of water in distribution system systems where chloramines are used. Numerous biofilm studies have been conducted under controlled conditions, such as pipe loops and annular reactors, but most of the studies have not considered taste-and-odor issues. The most comprehensive studies to date are those conducted by Khiari and her associates (2002) who evaluated taste-and-odor characteristics of water with biofilm development in pipe loops and annular reactors. Biofilms harvested from laboratory systems such as these and from actual distribution systems should be examined for actinomycetes and fungi that produce odors. The microbiological examination should be supplemented with some type of odor evaluation procedure such as triangle tests and Flavor Profile Analysis.
- Customer complaint records before and after conversion of a system to chloramines should be examined to determine if complaints decreased, and efforts should be made throughout the drinking water industry to involve water-treatment plant personnel in

the investigation of taste-and-odor complaints. Taste-and-odor complaints investigated by customer-service representatives are often lacking important details that can be provided only by laboratory personnel familiar with taste-and-odor evaluation procedures. Utilities that are forced to revert periodically from chloramines to chlorine in the distribution system to control water-quality problems associated with nitrification should be encouraged to include taste-and-odors evaluations as part of the overall data on which they base their decision to make the conversion. Likewise, increased taste-and-odor complaints following the conversion back to chlorine should be documented because they are likely to increase.

- Little is known about the formation and persistence of highly odorous iodonated and brominated compounds during chloramine reactions with bromide ion and chloride ion. Iodide ion primarily in groundwaters can react with free chlorine to produce hypoiodous acid, which can then form iodonated DBPs. Bromide ion is more prevalent in raw-water sources and brominated DBPs following chlorination have been studied. However, few studies were found in the literature that discussed brominated DBPs formed during chloramination. Some of the brominated species are also odorous. Questions regarding the formation of odorous iodonated and brominated DBPs by chloramine reactions can be answered by bench-scale experiments.
- Numerous studies have shown the benefits of changing the final disinfectant from chlorine to chloramines on DBP formation, but few of the excellent articles and reviews of this topic have considered the impacts on tastes and odors. The role of bromide ion in producing odorous DBPs in chloraminated water needs further investigation. Analyses of distribution-system samples for odorous DBPs following the conversion to chloramines would be costly and impractical, but useful data could be generated by bench-scale studies that focus both on chemical analyses and organoleptic evaluations.
- Only passing reference regarding color removal by chloramines has been made in the literature. Being weaker oxidants, chloramines are not expected to significantly reduce color caused by NOM, but the evidence for that assumption is scanty at best. Studies are needed to quantify the extent and rate of color reduction by chloramines and the role pH plays in the process. Highly colored natural waters should be used instead of laboratory-prepared waters containing commercial humic substances.

OZONE

Key Factors Affecting Water Quality Changes

The major factors affecting changes in color and tastes and odors during water treatment with ozone include: ozone dose and contact time, the ozone application point, the nature and concentration of NOM, pH, alkalinity, bromide ion concentration, ammonia concentration, and selection of final oxidant for distribution-system protection.

Positive Impacts on Water Quality

Theoretical

- Ozone, which is the strongest oxidant used for drinking water treatment, should effectively eliminate many odorous compounds from water (e.g. certain naturally occurring algal by-products and synthetic organic chemicals) that cannot be eliminated by treatment with chlorine, chloramines, or chlorine dioxide. Both ozonation and advanced oxidation processes (ozone-peroxide and ozone-UV) should eliminate odorous by-products of actinomycetes that are poorly oxidized by other oxidants.
- Molecular ozone is highly selective in the types of organic compounds it will oxidize, and conditions can be manipulated during the water treatment process to achieve different goals. Often, ozone's effectiveness can be enhanced by the creation of conditions that favor the formation of hydroxyl radical formation, which are even more powerful oxidants than ozone. Formation and destruction of odorous DBPs by ozonation are dependent on numerous conditions that have been reviewed in detail elsewhere (AwwaRF and Lyonnaise des Eaux 1995, Singer and Reckhow 1999).
- Another aesthetic property of drinking water, namely color, should be markedly reduced by ozonation because compounds causing color (usually fulvic acids) contain numerous carbon-to-carbon double bonds that are readily oxidized by ozone (Reckhow 1999b; Kirmeyer et al. 1993, Langlais et. al, 1991). Nickols and Varras (1992) noted that ozone is also highly reactive with some specific chromophoric groups, notably $-NH_2$ and $-OH$, and noted also that color can be reduced without a significant change in the organic concentration.
- Ozone can effectively oxidize reduced iron and manganese, both of which can create aesthetic problems in the distribution system.

Actual Experience

- Lalezary et. al, (1986) and Glaze et. al, (1990) demonstrated that ozone, both alone and in combination with hydrogen peroxide (H2O2) and ultraviolet (UV) irradiation (UV), effectively eliminated earthy and musty tastes and odors caused, respectively, by geosmin and 2-methlylisoborneol (MIB). Data presented by Glaze et. al, (1990) are summarized in Tables E.2 and E.3 and show that advanced oxidation treatment by AOP is more effective than ozone alone. In another pilot-plant study, Owen and Johnson (2002) demonstrated during an investigation at the Hillsborough River Water Treatment Plant that geosmin and 2-MIB could be effectively eliminated from raw water by ozone at dosages required for Cryptosporidium inactivation when 1.0 mg/L H_2O_2 was added following the required ozone-contact period. Lalezary et. al, (1986) also found that ozone was effective for removing another earthy-smelling compound, trichloroanisole.

Table E.2
Percent removals of 100 ng/L MIB and geosmin in Colorado River Water by ozone and advanced oxidation processes (Table 4 Glaze et. al, 1990)

Odor Compound	O_3 Only, 20 min contact time, % Removal		H_2O_2/O_3 (O_3 Dose, 0.2 mg/L·min^{-1}) 20 min contact time, % Removal		O_3/UV (O_3 Dose, 0.2 mg/L·min^{-1}) 20 min contact time, % Removal	
	0.1 O_3 mg/L·min^{-1}	0.2 O_3 mg/L·min^{-1}	H_2O_2 0.1 mg/L·min^{-1}	H_2O_2 0.2 mg/L·min^{-1}	UV Flux 0.19 Watts/L	UV Flux 0.56 Watts/L
Geosmin	38	86 & 92	95	81	87	99
MIB	40	73 & 82	89	79	71	83 & 75

Table E.3
Percent removals of 100 ng/L MIB and geosmin in California state project water by ozone alone and in combination with peroxide (Table 5, Glaze et. al, 1990)

Odor Compound	O_3 Only, 20 min contact time, % Removal		H_2O_2/O_3 (O_3 Dose, 0.2 mg/L/min), % Removal	
	0.1 O_3 Dose Mg/L·min^{-1}	0.2 O_3 Dose mg/L·min^{-1}	H_2O_2 0.1 mg/L·min^{-1} 15-min Contact	H_2O_2 0.2 mg/L·min^{-1} 10-min Contact
Geosmin	94	98	97	96
MIB	82	92	98	99

- Singer and Reckhow (1999) cited the work of Hoigne and his coworkers, who demonstrated the relative effectiveness of molecular ozone and AOP for eliminating synthetic organic chemicals (SOCs) from drinking water. Some of these compounds, such as certain pesticides and industrial chemicals, are readily oxidized by ozone while others are refractory. Certain pesticides, such as aldrin and heptachlor, cause musty and chlorinous odors but are readily oxidized by ozonation (AwwaRF and Lyonnaise des Eau 1995)

- Klimkina et. al, (1987), cited by Langlais et. al, (1991), reported color reductions ranging from 20 percent to 60 percent using high ozone dosages (8-13 mg/L).

- Wilczak et. al, (1993) found that particulate ferric hydroxide resulting from ferrous iron oxidation was readily removed by sedimentation, but oxidized manganese dioxide, which is colloidal, was not readily by removed by either sedimentation or filtration.

Potential Adverse Impacts on Water Quality

Theoretical

- Ozonation can generate odorous compounds and increase the concentration of assimilable organic carbon by partial oxidation of NOM (Khiari et. al 2002; Nickols and Varas (1992); AwwaRF and Lyonnaise des Eaux 1995 and 2002). Biodegradable by-products of ozonation include aldehydes, alcohols, ketones, keto-acids, and organic acids (Khiari et. al 2002; Reckhow 1996; Nickols and Varas 1992). However, taste-and-odor problems associated with these products have been focused mainly on aldehydes. Aldehyde production is greatest when ozonation is carried out under conditions that result in slower rates of ozone decomposition (Nickols and Varas 1992). Readily biodegradable organic compounds produced by ozonation of NOM can indirectly cause tastes and odors by enhancing growth of odor-producing bacteria and fungi in the bulk water and biofilms.
- Odorous DBPs may form during reactions between ozone, NOM, and inorganic ions such as iodide and bromide ion. Ozone can oxidize iodide ion and bromide ion and produce the corresponding hypohalous acids that can then react with NOM components to produce extremely odorous (usually medicinal) DBPs. Iodide ion in water supplies is rarer than bromide ion. Production of these DBPs is highly dependent on the concentration and type of NOM precursors, pH, temperature, and bromide and iodide ion concentrations. The formation of iodonated and brominated DBPs can be minimized by the addition of ammonia. Ammonia will form iodoamine and bromamine by reaction with hypoiodous acid and hypobromous acid, respectively, which form when ozone oxidizes the iodide and bromide ions.
- Manganese, whether it is in the +2 or +4 valence state. Overdosing with ozone, however, can oxidize the manganese to permanganate (+7), and, as a consequence, the treated water will develop a pink coloration.

Actual experience

- Aldehydes causing fruity odors have been found in drinking water following ozonation at moderate dosages and at less than optimal dosages (Anselme et. al, 1988; Glaze et. al, 1990). Other detailed accounts of aldehyde recoveries from ozonated water have been reviewed by Suffet et. al, (AwwaRF and Lyonnaise des Eaux 1995). Khiari et. al, (2002) reported that lower molecular-weight aldehydes often produced by ozonation have rancid or otherwise unpleasant odors, while the higher-molecular-weight aldehydes have pleasant odors, often fruity. Data from Glaze et. al, (1990) are presented in Table E.4.
- Bruchet et. al, (1992) demonstrated the potential for taste-and-odor problems caused by amino acids that are either created by ozonation or escape oxidation during the ozonation process. If chlorine is used as the final disinfectant, the amino acids may produce odorous aldehydes, some of which have very low OTCs.
- Crozes et. al, (1997) evaluated ozone for disinfection and taste-and-odor control at a lime softening plant in Wisconsin. Earthy, musty, and wet-hay odors were reduced in

intensity, but increasing the ozone dose unmasked fishy and swampy odors. At high ozone dosages, buttery and rancid odors were produced.

Table E.4
Percent removals of 100 ng/L of aldehydes in Colorado River water by ozone and advanced oxidation processes (Table 2 Glaze et. al, 1990)

Odor Compound	O_3 Only, 20 min contact time, % Removal		H_2O_2/O_3 (O_3, 0.2 mg/L·min^{-1}) 20 min contact time, % Removal	O_3/UV (O_3, 0.2 mg/L·min^{-1}) 20 min contact time, % Removal
	0.1 O_3 mg/L·min^{-1}	0.2 O_3 mg/L·min^{-1}	H_2O_2 0.2 mg/L·min^{-1}	UV Flux 0.56 Watts/L
1- Hexanal	- 450	42 & 41	73	42
1-Heptanal	- 610	14 & 10	41	40
2,4-Decadienal	75	99 & 99	97	> 99

Note: Negative percent removals indicate creation of the aldehyde during ozonation.

Extremely odorous iodonated THMs have been found at above-threshold levels in nitrified groundwater containing as little as 0.10 mg/L iodide ion at locations in France where ozonation was practiced, and the concentrations increased with increasing ozone dosages (AwwaRF and Lyonnaise des Eaux 1995).

- Singer and Reckhow (1999) stated that low concentrations of brominated DBPs may be expected following ozonation of bromide-containing waters, especially at lower pH and higher ozone dosages. Suffet et. al, (AwwaRF and Lyonnaise des Eaux 1995) pointed out, however, that the OTCs of these compounds have not been well-documented.
- While ferric hydroxide is readily removed by sedimentation, oxidized manganese is colloidal in nature and not readily removed by either sedimentation or filtration (Wilczak et. al, 1993). Over-dosing with ozone can create permanganate ion, which imparts a pink coloration to the water. Wilczak et. al, (1993) recommended that free chlorine be applied ahead of sand and anthracite filters to ensure that manganese is retained.

Unknown Issues/Information Gaps

Ozone as a treatment option has gained in favor in the past fifteen years, but the focus in the myriad of published studies has been primarily on disinfection and DBP reduction rather than on improvements in the aesthetic properties of water. The following are recommended to overcome this deficiency:

- Utilities should examine customer service records before and after ozonation was included in the treatment train and determine if the change affected the frequency of complaints, especially taste-and-odor complaints. Correlating the complaint records with the institution of ozonation may be difficult if the final disinfectant is also switched from chlorine to chloramines.

- Additional research is needed to determine the odor- and flavor threshold concentrations of odorous DBPs, especially brominated DBPs, formed during ozonation of natural waters.
- Additional research is also needed to determine the efficacy of ozone at different dosages and in waters with varying NOM concentrations for eliminating odorous aldehydes commonly produced during ozonation. Special attention should be given to the relationship between ozone dose and percent aldehyde removals following ozonation of waters containing high NOM concentrations.
- The relationships between ozone residual, hydrogen peroxide dosage, pH, and reaction time on taste-and-odor compounds are poorly understood. These relationships should be further explored because ozonation is likely to remain a treatment of choice for taste-and-odor control.
- Color reduction is an ancillary benefit of ozone that is usually not emphasized during pilot-scale and full-scale evaluations of ozone. More data are needed to clarify the conditions (e.g. pH, ozone dose, contact time) that produce acceptable treatment in highly colored waters.

CHLORINE DIOXIDE

Key Factors Affecting Water Quality Changes

Chlorine dioxide is most often added during water treatment as a preoxidant, though, occasionally, it is added either ahead of filtration or following filtration. Occasionally, it is added ahead of the clearwell so that the amount of chlorine or chloroamine needed for maintaining a distribution-system residual can be reduced. On some occasions, it may be used to prevent biological growths in open distribution-system reservoirs.

The increasing popularity of chlorine dioxide for water treatment is due in part to the fact that it oxidizes rather than chlorinates organic matter at pHs and dosages typically seen during drinking water treatment (Gordon and Rosenblatt 1995; Singer and Reckhow 1999). It has become more popular in the United States since 1998 when EPA established a relatively high MRDL (0.8 mg/L). The practical dose that can be applied, however, is limited primarily by constraints on the allowable concentration of its major by-product, the chlorite ion (MCL 1.0 mg/L), in finished water.

Factors affecting water-quality changes when chlorine dioxide is used include:

- Water temperature (affects chlorine dioxide reaction rates)
- TOC/DOC concentrations (affect the oxidant demand)
- Chlorine dioxide dose and contact time
- Exposure to sunlight (causes photodecomposition)
- Mixing intensity (increases loss of chlorine dioxide to the atmosphere because it is a dissolved gas)

- Application of other oxidants following chlorine dioxide addition, particularly chlorine and ozone.

Unlike chlorine, increasing pH does not impair the effectiveness of chlorine dioxide as an oxidant because, as a gas, it does not ionize in water as chlorine does.

Positive Impacts on Water Quality

Theoretical

- Chlorine dioxide provides an alternative to chlorine for pre-oxidation with none of the disadvantages normally associated with chlorination, primarily halogenated DBP formation. Few if any organic DBPs should be formed by NOM reactions with chlorine dioxide. Chlorine dioxide reacts with phenolic functional groups in NOM and with phenol itself without formation of chlorinated by-products that cause odors. The chemistry of chlorine dioxide suggests that it may not be effective for removing many algal and bacterial by-products that cause tastes and odors in drinking water. In fact, it, like ozone, may actually produce odorous compounds during reactions with NOM.
- Because it is a strong oxidant, chlorine dioxide should effectively oxidize iron and manganese, reduce or eliminate certain types of odors, and retard biological growths in treatment plant basins and distribution-system reservoirs. Ammonia is unreactive with chlorine dioxide and, therefore, does not increase the chlorine dioxide demand.
- As long as the MRDL is not exceeded, utilities should be able to effectively use chlorine dioxide for maintaining oxidant residual in the distribution system.
- Chlorine dioxide readily oxidizes both iron and manganese.

Actual Experience

- The earliest uses of chlorine dioxide during water treatment included were for reduction in odors caused by phenols, chlorophenols, and algal products (Aston 1947; Mounsey and Hagar 1946, Walker and Lee 1983). Numerous examples of successful uses of chlorine dioxide for reducing halogentated DBP formation have been published (Andrews et. al, 2001; Hulsey et. al, 2000; Routt 1989). One can infer from this information that the formation of odorous halogenated DBPs would be minimized, thus improving the organoleptic properties of the treated water.
- Chlorine dioxide was effective for controlling algae in open reservoirs by the Los Angeles Department of Water and Power distribution system (Stolarik and Liu 2000). Research has shown that some, but not all, sulfur-containing compounds that cause odors described as "fishy/swampy/grassy" can be effectively eliminated by chlorine dioxide treatment (Glaze et. al, 1990; Krasner et. al, 2004).
- The major inorganic by-product of chlorine dioxide reactions with NOM (chlorite ion) has been shown to reduce nitrification and biofilm growth in distribution systems at concentrations below the current MCL (McGuire et. al, 1999). The authors proposed that systems that use chloramines should consider adding small amounts of chlorite ion to finished water and thereby suppress, and perhaps eliminate,

nitrification. Utilities would probably be reluctant to add a regulated by-product (chlorite ion) to the drinking water, but chlorite ion would be present in water treated with chlorine dioxide unless steps were taken during treatment to remove it.

- Knocke et. al, (1990) found that manganese was rapidly oxidized by chlorine in low DOC waters within the pH range of 5.5 to 9.0. The reaction rate decreased with increasing pH. Iron was oxidized essentially instantaneously.

Potential Adverse Impacts on Water Quality

Theoretical

- Odor problems may develop in the distribution system if chlorite ion produced during chlorine dioxide treatment is not reduced to concentrations < 0.3-0.4 mg/L and the final disinfectant is free chlorine. Chlorine dioxide can regenerate in the distribution system when chlorine is used to provide a distribution-system residual.
- Chlorine dioxide may not eliminate many odorous by-products of microbial metabolism because it doesn't readily oxidize numerous types of organic compounds.

Actual Experience

- Chlorine dioxide is generally not effective in removing many biological products that cause odor tastes and odors in drinking water, especially aldehydes and alcohols. Glaze et. al, (1990) reported only marginal removals of 2,4-decadienal (fishy/swampy odor) and geosmin and 2-MIB (earthy and musty odors, respectively) from Colorado River water. Krasner et. al, (2004) reported that chlorine dioxide was relatively ineffective for eliminating other compounds that cause "fishy/swampy/grass" odors and the compound most commonly associated with cucumber odor (trans, 2-cis, 6-nonadienal). Dietrich et. al, (1995) used FPA to evaluate the effectiveness of low dosages of chlorine dioxide for reducing odors caused by a variety of compounds in water. They found that the oxidant was not effective in removing odors caused by β-cyclocitral (tobacco/grape odor), 2t,6c-nonadienal (cucumber odor), and phenethyl alcohol (floral/rose) but did eliminate the odor of linolenic acid (watermelon). Chlorine dioxide generated grassy odors from two organic acids (linoleic and linolenic).
- Hoehn et. al, (1990) reported that many utilities that use chlorine dioxide frequently receive customer complaints of odors that resemble kerosene or cat urine odors, especially if the final disinfectant is free chlorine. Their investigation showed that trace amounts of chlorine dioxide are regenerated in the distribution system by reactions between chlorite ion and free chlorine. When a customer's tap is opened, the chlorine dioxide is released and reacts with organic compounds already present in the air. The source of these organic compounds was identified most often as originating from new carpeting. If the final disinfectant is chloramine, the problems are generally eliminated because chloramines will not oxidize chlorite ion to chlorine dioxide. Routt (1989) concluded the same thing after studying the problem over a five-year period at a system in Kentucky. In another study, Hoehn et. al, (2002) found

that chlorine dioxide concentrations in the distribution system less than 0.1 mg/L were sufficient to cause customer complaints of kerosene- and cat-urine odors.

Unknown Issues/Information Gaps

- Not much information regarding color reduction by chlorine dioxide was found in the available literature. Additional studies are needed to evaluate the chlorine dioxide reduction of color caused by NOM, especially fulvic acids.
- More data are needed that identify odorous organic DBPs produced by chlorine dioxide treatment of NOM containing iodide and bromide ions.
- Research by Krasner and his associates indicate that chlorine dioxide is not very effective for reducing tastes and odors caused by aldehydes and other non-humic components of NOM in natural waters. Other studies of this nature are needed, perhaps with waters of different composition than Colorado River water and California State Project Water.
- Additional studies are needed to identify the chemicals that cause the kerosene- and cat-urine odors that are produced when chlorine dioxide reacts with substances released from new carpeting.

UV

Key Factors Affecting Water Quality Changes

Ultraviolet irradiation is not effective for taste-and-odor control unless it is used in combination with either ozone or peroxide. Mixing during UV irradiation is important, as are the concentrations and type of NOM compounds in the water at the point of UV application. Suspended solids and color are also important variables because they affect UV penetration in water. Ultraviolet irradiation "dosages" that are required for disinfection typically range from 10 to 100 millijoules per square centimeter (mJ/cm^2), but when UV is used for oxidation, usually in combination with hydrogen peroxide, dosages can as high as 5000 mJ/cm^2. "Low pressure" (LP) UV lamps, which are typically used for disinfection, emit UV irradiation primarily at 254 nm, whereas "medium pressure" (MP) lamps, which are used for oxidation, emit irradiation over a broader spectrum. When oxidation is the goal, the UV dose and hydrogen peroxide concentration are important variables, while if the treatment goal is taste-and-odor reduction, the type of odor compound becomes important.

Positive Impacts on Water Quality

Theoretical

- Ultraviolet irradiation alone should not be very effective for taste and odor removal. It should be effective, however, when used in combination with hydrogen peroxide or ozone because hydroxyl radicals, which are strong oxidants, are generated.
- Treatment with UV alone or in combination with peroxide or ozone may reduce the oxidant demand of the treated water and ultimately allow utilities to reduce the dosages of chlorine or chloramine that are required for maintaining a distribution-

system residual. The result would be a reduction in concentrations of halogenated DBPs and other by-products.

Actual Experience

- Linden et. al, (2002) showed that neither LP nor MP UV was effective for destroying MIB and geosmin. When used in combination with hydrogen peroxide (2 mg/L and 7 mg/L), MP UV was more effective than LP UV for degrading the two odor compounds. UV dosages required to achieve complete elimination of the MIB and geosmin were between 2,000 mJ/cm^2 and 5,000 mJ/cm^2. Romain et. al, (2003) evaluated geosmin destruction in pilot-scale and bench-scale studies using UV irradiation in combination with hydrogen peroxide and an aluminum-based catalyst, similar to what is known as "Fenton's Reaction." The average DOC in Lake Ontario raw-water during the study was only 2.1 mg/L. Irradiation was provided by a MP, high-intensity system. Geosmin spikes provided dosages ranging typically between approximately 80 and 120 ng/L. Results varied considerably, and the authors concluded that the UV/peroxide process (2.0-2.6 mg/L H_2O_2) would not reliably reduce geosmin concentrations to < 4 ng/L but could reliably reduce it to 10 ng/L on a consistent basis. Addition of coagulants (alum and PACl) as catalysts increased removals by only about 10-15 percent.
- Glaze et. al, (1990) reduced concentrations of odorous compounds in Colorado River water by treating it for 20 min with a combination of UV and hydrogen peroxide (0.56 watts/L and 0.2 mg/L·min, respectively). They reported the following reductions in compounds spiked into the raw water:

  - 1- heptanal, 74 percent
  - dimethyltrisulfide > 99 percent
  - 2,4-decadiental, 73 percent
  - geosmin, 40 percent
  - 2-MIB, 29 percent.

Potential Adverse Impacts on Water Quality

Theoretical

UV irradiation alone should not adversely affect water quality, but neither is it likely to oxidize taste-and-odor compounds nor otherwise improve the aesthetic properties of water. UV treatment supplemented with hydrogen peroxide may produce odorous, non-halogenated DBPs by formation of hydroxyl radicals that oxidize NOM but not at dosages normally used for drinking drinking water treatment.

Actual Experience

- Liu et. al, (2002) saw no significant changes in chlorinated DBP concentrations when water was treated first with UV at dosages < 500 mJ/cm^2 then with chlorine. When UV was applied at higher dosages and supplemented with high dosages of hydrogen peroxide (100 mg/L), chlorinated DBP concentrations actually decreased, but aldehydes and carboxylic acids were formed. The authors did not comment on the fact that the hydrogen peroxide dosages were much higher than those normally used during water treatment.

- Glaze et. al, (1990) found that concentrations of an odorous aldehyde (1-heptanal) increased by 27 percent following treatment of Colorado River water for 20 min with a combination of UV and hydrogen peroxide (0.56 watts/L and 0.2 mg/L·min, respectively).

Unknown Issues/Information Gaps

- Little is known about the impact of UV or UV/H_2O_2 on the aesthetic properties of drinking water, and the goals of UV treatment are not typically concerned with such issues. However, improvements in taste-and-odor and other aesthetic properties of water could easily be evaluated during UV studies designed for other purposes. Studies like those reported by Glaze et. al, (1990) could serve as a model for evaluating UV and UV/ H_2O_2 effects on other model odor compounds.
- No studies of color reduction by UV/H_2O_2 treatment were found in the literature. The subject should be investigated in pilot-plant-scale studies with a variety of high DOC natural waters.

BOOSTER CHLORINATION

Key Factors Affecting Water Quality Changes

It is not uncommon for utilities to boost chlorine and chloramine concentrations in the distribution system in order to offset potential problems caused by loss of oxidant residual. The impacts of this practice on tastes and odors in the distribution system are not well documented. Intuitively, one might suspect the key factors to include the oxidant demand, color, pH, DOC, ammonia concentration, and either chlorine or chloramine concentration in the system at the point where booster dosages are added.

Positive Impacts on Water Quality

Theoretical

Increasing the concentration of chlorine or chloramines in the distribution system at points where the residuals are low could prevent development of anaerobic conditions and microbial growth that lead to taste-and-odor problems.

Actual Experience

Wilczak et. al, (2003) surveyed utilities in the United States that boost chloramine concentrations in their distribution system. The major benefits were associated with preventing nitrification and reducing the dosage requirement at the treatment plant. Improvements in the organoleptic properties of water were not mentioned.

Potential Adverse Impacts on Water Quality

Theoretical

- Increasing the chlorine or chloramine residual may result in an increased number of customer complaints of chlorinous odors.
- If chlorine is added to water that already contains chloramines, the Cl_2:NH_3-N ratio will increase and cause odorous dichloramine formation. This issue was addressed in considerable detail by Barrett et. al, (1984) at Metropolitan Water District of Southern California.

Actual Experience

No information was found that specifically documented the impacts of booster chlorination on the organoleptic properties of water. Barrett et. al, (1984, 1985) stated, however, that taste-and-odor problems associated with dichloramine formation could be avoided if the blend ratio (chloraminated water with chlorinated water) is 5:1 or lower.

Unknown Issues/Information Gaps

The major deficiency in the current state of knowledge regarding the impacts of booster chlorination or chloramination is the lack of data on customer complaints associated with the practice. Customer complaint records should be reviewed to determine the nature of the complaint, the chlorine or chloramine concentrations in the distribution system at the point where the complaint was registered, and the association between the complaint and the addition of chlorine or chloramine at the booster station. Other data such as pH, temperature, HPC, etc. should also be noted if they exist.

SUMMARY

The positive and negative impacts of chloramines, ozone, chlorine dioxide, UV irradiation, and booster chlorination on the organoleptic properties of distributed water are summarized in Table E.5. The table also highlights gaps in the available information and suggests data that are needed to fill these gaps. The most glaring deficiency in the literature is the lack information about customers' reactions to changes in their drinking water following changes in disinfection practices at the treatment plant. Utilities could overcome this deficiency by examining customer-complaint records before and after the changes were made and determining whether the complaints were about tastes and odors, color, or turbidity.

Table E.5
Summary: positive and negative impacts of oxidants on organoleptic properties of drinking water

Oxidant	Positive Impacts	Negative Impacts	Additional Information Needed
Chloramines	• Reduced customer complaints of chlorinous odors • Residual persistence further into distributions system preventing anaerobic conditions, microbial growth • Formation of fewer odorous DBPs than chlorine	• Nitrification: loss of oxygen, biofilm growth • Potential formation of odorous dichloramine & trichloramines • Incomplete oxidation of odorous compounds (sulfides, algal products) • Aldehyde and nitrile formation • Masking of odors • Potential problems if chloraminated & chlorinated waters are blended.	• Impacts of biofilms on tastes-and-odors • Effects of conversion to chloramines on customer complaints, including chloramine concentrations tolerated by customers. • Odor thresholds of known DBPs formed by chloramine reactions with NOM • Chloramine's effects on color • Bromide role in odorous DBP formation. • Consequences of periodic reversion to chlorine on taste-and-odor complaints
Ozone	• Better color reduction than with chlorine or chlorine dioxide • More complete oxidation of odorous compounds than with other oxidants, especially in combination with UV and peroxide. • Readily oxidizes iron and manganese.	• Production of odorous by-products from NOM (e.g. aldehydes) • Oxidation of iodide & bromide to corresponding hypohalous acids that react with NOM to form odorous compounds. • Production of small molecular weight compounds that can form odorous compounds by reactions with chloramines or chlorine. • Potential for increased biofilm growth and T&O compounds produced by fungi, actinomycetes. • Overdosing Mn can produce MnO_4	• Documentation of customer satisfaction with water treated with ozone through analysis of customer complaint records. • Determination of odor- and flavor threshold concentrations of DBPs generated during ozonation, especially brominated forms. • Determination of O_3 dosages required to oxidize odorous DPBs produced from NOM. • Determination of O_3 dosages required for color removal at different pHs and types of NOM. • Determination of relationships between O_3 residual, hydrogen peroxide dosage, pH, and reaction time on T&O compounds.
Chlorine Dioxide	• Oxidizes but does not chlorinate organics to form odorous by-products.	• Customer complaints of kerosene-like and cat-urine-like odors in	• Color reduction by chlorine dioxide should be evaluated. • Production of odorous by-

Oxidant	Positive Impacts	Negative Impacts	Additional Information Needed
	• Oxidizes some but not all swampy/fishy/grassy/floral odors • Chlorite ion, is bacteriostatic in the distribution system, reducing nitrification and nitrifier biofilms. • Iron and manganese are readily oxidized in low DOC waters.	homes with new carpeting and other sources of indoor organic vapors. • Chlorine dioxide causes odor complaints such as "strong chlorinous" and "swimming pool." • Will not eliminate earthy/musty odors and certain others, such as grassy, fishy, swampy, and floral.	products by chlorine dioxide oxidation of NOM in water from a variety of locations should be studied. • Odor thresholds of known organic DBPs formed by oxidation of NOM with chlorine dioxide should be determined. • Compounds causing kerosene- and cat-urine odors in homes should be identified.
UV	• UV alone has no discernible effect on taste-and-odor removal. Some benefits have been demonstrated with UV/peroxide but only at higher UV dosages and peroxide concentrations. • UV/peroxide treatment may reduce the overall oxidant demand and thereby allow reduction of the final chlorine or chloramine dose.	• No negative impacts of UV treatment on tastes and odors were found. • Substitution of UV for chlorine, ozone, or chlorine dioxide for disinfection during treatment may result in higher halogenated DBP formation following final chlorination.	• While color and T&O reduction are not normally goals of UV treatment, future studies should examine the impacts of UV on these properties at varying UV dosages rates and with different types of lamps.
Booster Chlorination	• Booster chlorination may allow utilities to reduce the final dosage of chlorine or chloramine, thereby reducing the number of customer complaints of chlorinous odors.	• Booster chlorination of chloraminated water may create more dichloramine and thereby increase the number of customer complaints.	• Complaint records should be examined to determine the impact that booster chlorination/chloramination has had on the number and description of complaints.

APPENDIX F
AWWARF PROJECT NO. 2940
LONG-TERM EFFECTS OF DISINFECTION CHANGES ON WATER QUALITY

REPORT ON
PROJECT WORKSHOP

May 5 & 6, 2004

Held at the offices of
Tampa Bay Water
Clearwater, Florida

INTRODUCTION

The project workshop was held over a one and one half day period at the offices of Tampa Bay Water. The workshop participants included Project Team members, participating utilities, and other invited utilities as listed below.

Team Members

John Dyksen - Black & Veatch - Principal Investigator
Kevin Dixon - Black & Veatch
Bruce Long - Black & Veatch - Technical Advisory Group Chair
Dr. Michael McGuire - McGuire Environmental Consultants
Steve Harrison - Economic & Engineering Services
Dr. Robert Hoehn - Black & Veatch – author of issue paper on aesthetic changes
Dr. Anne Camper - Montana State University – author of issue paper on microbial changes
Dr. Philip Singer - University of North Carolina – author of issue paper on DBP changes
Jonathan Clement - Black & Veatch – author of issue paper on corrosion and metal release

Participating Utilities

Sylvia Barrett - Metropolitan Water District of Southern California
Christine Owen - Tampa Bay Water
John Trout - Tampa Bay Water
Robert Powell - Pinellas County (FL) Utilities
Eric Wert - Southern Nevada Water Authority
Manouchehr Boozapour - San Francisco Public Utilities Commission

Other Invited Utilities

Elizabeth Turner - Washington Aqueduct
John Yamona - PA-American Water Company
Christian Volk - IN-American Water Company
Luke Mulford - Hillsborough County, FL

OBJECTVES AND AGENDA

The objectives of the workshop were explained as follows:

- Obtain feedback on issue paper findings
- Obtain input from utility experiences
- Determine knowledge/data gaps
- Identify future research needs
- Determine any project course adjustments

The workshop agenda included four major blocks of time:

1. Review of project scope and objectives
2. Presentations of the four issue papers including a summary presentation
3. Presentations by the utilities on their experiences with disinfection changes
4. Breakout groups to review the presented material and to identify knowledge/data gaps and what is needed to fill those gaps

The full agenda is included in Appendix A (within this Appendix). The various PowerPoint presentations are contained in the following appendices (not included here):

Appendix B - Project Scope and Objectives
Appendix C - Issue Papers
Appendix D - Utility Experiences

Many questions and comments were raised during the presentations, and much of this was captured during the feedback in the breakout groups.

SUMMARY OF WATER QUALITY BENEFITS AND RISKS

Based on the findings of the four issue papers, the water quality benefits and risks associated with changing disinfection strategies were summarized and presented to the workshop participants, and are listed below:

Chlorine to Chloramines

Benefits

- May provide better biofilm control
- Result in lower DBPs
- Provide more consistent redox potential
- May be less reactive with Fe, Mn and organic matter
- Generate fewer chlorine T&Os and complaints

Risks

- Not as strong a disinfectant
- Nitrification can occur
- NDMA/nitrosamines can be formed
- IodoTHMs/dichloroacetaldehyde can be formed
- Nutrients are added to the water

Chlorine to Ozone

Benefits

- Greater disinfection capabilities
- No chlorinated DBPs formed
- Cl2 demand is reduced
- DO levels are increased
- Humics/fulvics are oxidized
- T&O is reduced

Risks

- Forms biodegradable material
- Forms bromate/brominated DBPs
- Forms chloral hydrate/chloropicrin
- Can adversely impact pH corrosion control
- Can cause fruity/rancid odors

Chlorine to Chlorine Dioxide

Benefits

- Greater disinfection capabilities
- No THM/HAA production
- Low corrosion/iron release potential

Risks

- Provides no residual
- Forms chlorate/chlorite
- Can produce kerosene/cat urine odors
- Provides poor control of earthy/musty and other types of odors

Chlorine to UV

Benefits

- Provides good disinfectant
- Does not produce DBPs
- Does not produce T&Os

Risks

- No residual provided
- No precursor reduction
- Nitrate can be converted to nitrite
- No T&O, color, Fe/Mn reduction

Booster Disinfection

Benefits

- Higher Cl2 residuals/may provide better biofilm control
- Possibly lower DBPs
- Minimize iron release
- Reduce microbially mediated tastes

Risks

- Disruption of biofilm
- May cause increase in HPCs, T&O, and color because of biofilm changes
- Possibly increase DBPs
- Increase chlorinous T&Os

RESULTS OF BREAKOUT GROUPS

During the morning of the second day, the workshop participants were divided into two breakout groups as follows:

Chloramination and Booster Disinfection

Ozone, Chlorine Dioxide, and UV

Each group was asked to provide any comments on the presentations, to identify knowledge/data gaps, and to identify potential projects or means to obtain the data. In general, the groups focused on data gaps and how to fill those gaps. The results of the breakout groups may be summarized as follows:

Data Gaps/Issues Raised Relating to All Disinfectants

- What impacts will a switch in disinfectant have on the Candidate Contaminant List (CCL) organisms?
- How important is good corrosion control? (consensus was that it is very important)

Additional Questions to Ask the Participating Utilities

1. When changing to booster chlorination, is there a significant change in DBPs – both levels and composition?
2. If one has an existing chlorine booster station and then switches to chloramines, what is done with the booster station? Might it be kept for flushing or for emergency response?
3. Has the utility seen any anomalies when switching to another disinfectant? (Some utilities in Florida have experienced green water when switching to chloramines)
4. Has the utility used annular reactors to evaluate biofilms or biofilm control?
5. How long do transient effects last?
6. Have utilities noticed any change in the morphology/ecology of bacteria in the distribution system? (even anecdotal information) Any change in the biofilm?
7. Any change in AOC/BDOC?
8. What is the water age in the system?
9. What is the change in disinfectant residual - % of samples greater than 0.2 mg/l?
10. Have they used annular reactors to study distribution system changes?

Addition to Scope of Project

- Develop a list of criteria that a utility can use to determine whether a switch from chlorine to another disinfectant will not result in major risks – what are some red flags that utilities should be aware of?

Future Research Needs to Fill Knowledge/Data Gaps

Chloramines

- Determine an optimum flushing regime when switching from a free chlorine residual to a chloramines residual
- Determine chloramine levels that are needed to provide equivalent disinfection to chlorine dosages
- Evaluate the nature of DBPs formed by chloramines; conduct occurrence survey of DBPs identified in Richardson report
- Study redox conditions and nitrification effects throughout distribution systems; determine pH impact on lead and copper species from nitrification
- Evaluate the change in tastes/odors when using chloramines
- Investigate the occurrence of low chloramine residuals and organic chloramines (active/inactive)
- Evaluate the impacts of switching to chloramines on Pb/Cu release; on brass water meters; on pipes (is the water more corrosive)
- Investigate the control of nitrification using chlorite ion
- Evaluate the effect of using pulsed disinfection on corrosion
- Describe and quantify the processes occurring at pipe surface vs. bulk water – include development of methods that can be taken outside the lab
- Investigate and develop methods to assess regrowth potential of the water

- Determine the ramifications of nitrification on corrosion, red water, Pb/Cu, T&O, and NDMA; determine how to return the distribution system back to pre-nitrification conditions

Ozone

- Develop better methods to determine NOM
- Develop understanding of how utilities can control microbial growth caused by ozonated NOM
- Investigate the nature and properties of oxidation by-products
- Evaluate methods of bromate control (treatment issue)
- Evaluate the impacts of ozone on particulate and inorganic molecules
- Develop odor and flavor profiles and thresholds for ozonated water
- Conduct a study to determine: (treatment issues)
- How many plants follow ozone with biological filtration
- How many utilities avoid biological filtration because of manganese concerns
- Reasons why utilities have discontinued biologically active filters
- Impacts of temperature on biological filter ecology
- Pipe loop study to determine whether there is a minimum secondary disinfectant level that is necessary to offset AOC/BDOC produced by ozone w/o biological filtration – example is Taylor's ongoing AwwaRF project
- Research project to answer question of how much BDOC/AOC is generated per unit of ozone applied or per mg of carbon ozonated – real world data, data and characterization of system, their water quality and nature of organics – goal is to develop a predictive tool
- Project to survey Florida utilities to better understand oxidative sequencing of sulfide and what organisms it can support. Would involve lab analyses.
- Research project to develop good procedures to measure AOC and BDOC or development of single surrogate.
- "True" occurrence survey that supplements the Richardson-Weinberg study – could be part of larger survey.

Chlorine Dioxide

- Determine possible selective pressure for opportunistic pathogens in distribution systems and health care facilities
- Study interactions of chlorine dioxide with natural organics and formation of by-products
- Study chlorine dioxide impacts on corrosion and particle transformations
- Identify tastes and odors created by chlorine dioxide with different waters
- International study to determine why some systems can successfully use chlorine dioxide as a secondary disinfectant and without odor problems. Also, what conditions in a customers' home environment cause odor problems
- Analytical study to determine if LGB kits are equivalent to wet chemical methods

- Study of utilities to determine how does the microbial environment shift because of chlorine dioxide including documenting nitrification inhibition.
- Study to identify the compounds that cause kerosene and cat urine odors and their associated health effects
- Re-analyses of the ICR data to determine the relationship between chlorite's distribution and chlorine dioxide usage

UV

- Study to determine if UV increases bio-availability of NOM
- Study to determine how effective UV can be in controlling DBP formation when integrated with post-disinfection practices
- Investigation of potential synergies of UV with other disinfectants
- Study to determine under what conditions UV reduces inorganic compounds
- Study to determine if there are any taste or odors associated with addition of UV
- Questionnaire about risk of mercury release from UV lamps
- Questionnaire to determine extent of increase in nitrite with UV useage
- Study to confirm that DNA repair of UV-injured organisms is not a significant issue

Booster Chlorination

- Studies illustrating the response of model or actual distribution systems with established biofilms to both sudden and gradual changes in free chlorine residuals
- Studies using distribution system models to predict how DBP concentrations and locations of concern would change with booster chlorination. Studies should be verified with real data
- Collection of customer data regarding taste and odor associated with booster chlorination

APPENDIX A
WORKSHOP AGENDA

Wednesday - May 5, 2004

8:30 am - 9:00 am	Welcome/Introductions/Review project goals and status (B&V)
9:00 am - 9:15 am	Workshop Agenda/Goals (B&V)
9:15 am - 9:50 am	Changes in Microbial Quality (Camper)
9:50 am - 10:10 am	BREAK
10:10 am - 10:45 am	Changes in Chemical Quality – Corrosion, Metal Release, Particle Changes (Clement)
10:45 am - 11:20 am	Changes in Chemical Quality – DBPs (Singer)
11:20 am - 11:55 pm	Changes in Aesthetic Quality (Hoehn)
11:55 am - 12:10 pm	Summary of Water Quality Impacts (B&V)
12:10 pm - 1:00 pm	LUNCH
1:00 pm - 6:00 pm	Participating Utility presentations PA-American Water (Yamona) Washington Aqueduct (Turner) Southern Nevada Water Authority (Wert) Pinellas County Utilities (Powell) MWDSC (Barrett) IN-American Water (Volk)
7:00 pm	DINNER

Thursday - May 6, 2004

8:00 am - 8:15 am	Recap of Day 1 and Review Day 2 Agenda (B&V)
8:15 am - 11:30 am	Breakout Groups – discuss knowledge and data gaps (what we don't know)
11:30 pm - 12:30 pm	Reports back from Breakout Groups
12:30 pm - 1:00 pm	Closing Comments/Adjourn (B&V)
1:00 pm	LUNCH

APPENDIX G
UTILITY QUESTIONNAIRE

Directions:

1. Complete Parts 1 through 5 of the Questionnaire (see tabs at the bottom of this worksheet). For treatment plants/well(s) where a disinfectant change has not yet been made, it is not necessary to fill out parts 4 and 5.
2. Complete Part 2 of the Questionnaire for each Water Treatment Plant or Well(s) where you have made a change or plan to make a change in the disinfection process. One "WTP" or "well(s)" is defined as that which produces one source of finished water at an entry point to the water distribution system.
3. Email completed questionnaire to John Dyksen at dyksenj@bv.com or Mail to:

    John Dyksen
    Black & Veatch
    475 Kinderkamack Road
    Oradell, NJ 07649-0160
4. Based on the results from parts 4 and 5 of the questionnaire, email or mail water quality and customer complaint data to John Dyksen (see addresses above). Clearly document each dataset (including the name of each parameter, date and time of sample collection, sampling location, data, and units). Delete any information that is not specifically requested. If we need more information we will request it. Excel spreadsheet is preferred. Please do not email files greater than 3 MB.
5. Please direct any questions to John Dyksen at 201-261-3627.

1.0 Utility Information

1.1 Please list the Utility's contact information below.

Utility Name	
Address	
Address	
City, State, Zip	
Contact Name	
Phone	
Fax	
Email	

2.0 Water Treatment Plant or Well(s) -

Complete a WTP/Well Questionnaire for each Water Treatment Plant or Well(s) where you have made a change or plan to make a change in the disinfection process. One "WTP" or "well(s)" is defined as that which produces one source of finished water entering the water distribution system.

2.1 Plant Flows – Please list the following plant flows.

2.1.A	Design Capacity		**mgd**
2.1.B	Average Daily Flow		**mgd**
2.1.C	Maximum Daily Flow		**mgd**

2.2 List source water supply and estimated percent of total supply capacity (if applicable)

		Name of Source(s)	**Estimated Percent of Total Supply Capacity**	
2.2.A	Ground water			%
2.2.B	Ground water under the influence of surface water			%
2.2.C	Surface water (impoundment)			%
2.2.D	Surface Water (free flowing stream)			%

2.3 Raw Water Quality – Please check appropriate range

		Average	**Maximum**
2.3.A	Turbidity	☐ <5 NTU	☐ <5 NTU
		☐ 6-10 NTU	☐ 6-10 NTU
		☐ 11-20 NTU	☐ 11-20 NTU
		☐ 21-50 NTU	☐ 21-50 NTU
		☐ >50 NTU	☐ >50 NTU
2.3.B	Alkalinity	☐ <25 mg/L	☐ <25 mg/L
		☐ 26-50 mg/L	☐ 26-50 mg/L
		☐ 51-75 mg/L	☐ 51-75 mg/L
		☐ 76-100 mg/L	☐ 76-100 mg/L
		☐ >100 mg/L	☐ >100 mg/L
2.3	Total Hardness	☐ <50 mg/L	☐ <50 mg/L
		☐ 51-100 mg/L	☐ 51-100 mg/L
		☐ 101-150 mg/L	☐ 101-150 mg/L
		☐ 151-200 mg/L	☐ 151-200 mg/L
		☐ >200 mg/L	☐ >200 mg/L
2.3.D	pH	☐ <6.0	☐ <6.0
		☐ 6.1-6/5	☐ 6.1-6/5
		☐ 6.6-7.0	☐ 6.6-7.0
		☐ 7.1-7.5	☐ 7.1-7.5
		☐ 7.6-8.0	☐ 7.6-8.0
		☐ >8.0	☐ >8.0

		Average	Maximum
2.3.E	TOC	☐ <2 mg/L	☐ <2 mg/L
		☐ 2.1-4.0 mg/L	☐ 2.1-4.0 mg/L
		☐ 4.1-6.0 mg/L	☐ 4.1-6.0 mg/L
		☐ >6.0 mg/L	☐ >6.0 mg/L
2.3.F	UV 254	☐ <0.02 cm $^{-1}$	☐ <0.02 cm $^{-1}$
		☐ 0.021-0.05 cm $^{-1}$	☐ 0.021-0.05 cm $^{-1}$
		☐ 0.051-0.1 cm $^{-1}$	☐ 0.051-0.1 cm $^{-1}$
		☐ 0.11-0.5 cm $^{-1}$	☐ 0.11-0.5 cm $^{-1}$
		☐ 0.51-1.0 cm^{-1}	☐ 0.51-1.0 cm^{-1}
		☐ >1.0 cm $^{-1}$	☐ >1.0 cm $^{-1}$
2.3.G	Iron	☐ <0.2 mg/L	☐ <0.2 mg/L
		☐ 0.21-0.4 mg/L	☐ 0.21-0.4 mg/L
		☐ 0.41-.06 mg/L	☐ 0.41-.06 mg/L
		☐ 0.61-1.0 mg/L	☐ 0.61-1.0 mg/L
		☐ >1.0 mg/L	☐ >1.0 mg/L
2.3.H	Manganese	☐ <0.01 mg/L	☐ <0.01 mg/L
		☐ 0.011-0.04 mg/L	☐ 0.011-0.04 mg/L
		☐ 0.041-0.06 mg/L	☐ 0.041-0.06 mg/L
		☐ 0.061-0.1 mg/L	☐ 0.061-0.1 mg/L
		☐ >0.1 mg/L	☐ >0.1 mg/L
2.3.1	Bromide	☐ <0.01 mg/L	☐ <0.01 mg/L
		☐ 0.011-0.03 mg/L	☐ 0.011-0.03 mg/L
		☐ 0.031-0.06 mg/L	☐ 0.031-0.06 mg/L
		☐ 0.061-.1 mg/L	☐ 0.061-.1 mg/L
		☐ 0.11-0.5 mg/L	☐ >0.11-0.5 mg/L
		☐ >0.5 mg/L	☐ >0.5 mg/L
2.3.J	Color	☐ <10	☐ <10
		☐ 11-20	☐ 11-20
		☐ 21-40	☐ 21-40
		☐ 41-60	☐ 41-60
		☐ >60	☐ >60

2.4	**Finished Water Quality – Please check appropriate range.**	**Average**	**Maximum**
2.4.A	Turbidity	☐ <0.05 NTU	☐ <0.05 NTU
		☐ 0.05-0.1 NTU	☐ 0.05-0.1 NTU
		☐ 0.11-1.21 NTU	☐ 0.11-1.21 NTU
		☐ 0.21-0.3 NTU	☐ 0.21-0.3 NTU
		☐ >0.31 NTU	☐ >0.31 NTU
2.4.B	Alkalinity	☐ <25 mg/L	☐ <25 mg/L
		☐ 26-50 mg/L	☐ 26-50 mg/L
		☐ 51-75 mg/L	☐ 51-75 mg/L
		☐ 76-100 mg/L	☐ 76-100 mg/L
		☐ >100 mg/L	☐ >100 mg/L
2.4.C	Total Hardness	☐ <50 mg/L	☐ <50 mg/L
		☐ 51-100 mg/L	☐ 51-100 mg/L
		☐ 101-150 mg/L	☐ 101-150 mg/L
		☐ 151-200 mg/L	☐ 151-200 mg/L
		☐ >200 mg/L	☐ >200 mg/L
2.4.D	pH	☐ <6.5	☐ <6.5
		☐ 6.6-7.0	☐ 6.6-7.0
		☐ 7.1-7.5	☐ 7.1-7.5
		☐ 7.6-8.0	☐ 7.6-8.0
		☐ >8.5	☐ >8.5
2.4.E	TOC	☐ <2 mg/L	☐ <2 mg/L
		☐ 2.1-4.0 mg/L	☐ 2.1-4.0 mg/L
		☐ 4.1-6.0 mg/L	☐ 4.1-6.0 mg/L
		☐ >6.0 mg/L	☐ >6.0 mg/L
2.4.F	UV 254	☐ <0.02 cm^{-1}	☐ <0.02 cm^{-1}
		☐ 0.021-0.05 cm^{-1}	☐ 0.021-0.05 cm^{-1}
		☐ 0.051-0.1 cm^{-1}	☐ 0.051-0.1 cm^{-1}
		☐ 0.11-0.5 cm^{-1}	☐ 0.11-0.5 cm^{-1}
		☐ 0.51-1.0 cm^{-1}	☐ 0.51-1.0 cm^{-1}
		☐ >1.0 cm^{-1}	☐ >1.0 cm^{-1}
2.4.G	Iron	☐ <0.05 mg/L	☐ <0.05 mg/L
		☐ 0.06-0.1 mg/L	☐ 0.06-0.1 mg/L
		☐ 0.11-0.2 mg/L	☐ 0.11-0.2 mg/L
		☐ 0.21-0.3 mg/L	☐ 0.21-0.3 mg/L

		Average	Maximum
		☐ >0.3 mg/L	☐ >0.3 mg/L
2.4.H	Manganese	☐ <0.001 mg/L	☐ <0.001 mg/L
		☐ 0.0011-0.004 mg/L	☐ 0.0011-0.004 mg/L
		☐ 0.0041-0.006 mg/L	☐ 0.0041-0.006 mg/L
		☐ 0.0061-0.01 mg/L	☐ 0.0061-0.01 mg/L
		☐ >0.01 mg/L	☐ >0.01 mg/L
2.4.1	Bromate	☐ <0.002 mg/L	☐ <0.002 mg/L
		☐ 0.0021-0.004 mg/L	☐ 0.0021-0.004 mg/L
		☐ 0.0041-0.006 mg/L	☐ 0.0041-0.006 mg/L
		☐ 0.0061-0.008 mg/L	☐ 0.0061-0.008 mg/L
		☐ 0.0081-0.01 mg/L	☐ 0.0081-0.01 mg/L
		☐ >0.01 mg/L	☐ >0.01 mg/L
2.4.J	Color	☐ <5	☐ <5
		☐ 6-10	☐ 6-10
		☐ 11-15	☐ 11-15
		☐ 15	☐ 15
2.4.K	Total THMs	☐ <20 ug/L	☐ <20 ug/L
		☐ 21-40 ug/L	☐ 21-40 ug/L
		☐ 41-60 ug/L	☐ 41-60 ug/L
		☐ 61-80 ug/L	☐ 61-80 ug/L
		☐ >80 ug/L	☐ >80 ug/L
		☐ Don't know	☐ Don't know
2.4.L	HAA5	☐ <15 ug/L	☐ <15 ug/L
		☐ 16-30 ug/L	☐ 16-30 ug/L
		☐ 31-45 ug/L	☐ 31-45 ug/L
		☐ 45-60 ug/L	☐ 45-60 ug/L
		☐ >60 ug/L	☐ >60 ug/L
		☐ Don't know	☐ Don't know

2.5 Treatment Process

2.4.A Please check appropriate treatment process

☐ **Conventional Treatment –** chemical addition, flocculation, sedimentation, filtration

☐ **Direct Filtration** - coagulation, flocculation, filtration

☐ **Iron Removal** - oxidation, filtration

☐ **VOC Removal** - airstripping or GAC

☐ **Chemical Softening**

☐ Disinfection only

2.5B 2.	**Do you use a corrosion inhibitor or sequestering agent?**	☐ **Yes**	☐ **No**
2.6	Has a disinfectant change been made at this treatment plant/well?	☐ **Yes**	☐ **No**

2.7 Please fill in the following table regarding treatment plant disinfection.

If YES to question 2.6, please describe your disinfection practices BEFORE the disinfectant change.
If NO to question 2.6, please describe your current disinfection practices.

		Disinfectant	Dose Range
2.7.A	Pre-disinfection	☐ Chorine	☐ <0.3 mg/L
		☐ Chloramines	☐ 0.31-0.5 mg/L
		☐ Ozone	☐ 0.51-1.0 mg/L
		☐ Chlorine dioxide	☐ 1.1-2.0 mg/L
		☐ UV	☐ >2.0 mg/L
		Where is the disinfectant added:	☐
2.7.B	Primary	**Disinfectant**	**Dose Range**
		☐ Chorine	☐ <0.3 mg/L
		☐ Chloramines	☐ 0.31-0.5 mg/L
		☐ Ozone	☐ 0.51-1.0 mg/L
		☐ Chlorine dioxide	☐ 1.1-2.0 mg/L
		☐ UV	☐ >2.0 mg/L
		Where is the disinfectant added:	☐
2.7.C	Secondary	**Disinfectant**	**Dose Range**
		☐ Chorine	☐ <0.3 mg/L
		☐ Chloramines	☐ 0.31-0.5 mg/L
		☐ Ozone	☐ 0.51-1.0 mg/L
		☐ Chlorine dioxide	☐ 1.1-2.0 mg/L
		☐ UV	☐ >2.0 mg/L
		Where is the disinfectant added:	☐
2.7.D	Booster chlorination	**Disinfectant**	**Dose Range**
		☐ Chorine	☐ <0.3 mg/L
		☐ Chloramines	☐ 0.31-0.5 mg/L
		☐ Ozone	☐ 0.51-1.0 mg/L
		☐ Chlorine dioxide	☐ 1.1-2.0 mg/L
		☐ UV	☐ >2.0 mg/L
		Where is the disinfectant added:	☐

2.8 Please fill in the following table regarding treatment plant disinfection.

If YES to question 2.6, please describe your disinfection practices AFTER the disinfectant change.
If NO to question 2.6, please describe your planned disinfection practices.

2.8.A Pre-disinfection

Disinfectant
☐ Chorine
☐ Chloramines
☐ Ozone
☐ Chlorine dioxide
☐ UV

Dose Range
☐ <0.3 mg/L
☐ 0.31-0.5 mg/L
☐ 0.51-1.0 mg/L
☐ 1.1-2.0 mg/L
☐ >2.0 mg/L

Where is the disinfectant added: ☐

2.8.B Primary

Disinfectant
☐ Chorine
☐ Chloramines
☐ Ozone
☐ Chlorine dioxide
☐ UV

Dose Range
☐ <0.3 mg/L
☐ 0.31-0.5 mg/L
☐ 0.51-1.0 mg/L
☐ 1.1-2.0 mg/L
☐ >2.0 mg/L

Where is the disinfectant added: ☐

2.8.C Secondary

Disinfectant
☐ Chorine
☐ Chloramines
☐ Ozone
☐ Chlorine dioxide
☐ UV

Dose Range
☐ <0.3 mg/L
☐ 0.31-0.5 mg/L
☐ 0.51-1.0 mg/L
☐ 1.1-2.0 mg/L
☐ >2.0 mg/L

Where is the disinfectant added: ☐

2.8.C Booster chlorination

Disinfectant
☐ Chorine
☐ Chloramines
☐ Ozone
☐ Chlorine dioxide
☐ UV

Dose Range
☐ <0.3 mg/L
☐ 0.31-0.5 mg/L
☐ 0.51-1.0 mg/L
☐ 1.1-2.0 mg/L
☐ >2.0 mg/L

Where is the disinfectant added: ☐

2.9 **Upgrades/Modifications - Starting from the most recent, please provide information of any major upgrades/modifications to the disinfection process at this WTP or well(s) over the past 10 years.**

Year	Upgrade/Modification Description

APPENDIX H
DETAILED INFORMATION FROM PARTICIPATING UTILITIES THAT SWITCHED FROM CHLORINE TO CHLORAMINES

CHLORINE TO CHLORAMINES

Detailed information on the water quality impacts resulting from the change from using chlorine as a secondary disinfectant to using chloramines as a secondary disinfectant are described in this appendix. The data from all the participating utilities are included except for Brick Township MUA, San Francisco PUC, and Washington Aqueduct which are included as case studies in Chapter 3.

MICROBIAL QUALITY

Concord, NH

Concord collects samples monthly from fourteen distribution sites and analyzes water quality for coliform, heterotrophic plate counts (HPCs) using the R2A method and general chemical quality. Prior to the conversion to chloramines in May, 1996, some of the low flow sites in Concord consistently had HPC numbers in excess of 6,000 colony forming units per milliliter (cfu/ml)(Figure H.1) with the average for the distribution system at about 1,200 cfu/ml. Within three months after the conversion, none of the sites had HPCs in excess of 3,000 cfu/ml and in the two years after chloramine conversion, the heterotrophic plate count average had dropped to 65 cfu/ml. At three sites, temporary increases in HPCs were observed, possibly from the displacement of established biofilms. At these sites, the microbial numbers returned to levels seen during chlorination within a year or two after the conversion. No coliforms were detected either before the conversion to chloramines or after the conversion to chloramines.

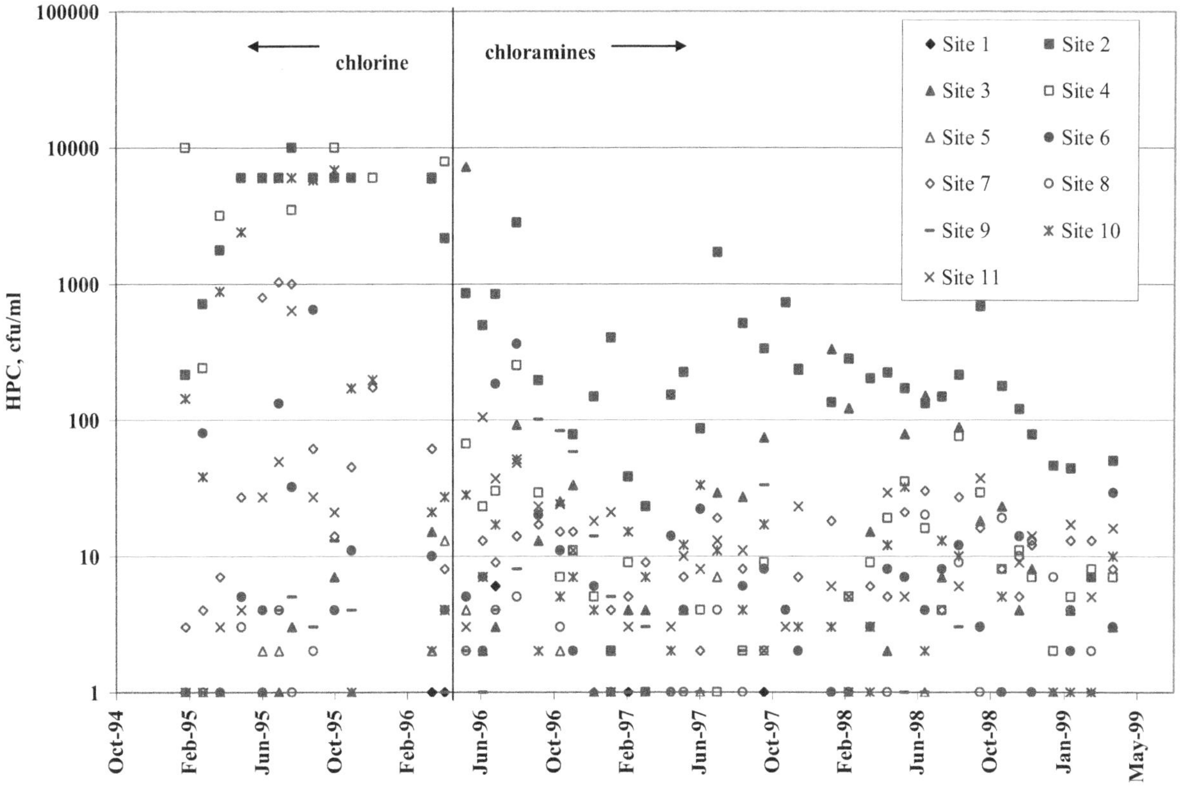

Figure H.1 Concord, NH heterotrophic plate count data

IN-American Water Co.

In the three years prior to their conversion to chloramines, IN-American experienced HPC levels averaging 4.3 cfu/ml when using SPC agar. There was no significant change during the nine months after conversion to chloramines with the average being 3.3 cfu/ml. HPC levels using R2A were much higher with average concentrations of around 420 cfu/ml before chloramination. After conversion to chloramines, the average dropped to 31 cfu/ml. They note that during a subsequent period of planned flushing with free chlorine, R2A counts increased to 102 cfu/ml.

During the three years prior to the conversion, 24 of 4,046 (0.60 percent) distribution system samples were positive for coliforms. After the conversion, three of 1,868 (0.16 percent) of the samples were coliform-positive. During 1994-95, only 11 of 3,008 (0.36 percent) of the distribution system samples were coliform-positive.

IN-American also found that disinfectant residuals throughout the distribution system were higher after the conversion to chloramines. When comparable disinfectant dosages were used, 22-24 percent of the chlorinated samples had residuals of less than 1.0 mg/L, while only 5.5 percent of the chlorinated samples had residuals of less than 1.0 mg/L. At one site where it was historically difficult to maintain free chlorine residuals (range of 0.00 – 0.37 mg/L), the chloramine residual has ranged from 0.21 – 1.8 mg/L.

MWDSC, CA

The Metropolitan Water District of Southern California (MWDSC) switched from free chlorine to chloramines disinfection in June 1985 as the most cost-effective alternative to meet trihalomethans regulations. A comprehensive monitoring program was initiated to assess changes in the bacteriological and chemical quality of MWDSC's water as a result of the changeover. The monitoring program consisted of weekly anslyses of chemical, physical, and bacteriological parameters from 52 sites within MWDSC's distribution system one year before the changeover (June 1984 through May 1985) and 1 year after it (June 1985 through May 1986).

Sample locations were selected to represent a variety of common conditions in the distribution system—e.g., high and low flow rates, short and long detention times, and treatment plant effluents. The chemical and physical measurements recorded included temperature, electroconductivity, turbidity, threshold odor number, flavor-profile analysis, free and total ammonia, and free chlorine and chloramines concentrations. The bacteriological analyses included quantifying total coliforms and heterotrophic plate count (HPC) bacteria. In addition, bacterial isolates from 10 of the sites were identified to genus to determine qualitative changes in the bacterial quality of the water as a result of the changeover.

The overall numbers of coliforms in MWDSC's water before and after the change to chloramines were equivalent. No coliforms were detected in 2123 samples collected before the changeover, and only one coliform was detected in 2359 samples analyzed during chloramines use. Similarly, no change was observed in HPC levels. The geometric mean of the m-R2A counts before and after the changeover increased from 3 to 5 CFU/mL, respectively. Most of this increase occurred in the summer months.

Another impact of the switchover from chlorine to chloramines was the occurrence of nitrification. In 1987, nitrification occurred in a finished water reservoir. The first indication of nitrification occurred in June. The ammonia-oxidizing bacteria (AOB) levels increased from <0.2 (detection limit) to 3.0 MPN/mL; however, nitrite was not detected at this time. It was not until early August that nitrite levels rapidly increased (within six days) to 0.030 mg/L. The chloramines level decreased from 1.5 to less than 1.0 mg/L, and the AOB levels increased to 22 MPN/mL.

In the summer of 1994, a southland water utility experienced nitrification within its distribution system. At some of the locations, the nitrite-nitrogren levels exceeded 0.1 mg/L, and the chloramine levels were less than 0.1 mg/L. The utility's water supply came from a blend of groundwater and MWDSC's surface water. The finished surface water typically had a chloramines residual of 1.8 mg/L at the point of entering the utility's distribution system. The groundwater was highly colored and had levels of high organic matter. This water was treated, then chloraminated. The total chloramines level leaving the treatment plant ranged between 2.2 and 3.5 mg/L.

The results indicated that the nitrifiers did not originate from the source waters or the treatment plant effluent. The AOB was only detected from the distribution system. The results, from 10 distibution sites representing areas with the highest nitrite levels, indicated that nitrification was well established. From 10 sites on a single day, the average AOB level was 99 MPN/mL (range, 0.84 to 570 MPN//mL), the average nitrite level was 0.014 mg/L (range, 0.006 to 0.030 mg/L), and the average chloramines level was 1.1 mg/L (range, 0.5-2.5 mg/L). Previous work performed by MWDSC indicated that the onset of nitrification occurred in their system when AOB levels were between 5 to 30 MPN/mL providing supporting evidence that the AOB were well established in the southland's system. Coliforms in this southland utility were not detected anywhere in the distribution system during this time period, and the HPC bacteria were very low (<33 CFU/mL).

The results of the two nitrification episodes indicated that when nitrite is detected, the AOB are typically well established in biofilms attached on the surfaces of materials. Therefore, once the AOB are established, it is conceivable that nitrite levels can increase noticeably within hours. For example, in one of MWDSC's reservoirs, the nitrite levels (as nitrogen) increased from 0.015 to 0.020 mg/L in approximately 4 hours.

PA – American Water Co.

PA-American converted two connected systems, Brownell and Fallbrook, from free chlorine to chloramines from May, 2002 through August, 2003. Free chlorine was the secondary disinfectant prior to chloramine treatment and from September, 2003 through late April, 2004. Heterotrophic plate counts with free chlorine as the disinfectant resulted in average counts of 19 cfu/ml at free chlorine residuals of 1.2 mg/L. During the chloramine conversion, heterotrophic plate count numbers rose to an average of 142 cfu/ml. The total chlorine residual during this time was 1.5 mg/L. During the reversion to free chlorine, heterotrophic plate count numbers averaged 77 cfu/ml and the free chlorine residual was 1.1 mg/L. The use of chloramines as the distribution system disinfectant was reestablished in April, 2004. Monthly minima, maxima, and averages of heterotrophic plate count numbers from May through December 2004 are presented in Figure H.2. The average heterotrophic plate counts were 13 cfu/ml, less than those observed during free chlorination, even through the warm water months.

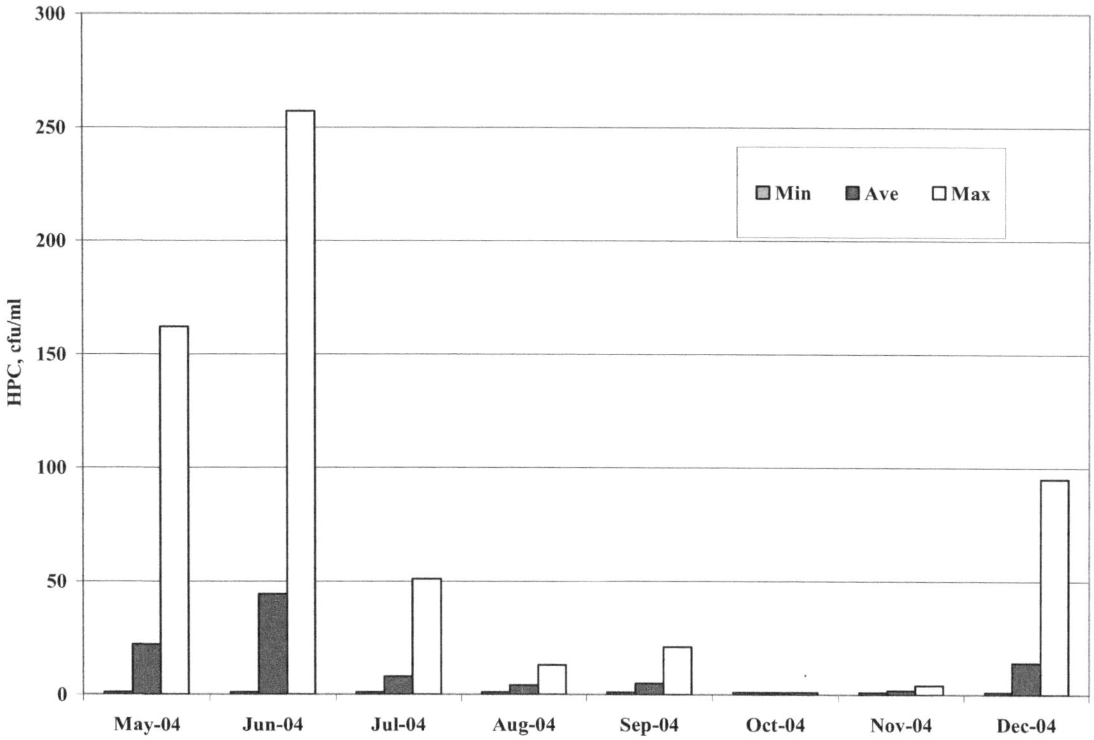

Figure H.2 PA - American HPC data for two chloraminated systems

Philadelphia, PA

The Philadelphia Water Department (PWD) has been using chloramines for secondary disinfection at its three water treatment plants since the 1975. The move to chloramines was motivated in part by summertime die-off of the chlorine residual, especially in open finished water reservoirs. Today, PWD maintains a chloramine residual at about 2.0 mg/L at the entry point to the distribution system. The three treatment plants aim to keep this residual from dropping below 1.5 mg/L or exceeding 2.7 mg/L. Out in the distribution system, the goal is to maintain the residual at or above 1.0 mg/L and to take action when it drops below 0.5 mg/L. The average percent die-off in chloramine residual ranges from <10 percent during the winter (water temperatures below 10°C) to 20-40 percent. PWD reports that the switch to chloramines has solved the problems with maintaining a total chlorine residual throughout the system especially in storage tanks and reservoirs during the summer (water temperatures above 20°C) based on actual monitoring data.

PWD's distribution system monitoring program includes over 400 samples each month. Based on their monitoring program, biofilms have not played a role in driving distribution system practices. In addition, PWD has found that the switch to chloramines in the 1970s has not had a negative effect on distribution system bacterial occurrences. Out of 5478 routine monitoring samples from the distribution system in 2001, only 0.3 percent exceeded 100 cfu/ml in HPC (occurrences were during May-September) and only 0.2 percent were positive for total coliforms (which appeared during June-November).

Tampa Bay Water, FL

Tampa Bay Water did not provide microbiological data.

CHEMICAL QUALITY - DBPs

Chesterfield County, Virginia

Chesterfield County made two major treatment changes in the span of two years. In 1998, they converted from chlorine to chloramines as the secondary disinfectant and the following year changed from alum to ferric sulfate for coagulation. The effect of both of these changes was to lower the trihalomethane levels in the distribution system significantly thereby allowing Chesterfield to comply with the 80 μg/L trihalomethane limit of the Disinfectant/Disinfection By-Product Rules. Figure H.3 presents these data. Chesterfield is also in compliance with the haloacetic acid limit of 60 μg/L, though no data are available prior to 2001.

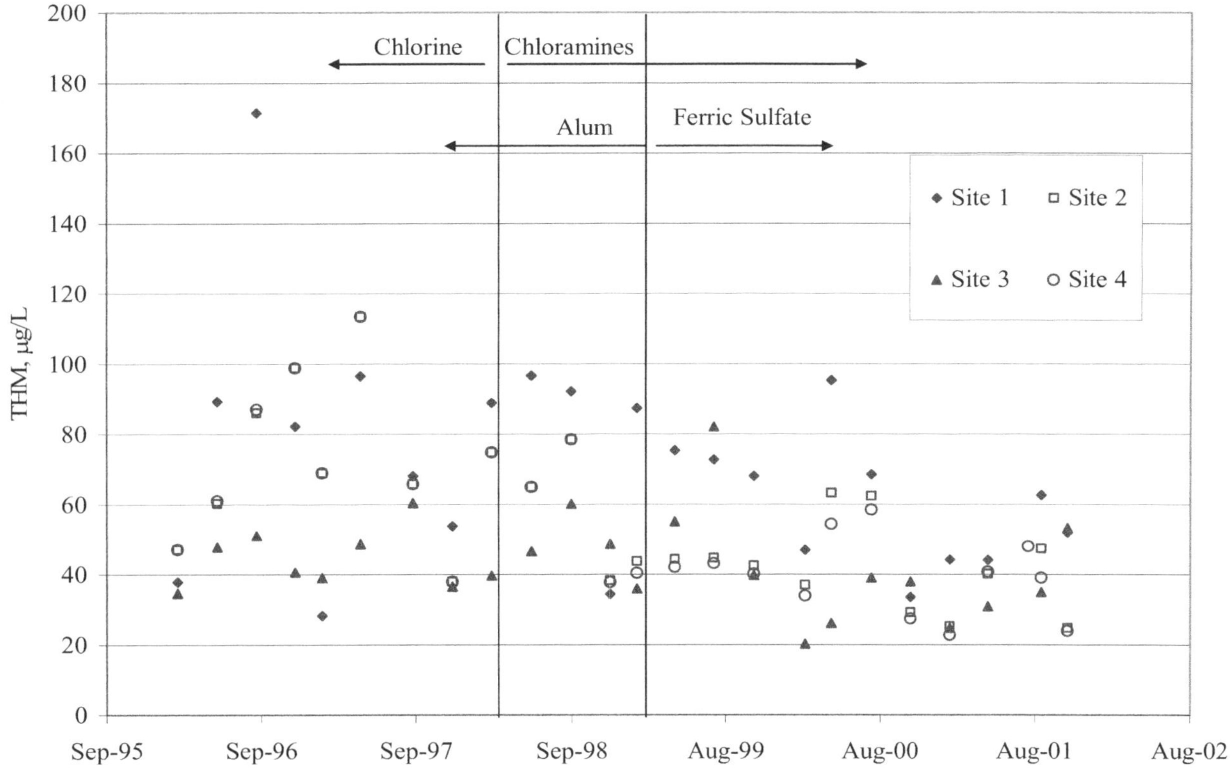

Figure H.3 Chesterfield County TTHM data

Concord, NH

Concord sampled for trihalomethanes at four sites in the distribution system, before and after the conversion to chloramines. Before the conversion to chloramines, trihalomethane concentrations were elevated in August, typically the season of highest water temperatures and greatest chlorine demand. One value exceeded the current standard of 80 μg/L. After the conversion to chloramines in May 1996, trihalomethane concentrations were lower with no values exceeding 60 μg/L. Figure H.4 presents the trihalomethane data from the four sampling sites.

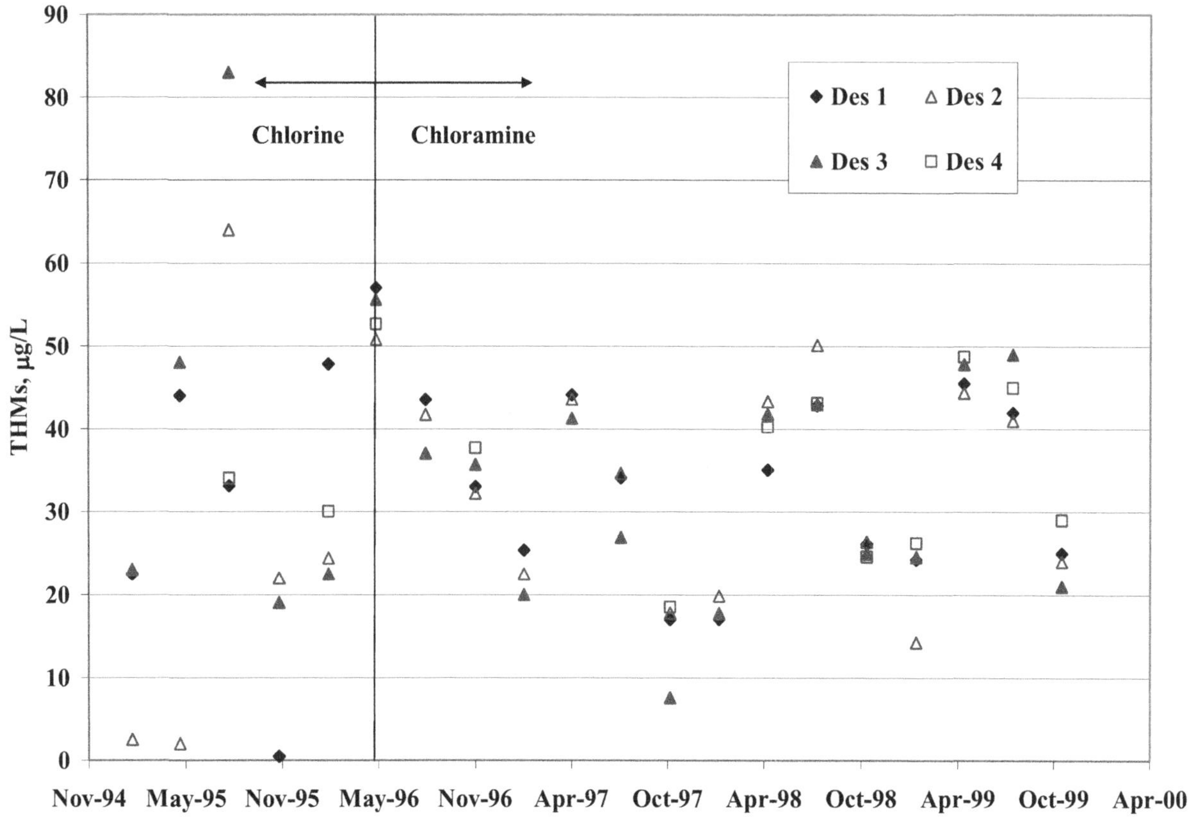

Figure H.4 Concord, NH TTHM data

IN-American Water Co.

In the years prior to their conversion to chloramines, IN-American increased their free chlorine residual because of coliform occurrences. As a result, TTHM concentrations averaged over 100 μg/L just prior to the conversion compared to averaging 77 μg/L before the increase in free chlorine residual. Following the conversion to chloramines, TTHM concentrations averaged 63 μg/L. HAA5 concentrations during free chlorination averaged 88 μg/L, and after conversion to chloramines dropped to 51 μg/L. Averages for the total remaining organic DBPs that were analyzed decreased from 52 to 19 μg/L after conversion to chloramines.

Newport News, VA

Newport News has two water treatment plants – Harwood Mills and Lee Hall. Chloramines were initiated as the secondary disinfectant at both plants in 1998. Since the conversion to chloramines, Newport News has observed a decrease in trihalomethane and haloacetic acid concentrations even in 1999 when Hurricane Floyd caused the raw water TOC concentration to exceed 20 mg/L, well above the average value of 4 to 6 mg/L. Figures H.5 and H.6 present disinfection by-product data for the period from 1997 through 2001.

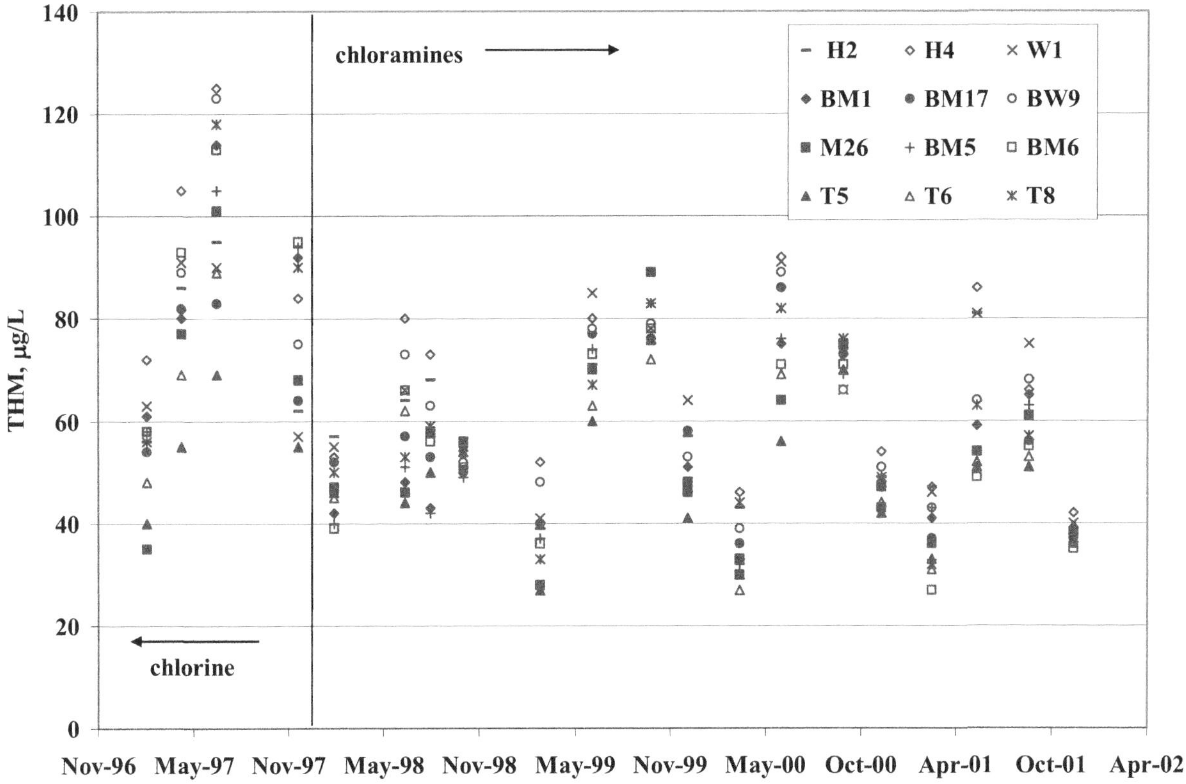

Figure H.5 Newport News, VA TTHM data

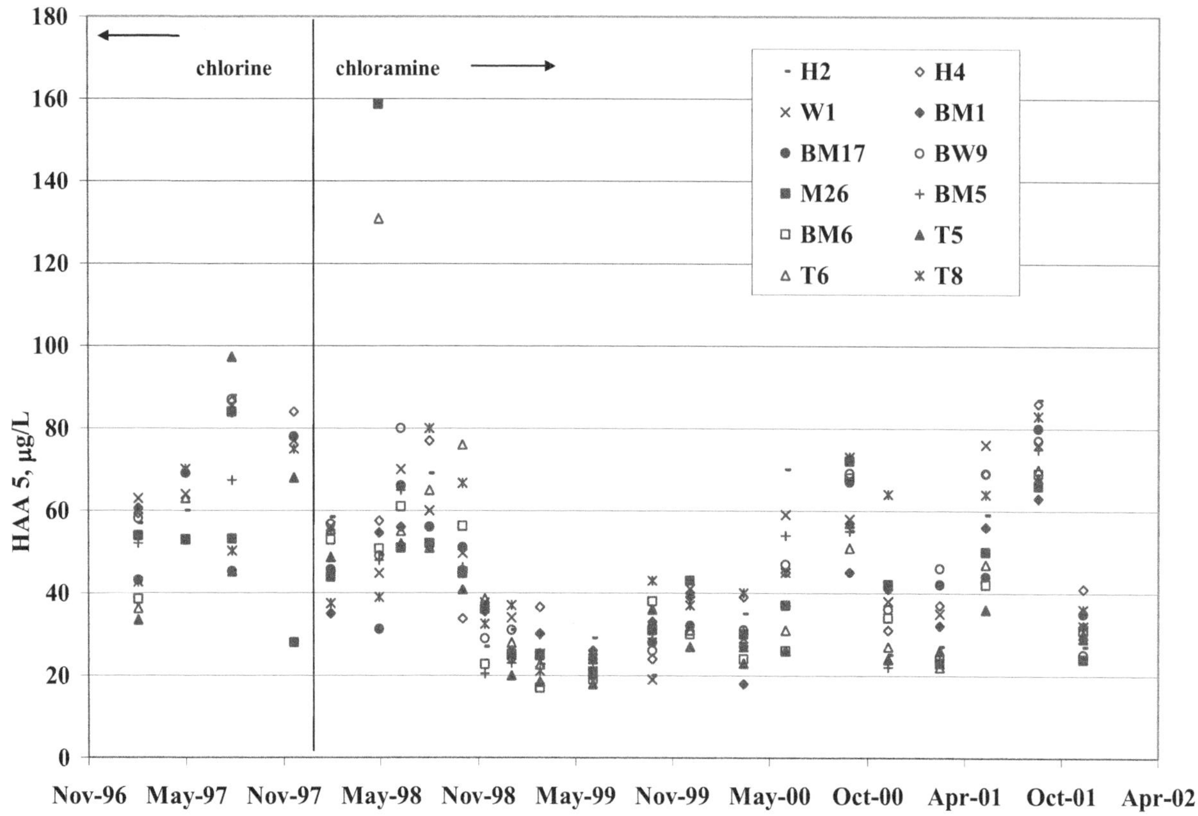

Figure H.6 Newport News, VA HAA5 data

Philadelphia, PA

With the use of chloramines, DBP levels in the PWD distribution system have been relatively low. There are no data on DBPs before the conversion to chloramines. In 2002, PWD reported running annual average (RAA) TTHMs of 40-50 ug/L and RAA HAAs of 25-35 ug/L. The range of values at individual locations was reported to be 19-92 ug/L for TTHMs and 18-52 ug/L for HAA5s. Cyanogen chloride is a disinfectant by-product that is expected to occur at higher concentrations in systems that use chloramines. PWD conducted monitoring for 18 months under the Information Collection Rule (1997-1998). Overall, for three plant effluents (entry points) and three distribution points representative of maximum travel times, the cyanogen chloride ranged from 2-8 µg/L, with the highest concentrations occurring at the maximum travel time locations.

PA – American Water Co.

While disinfection by-product levels were low for the Brownell and Fallbrook combined systems before the change to chloramination, the use of chloramines reduced the levels of trihalomethanes and the five regulated haloacetic acids even further. Figure H.7 presents disinfection by-product data for the systems starting in 2001, prior to converting to chloramines in 2002 and 2003, as well as through the remainder of 2003 and early 2004 before chloramination was reestablished in April 2004.

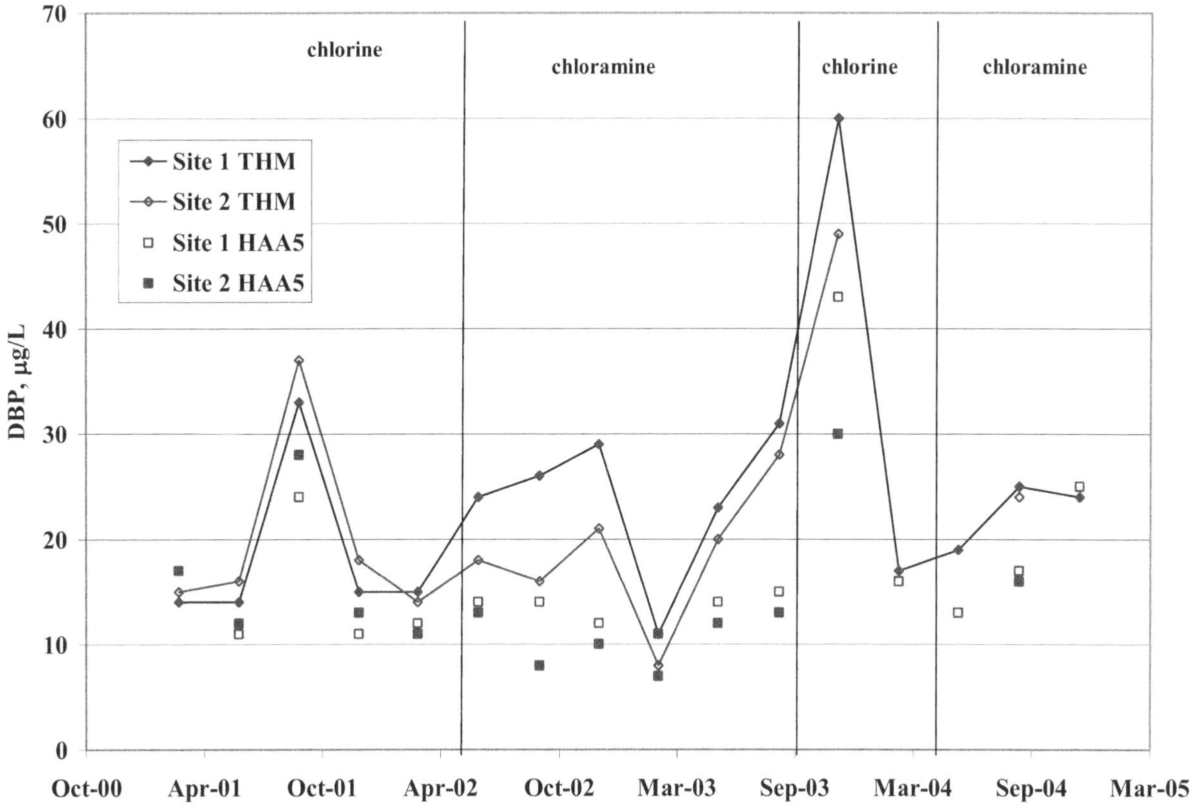

Figure H.7 PA-American disinfection by-product data for systems 1 and 2

Tampa Bay Water, FL

Tampa Bay Water was created in 1998 to coordinate and provide water supply to six governmental entities in west central Florida, a region of about 2,000,000 people. Tampa Bay Water serves Hillsborough County, Pasco County, Pinellas County, New Port Richey, St. Petersburg, and Tampa from 13 wellfields, one treated fresh water plant and a desalinization plant withdrawing from Tampa Bay.

Since 2002, the region has had to rely more on surface water and desalinization as primary sources and decrease pressure on the wellfields. The 66 mgd surface water treatment plant came on-line in September, 2002. It is a conventional plant using the high rate Actiflo clarification process followed by ozone oxidation, dual media filtration with granular activated carbon filter caps, and chloramines for secondary disinfection. The 25 mgd Tampa Bay desalinization plant came on-line in March, 2003.

The City of New Port Richey is served by Tampa Bay Water using groundwater from the Starkey/North Pasco aquifer. The raw groundwater which has high levels of color, alkalinity (>100 mg/l as $CaCO_3$), and hardness (> 200 mg/L) receives disinfection before it is distributed to the 29,000 residents of New Port Richey. In May, 2004, the gas chlorine system was changed to sodium hypochlorite and an ammonia feed system was added to form monochloramines for secondary disinfection. New Port Richey saw a significant reduction in TTHM levels as a result of the conversion, as shown on Figure H.8.

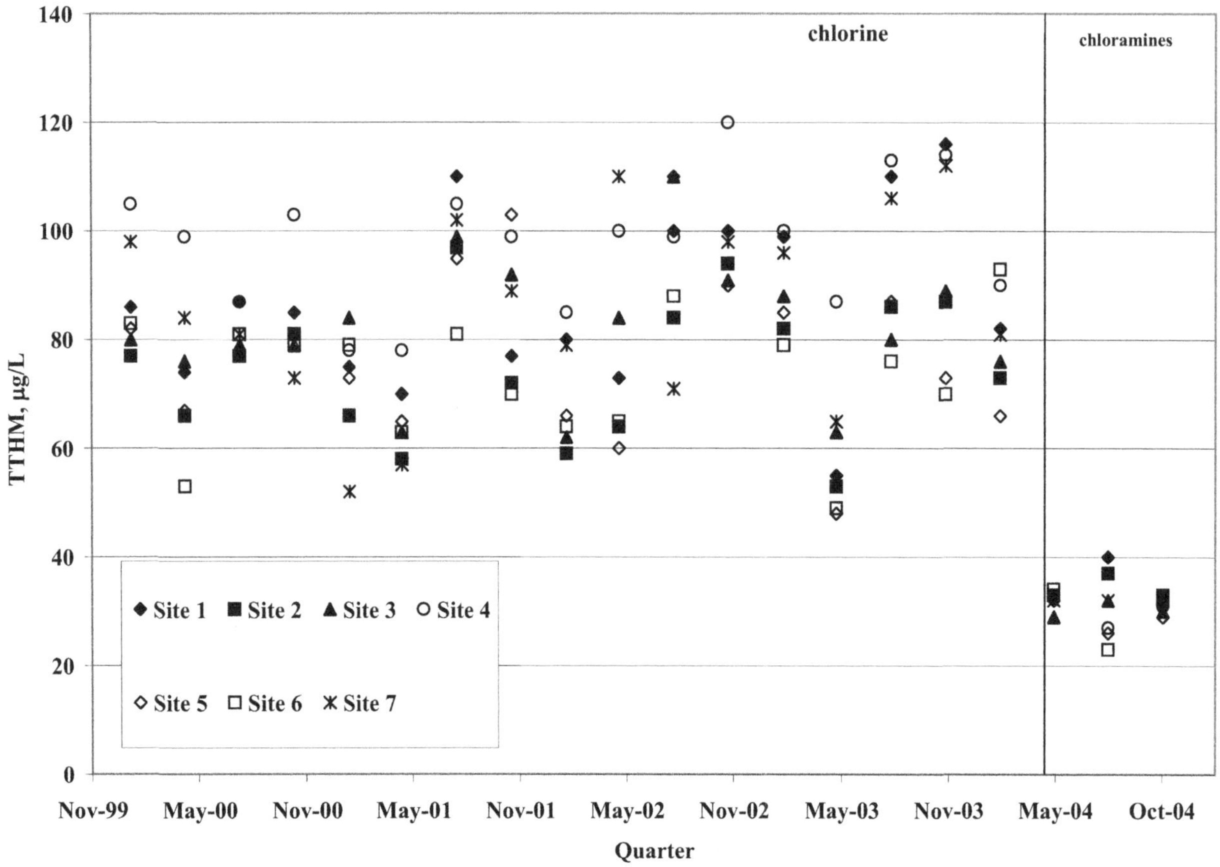

Figure H.8 New Port Richey TTHM data

CHEMICAL QUALITY – CORROSION/METAL RELEASE

Concord, NH

Concord had exceeded the lead "action level" in the first two sampling rounds conducted in 1992. Concurrent with the conversion to chloramines, Concord initiated improvements to their corrosion control to create a more consistent finished water quality and reduce lead solubility. Both the finished water pH and alkalinity targets were raised to 9.50 and 30mg/L as $CaCO_3$ respectively and the phosphate inhibitor feed was stopped. As the finished water quality became more consistent, lead and copper levels decreased and since October 1997, Concord has been in compliance with the lead "action level and has low levels of first draw copper. Graphs of the pH change and the lead and copper 90th percentile data are presented below (Figures H.9 and H.10).

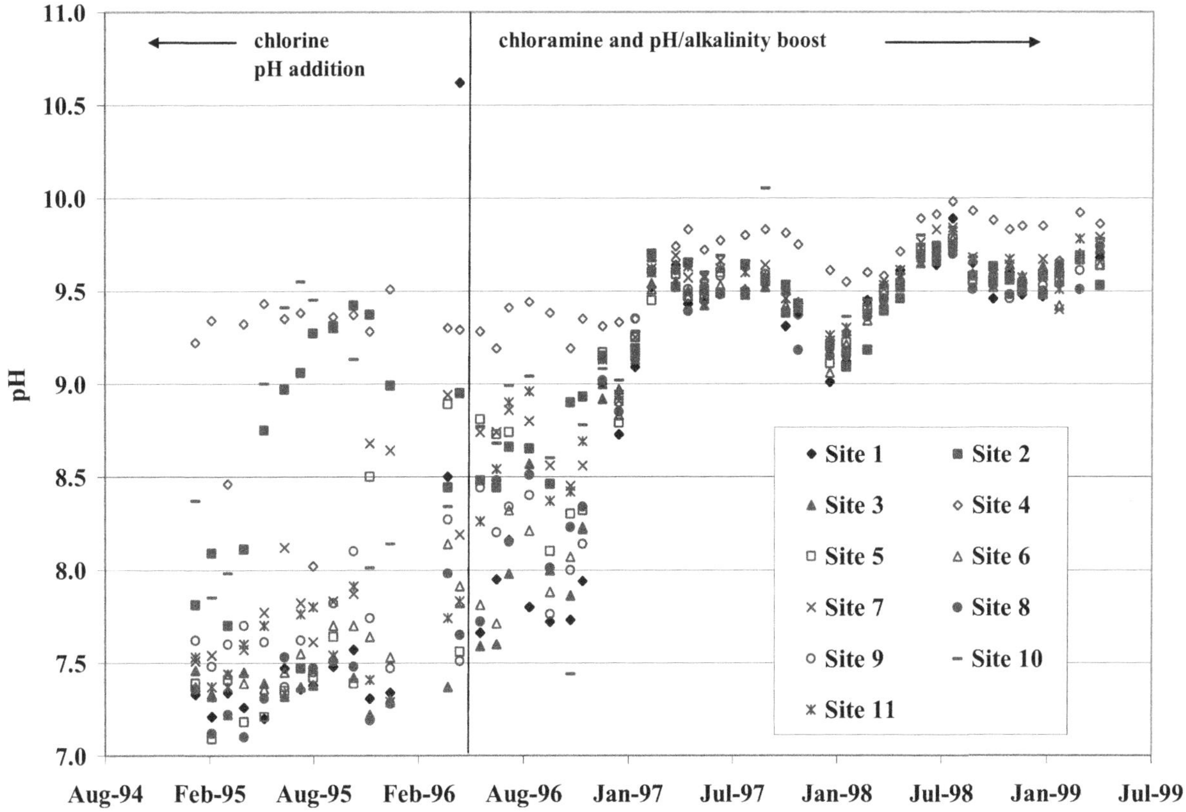

Figure H.9 Concord, NH pH data

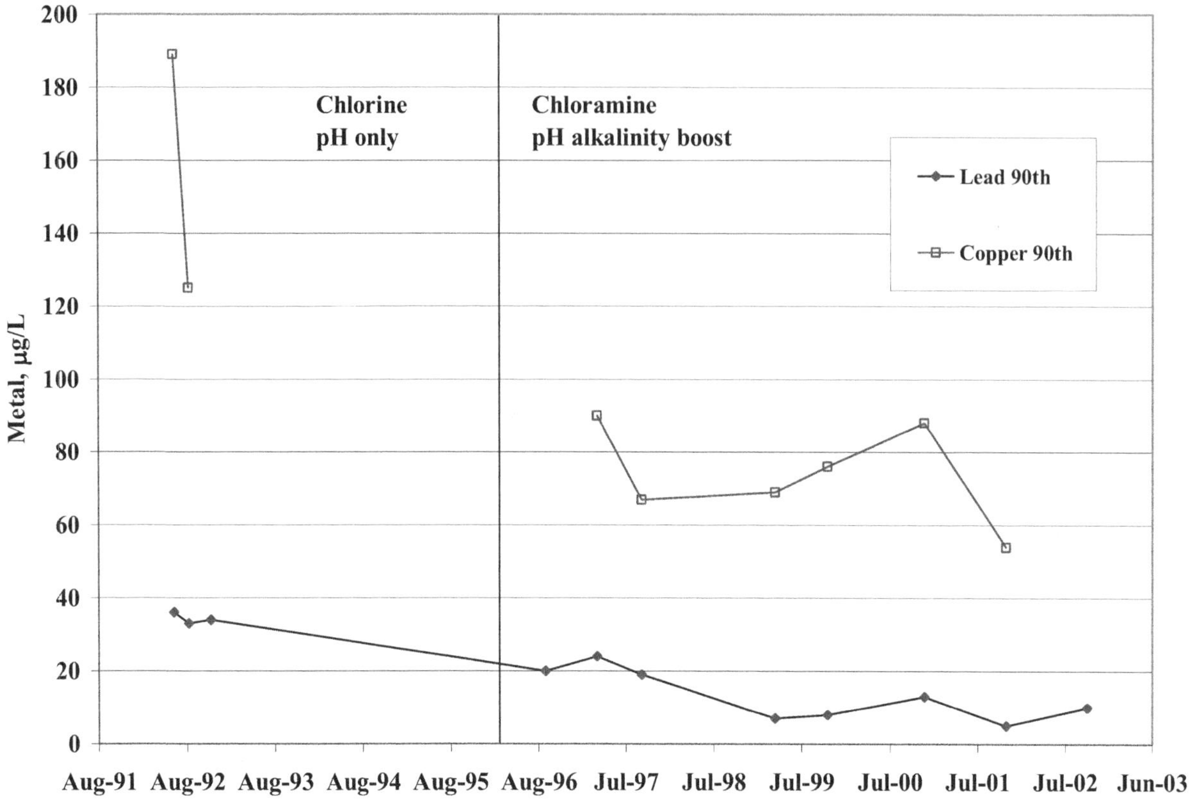

Figure H.10 Concord, NH 90th percentile lead and copper values

Newport News, VA

Concurrent with the conversion to chloramines in 1998, Newport News also raised the pH of the finished water to a target value of 7.5 leaving the two water treatment plants. Figure H.11 presents pH data for both of the water treatment plant effluents and values from the distribution system. There is quiet a bit of variability in the distribution system pH with values ranging from 7.5 to 9.

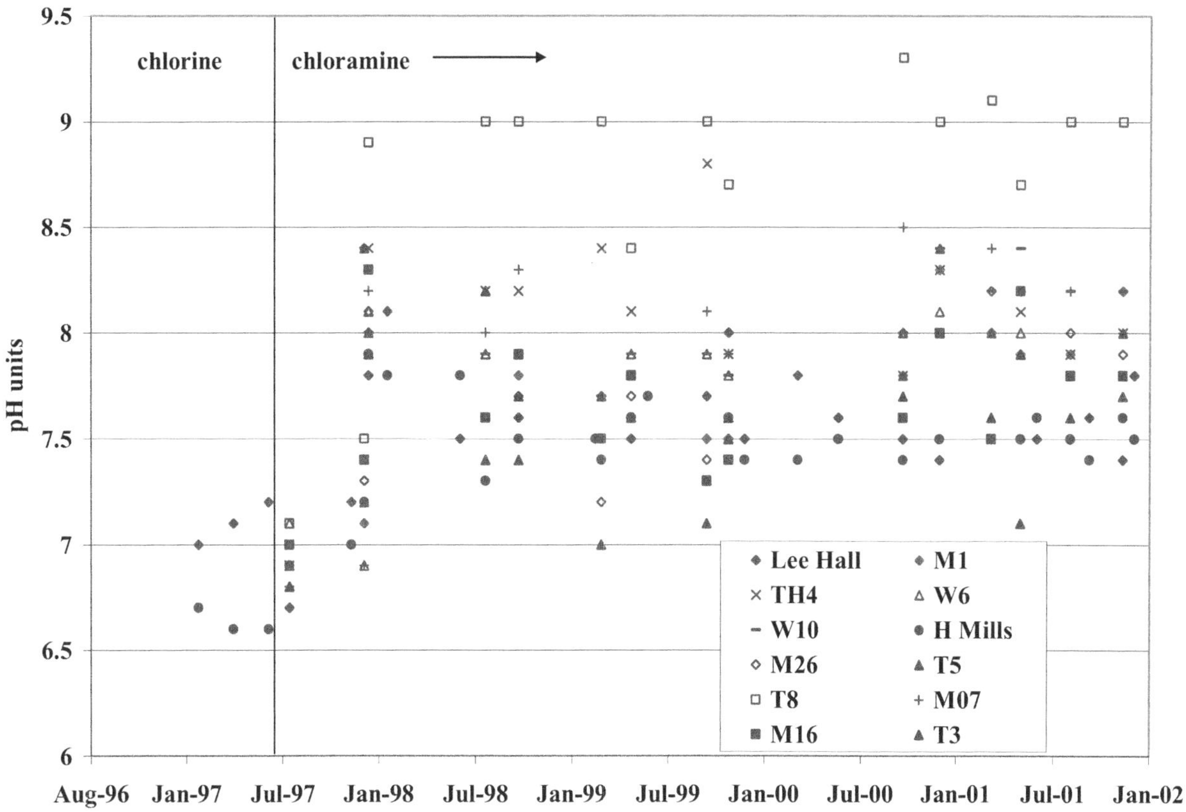

Figure H.11 Newport News, VA distribution system pH data

Philadelphia, PA

Since 1992, the 90th percentile lead levels in the PWD system have ranged from 0.009-0.015 mg/L with the highest levels occurring in the early 90s. The 90th percentile copper levels have ranged from 0.3-0.9 mg/L. Overall, PWD reports that the switch to chloramines has solved problems with rusty and red water complaints.

Tampa Bay Water, FL

Tampa Bay Water did not provide any corrosion or particle data.

AESTHETIC QUALITY – T&O

Concord, NH

Concord records date, location and type of water quality complaint and has a program to sample and resolve complaints. Before conversion to chloramines in 1996, the predominant water quality complaint was taste and odor, particularly chlorinous tastes though metallic and bitter tastes were also reported. Red water complaints were second most common issue with consumers. After the conversion to chloramines, the reports of chlorinous tastes decreased. Reports of discolored water increased for a few years after the conversion to chloramines and the concurrent increase in pH and alkalinity. Concord noted that they detected chlorine residuals in places that had no detectable residual before the conversion. Also, in the summer of 1999, vandals opened a valve which emptied a storage tank and reversed flow to the distribution system, resulting in a rash of discolored water complaints. The appearance of white particles in the water correlated to the elevation in finished water pH above 9.5 so Concord revised the finished water pH target to 9.3. The data are presented on Figure H.12.

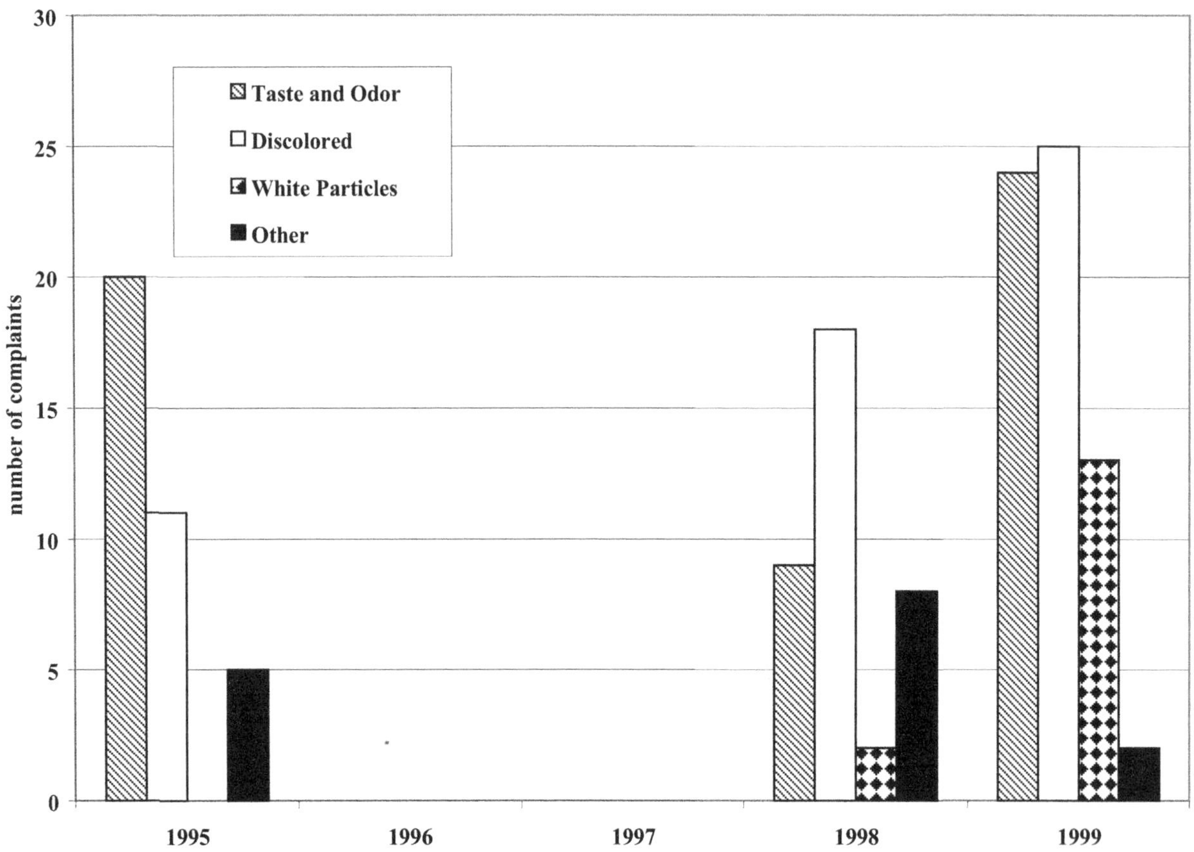

Figure H.12 Concord, NH customer complaint data

IN-American Water Co.

While Indiana American cited a decrease in taste and odor complaints after the change to chloramines in 1992, they no longer have records of these complaints.

MWDSC, CA

As part of the monitoring program implemented by MWDSC, flavor-profile analysis (FPA) was run. The FPA technique is an objective sensory method that can provide reproducible data. FPA indicates that monochloramine is not as good as free chlorine in oxidizing certain off-odors in the water. This reduced oxidation capability was evidenced by an increase in the occurrence of fishy/swampy flavors after the changeover to chloramines. To provide for better oxidation of the fishy/swampy taste-and-odor compounds in the source waters, MWDSC applies chlorine before ammonia in the treatment plants. It was interesting to note that chlorine and bleach flavors did not occur as much when MWDSC was using chloramines as when it was using free chlorine. There was no other significant change in the flavor of the water after MWDSC switched disinfectants from free chlorine to chloramines.

Newport News, VA

After the conversion to chloramines for secondary disinfection in 1998, Newport News detected a decrease in customer complaints citing a "chemical taste". Earthy, musty tastes and odors also appeared to decrease, though Hurricane Floyd may have contributed to the September 1999 rise in those complaints as it washed nutrients, microbes, and organic material into the reservoirs. The data are presented on Figure H.13.

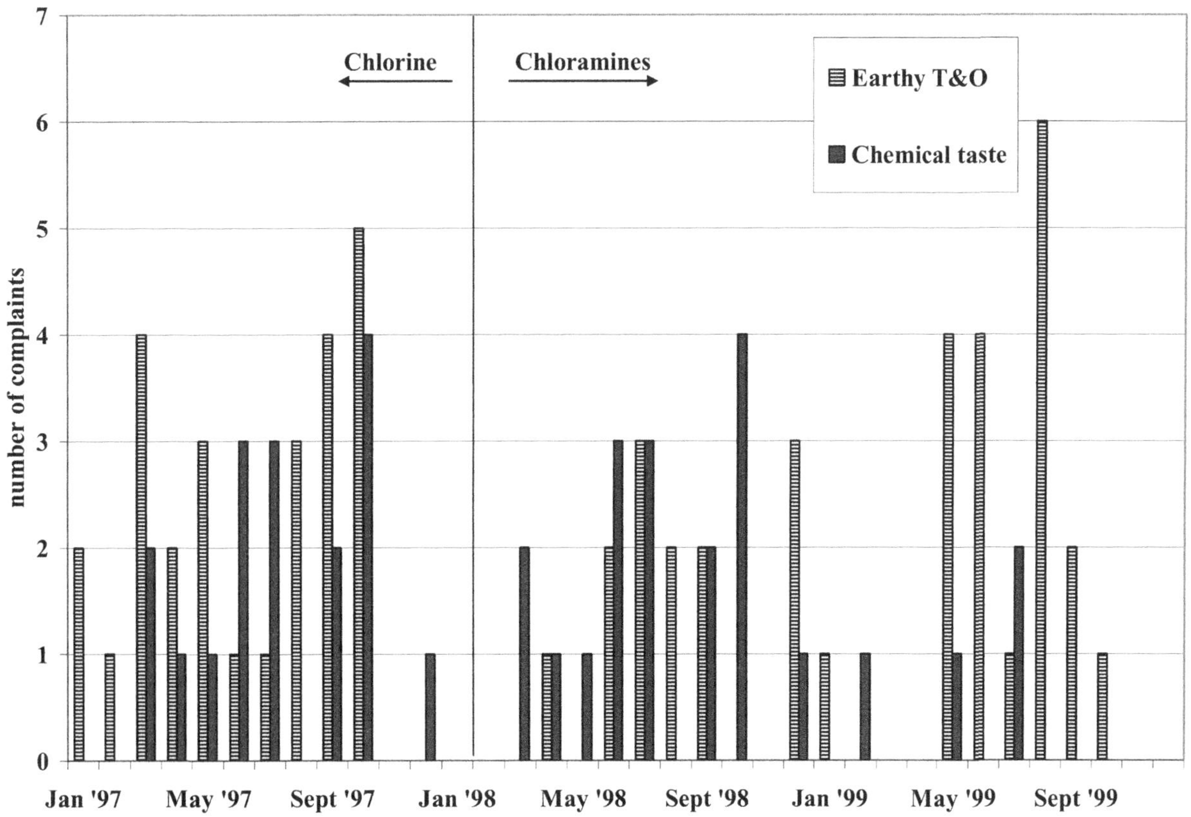

Figure H.13 Newport News taste and odor complaint data

PA – American Water Co.

While the Brownell/Fallbrook system had relatively low levels of taste and odor complaints, there were intermittent complaints of discolored water or particles in the water. Within a few months after the conversion to chloramines, the number of discolored water complaints had decreased and customers reported fewer chlorinous taste and odor complaints. The data are presented on Figure H.14.

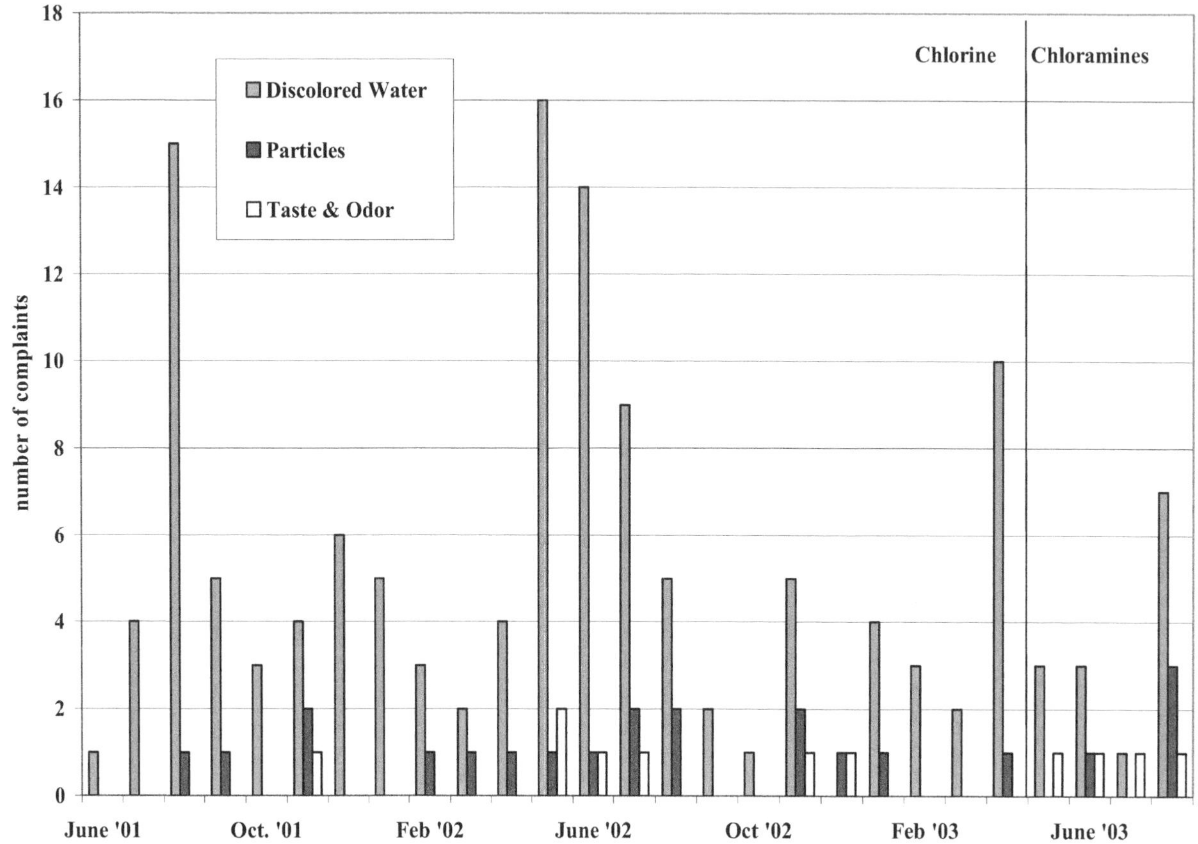

Figure H.14 PA - American customer complaint data

Philadelphia, PA

One of the key motivations for PWD to convert to chloramines was to reduce the chlorinous flavor of the water. Flavor Profile Analysis (FPA) work at PWD during the 1980s found that the dose-response curve (flavor intensity vs concentration) for monochloramine was less steep than for free chlorine. They found that:(1) it takes more chloramine than chlorine to produce a noticeable flavor in water; and (2) typical fluctuations in the chloramine residual should not be detected by most customers. Thus, the switch from free chlorine to monochloramine was expected to have a benefit on customer attitudes. PWD explored its customers' attitudes during the 1980s (Burlingame *et. al,* 2003). A taste test and survey was done in which 400 people tasted four waters and provided their perceptions. The results showed that: customers were not highly sensitive to differences between the taste of different waters (random selection would give 25 percent for each water, and the difference between the least preferred and most preferred waters was 18 vs 34 percent; and the preference for water trended with the level of the chloramine residual. Flavor Profile Analysis (APHA, 1998) described Philadelphia tap water as having a weak chlorinous and musty flavor. Sensitive consumers used the following descriptors more frequently when describing Philadelphia tap water than when describing the other less chlorinated waters: chlorinous, aftertaste, swimming pool, chemical, and metallic.

The data collected in the public perception studies indicated that: 1) consumers have a negative perception of chlorinous tastes and odors in general; 2) perception of such "off" tastes can negatively influence their perception of tap water healthiness and encourage purchasing of tap water alternatives; 3) many consumers are insensitive to monochloramine at the levels typically used in the distribution system. However, the studies noted that many consumers may have been misidentifying other off flavors as "chlorinous," so a negative perception may persist despite the fact that they cannot taste the residual. By keeping the chloramine residual at or below 2.0 mg/L, Philadelphia has benefited from more positive customer attitudes and a higher level of consumer confidence in tap water safety and healthfulness. This level is below the sensitivity threshold of most people as long as the dichloramine concentration is kept to a minimum.

Tampa Bay Water, FL

The City of New Port Richey gets few taste or odor complaints, typically less than 1 per 10,000 people. After the conversion to chloramines, the already low number of taste, odor, or combined taste and odor complaints appeared to decrease. Data for the five months preceeding the conversion and the seven months after the conversion are presented in Figure H.15.

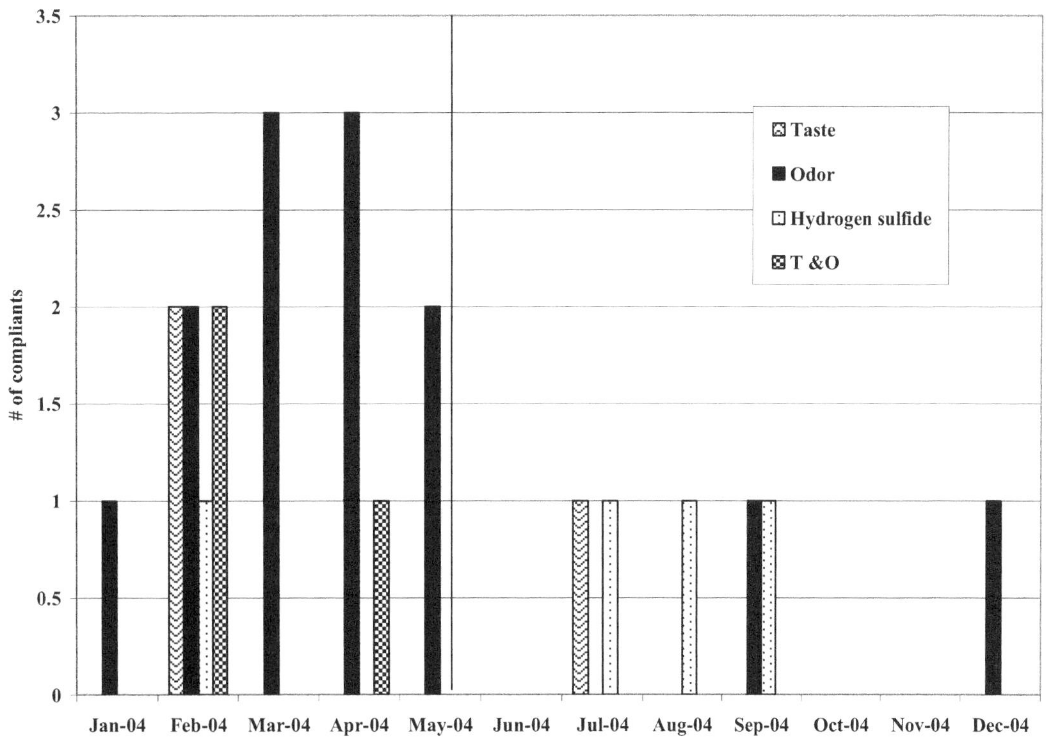

Figure H.15 New Port Richey customer taste and odor complaint data

APPENDIX I
DETAILED INFORMATION FROM PARTICIPATING UTILITIES THAT SWITCHED FROM CHLORINE TO OZONE

CHLORINE TO OZONE

Detailed information on the water quality impacts resulting from the change from using chlorine as a primary disinfectant to using ozone as a primary disinfectant are described in this appendix. The data from all the participating utilities are included except for Milwaukee Water Works and Greater Vancouver Regional District, which are included as case studies in Chapter 4.

MICROBIAL QUALITY

Lincoln Water System, NE

The Lincoln Water System (LWS) has two water treatment plants. The West Plant has a capacity of 60 mgd and uses aeration as a pre-treatment process, chlorine for primary disinfection, and chloramines for secondary disinfection. The East Plant was added to the system in 1994 and includes pre-chlorination, ozone for primary disinfection, and chloramines for secondary disinfection. Following the start-up of the East Plant using ozone, AOC levels leaving the plant were higher than what is currently considered to be normal for entering the distribution system. However, LWS did not observe any change in microbial conditions in the distribution system. There were no changes in HPC levels or coliform positives. Also, no loss in chloramine residual was observed as a result of the East Plant coming on line.

MWRA, MA

MWRA provides treated but not filtered water to 46 communities in the greater Boston area. Treatment consisted of pre-chlorination, chloramination, and pH and alkalinity adjustment before the new ozonation facility at Walnut Hill started up in July, 2005. Current treatment consists of ozonation, chloramination, and pH and alkalinity adjustment. In addition to treatment improvements, MWRA has constructed covered storage to replace all five open finished water storage reservoirs and has built new transmission conduits so that the older tunnels could be inspected and repaired. The work started in 1997 and was completed in May, 2004 with the commissioning of the Metrowest tunnel.

MWRA collects approximately 600 microbial samples from their transmission system each month and analyzes microbial samples from 37 of the 46 communities that they serve. Total coliform data from the MWRA samples and from the community samples are plotted in Figure I.1. The conversion to ozone as the primary disinfectant has improved *Giardia* and *Cryptosporidium* inactivation significantly.

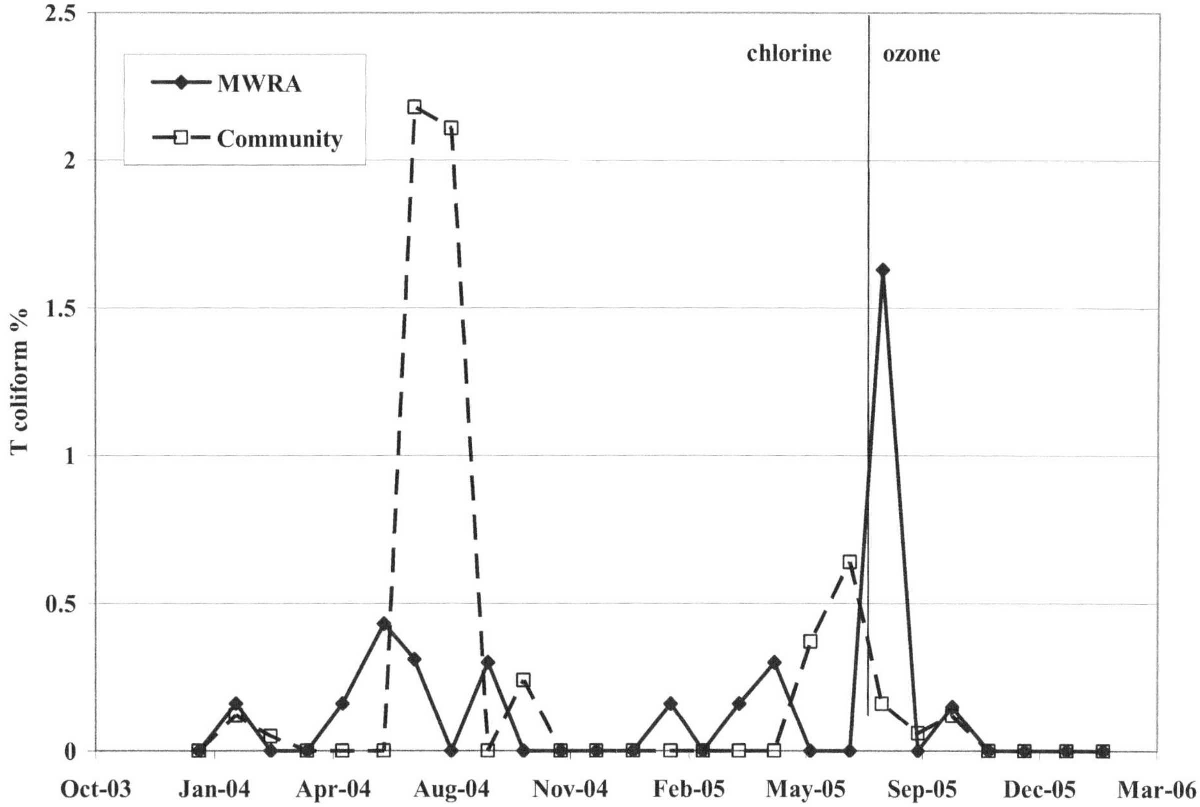

Figure I.1 Total coliform data for MWRA

Heterotrophic plate counts are obtained at the entry point and at the center of the distribution system. HPC data are presented in Figure I.2.

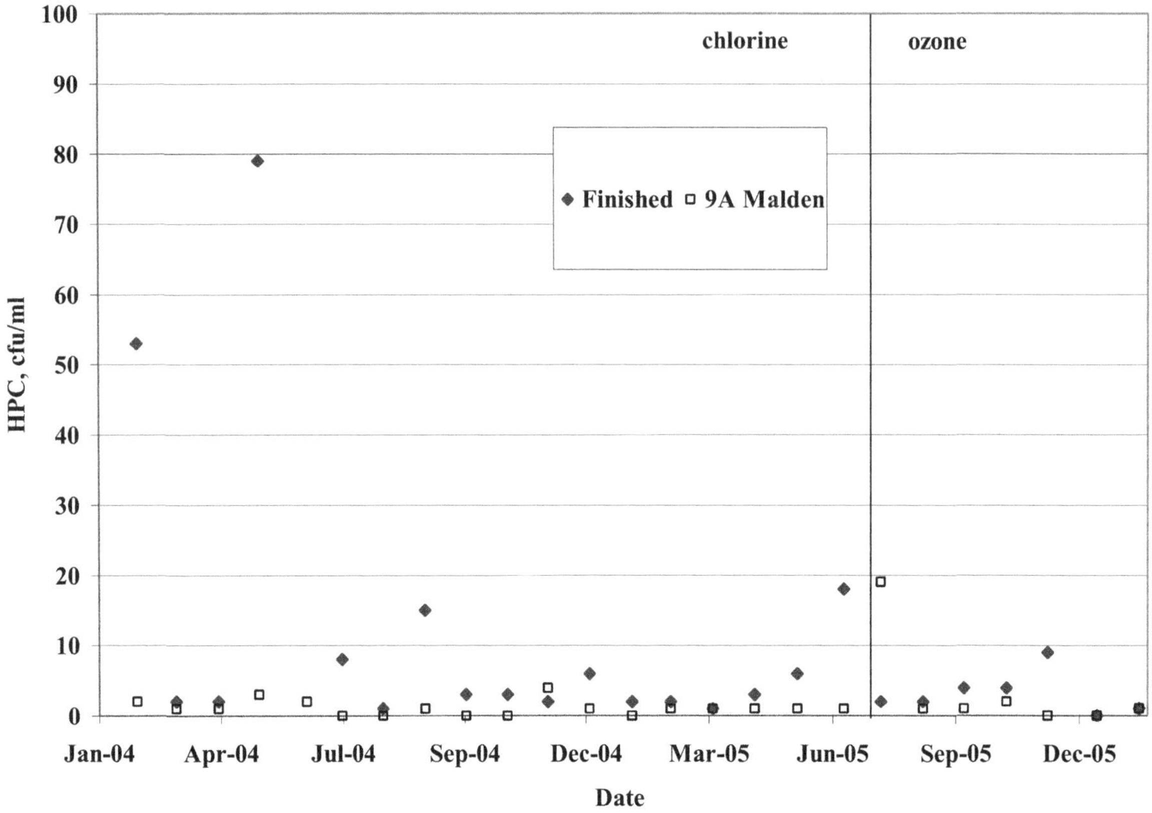

Figure I.2 HPC data for MWRA

Newport News, VA

Newport News converted the Harwood Mills plant to ozone for primary oxidation in 2002. The other plant is in the process of renovation and conversion to ozone for primary disinfection. Newport News has not observed any detectable change in total coliform or heterotrophic plate count numbers in the distribution system since the change to ozone.

Southern Nevada Water Authority (SNWA)

SNWA utilizes Lake Mead and the Colorado River as water supplies. Water is treated at two plants – Alfred Merritt Smith and River Mountains. From 1972 until July 2003, they used chlorine for both primary and secondary disinfection. In July 2003, they switched to using ozone for primary disinfection and continued to use chlorine for secondary disinfection. As a result of the change in disinfection strategy, SNWA experienced increases in assimilable organic carbon (AOC), chlorine decay rates, and HPC counts. Figure I.3 illustrates the increase in AOC from around 150 μg/L with chlorine to about 300 μg/L with ozone. Figure I.4 shows the increase in HPC levels for one part of their distribution system. HPC levels increased from around 10 cfu/ml with chlorine to around 50 cfu/ml with ozone. Figure I.5 illustrates the increase in chlorine decay.

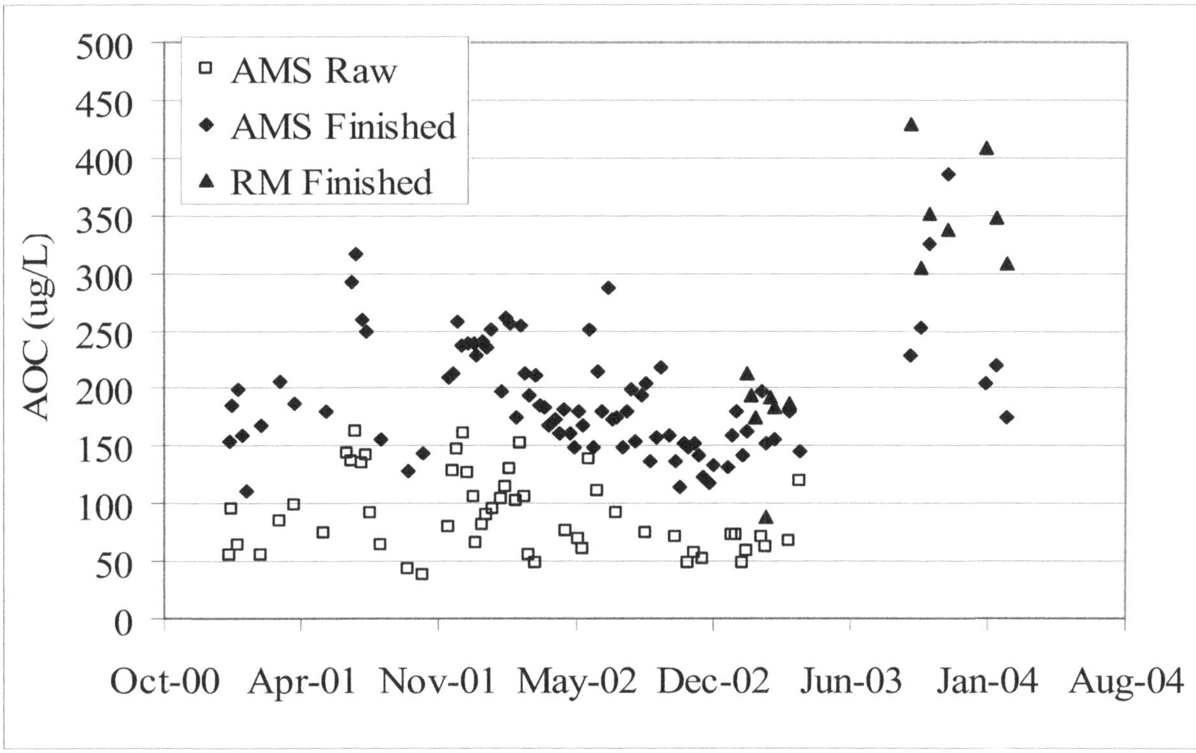

Figure I.3 SNWA assimilable organic carbon data

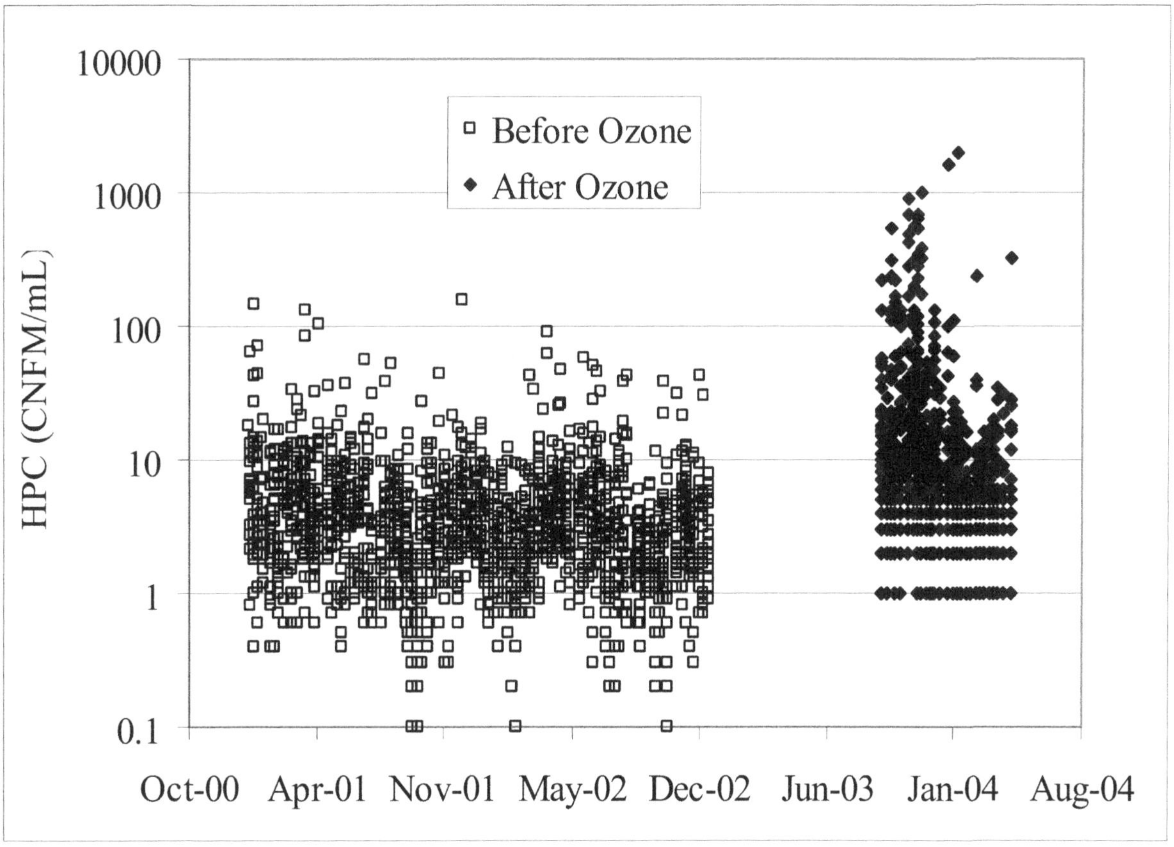

Figure I.4 SNWA heterotrophic plate count data

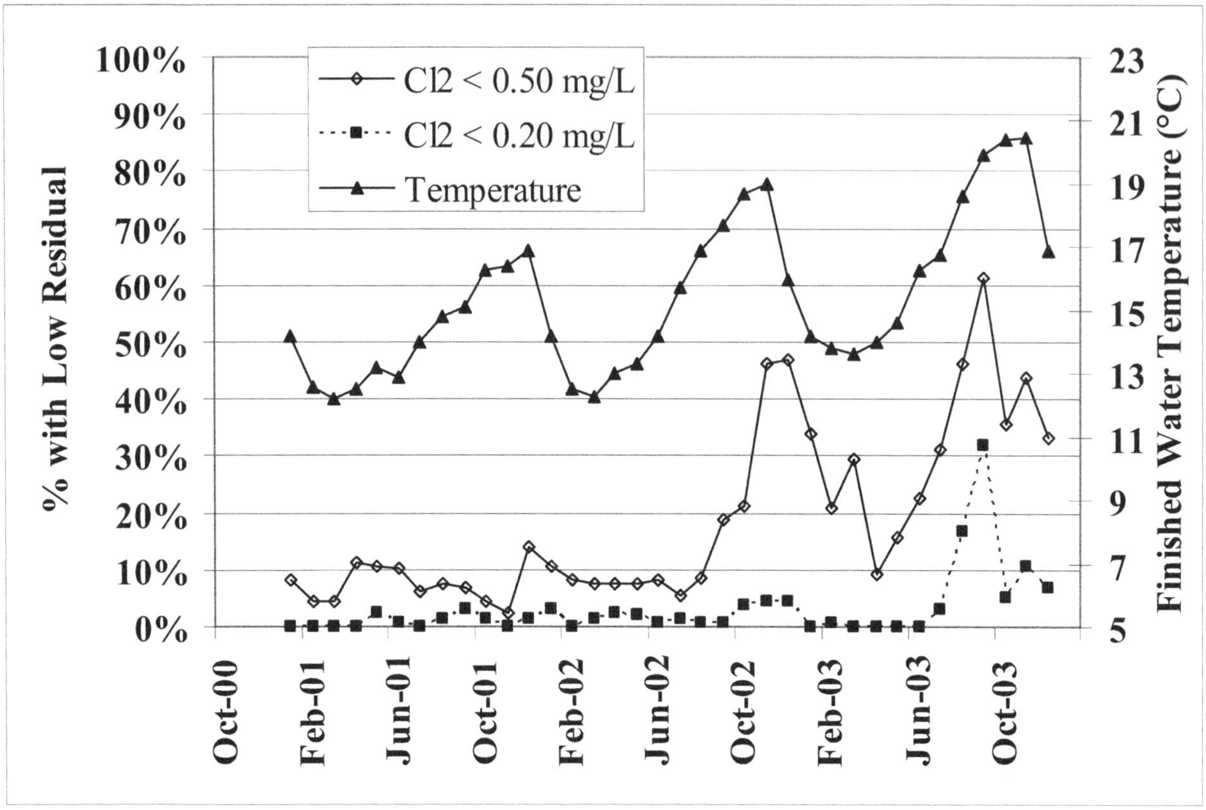

Figure I.5 SNWA chlorine decay data

CHEMICAL QUALITY - DBPs

Lincoln Water System, NE

After LWS placed their East Plant with ozone on line, TTHM levels did not change and remained low. Although they did not have HAA5 data from before the East Plant coming on line, HAA5 levels have been very low.

Massachusetts Water Resources Authority (MWRA)

In anticipation of the startup of the new Walnut Hill facility, MWRA conducted research to determine the effects of ozone on disinfection by-product formation before the full-scale facility went on-line. Parallel pilot scale treatment trains were operated – one with ozone addition and contact, one without but both received free chlorine addition and contact, ammonia addition to form chloramines and pH and aklalinity adjustment before passage through a model distribution system. Comparison of ozonated water versus non-ozonated water for regulated disinfection by-product formation revealed that ozonation resulted in reductions of 75 to 80% of THM levels and 75% of the HAA levels compared to non-ozonated water levels. Since the Ozone Plant went on-line in July, 2005, MWRA has observed the TTHM running average drop from 77 μg/l (2nd quarter of 2005) to 6 μg/L (2nd quarter of 2006) and HAA5 levels have dropped as dramatically. HAA5 are now in the range of 7 to 12 μg/l.

In addition to the regulated DBPs of trihalomethanes and haloacetic acids, analyses for oxygenated by-products such as aldehydes and keto-acids were performed for MWRA by the University of Massachusetts. This study concentrated on evaluating the formation and persistence of the most common aldehydes and keto-acids; formaldehyde, acetaldehyde, glyoxal, and methyl glyoxal for the aldehyde group, and glyoxalic acid and pyruvic acid for the keto acids.

These compounds are highly biodegradable and have the potential to support regrowth in filters or in the distribution system. The ozonated water did have significantly higher levels of aldehydes than the raw water and the higher levels persisted through subsequent chlorination of the water when compared to non-chlorinated water. The aldehydes are shown in Figure I.6.

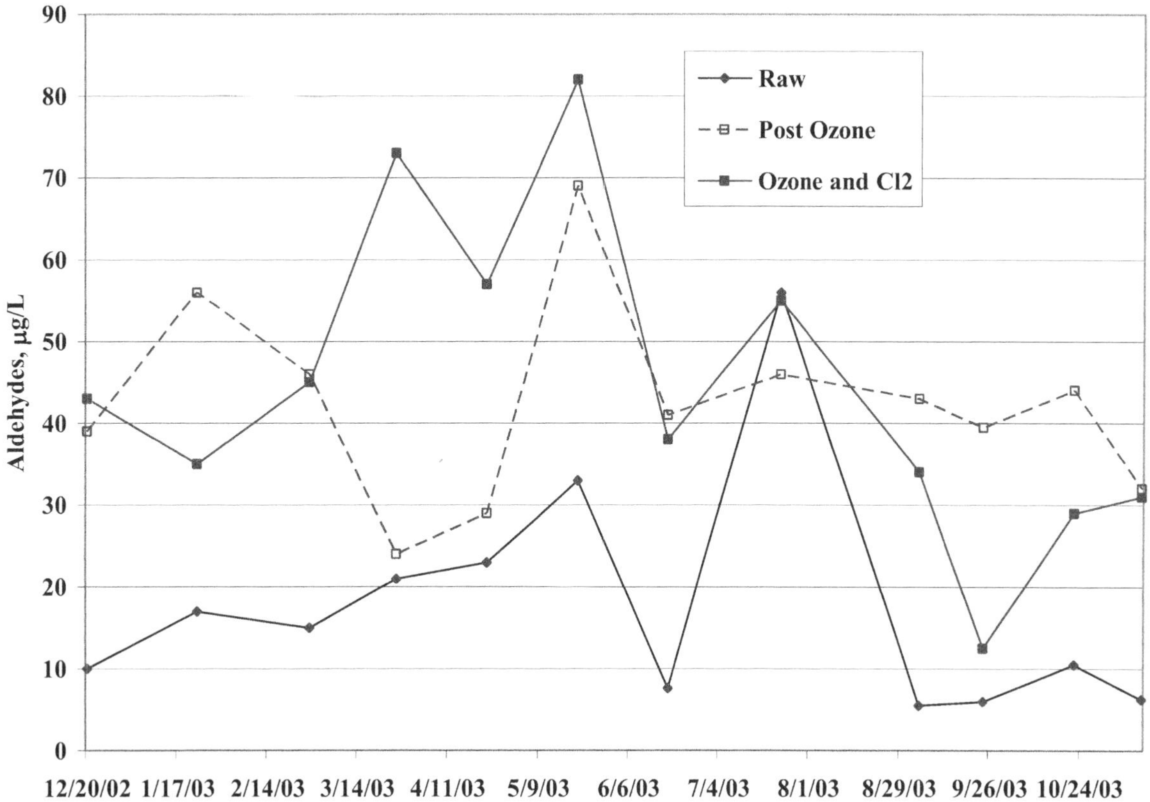

Figure I.6 MWRA aldehyde data

Ozonation also creates substantial levels of keto-acids in the ozonated water. However, the keto-acids appear to be less stable than aldehydes and the levels of keto-acids after ozonation, subsequent chlorinations and chlorine contact in the clearwell are similar to the levels in non-ozonated chlorinated water after chlorine contact. Figure I.7 presents the keto acid data for raw, ozonated and ozonated and non-ozonated, post-chlorinated clearwell water.

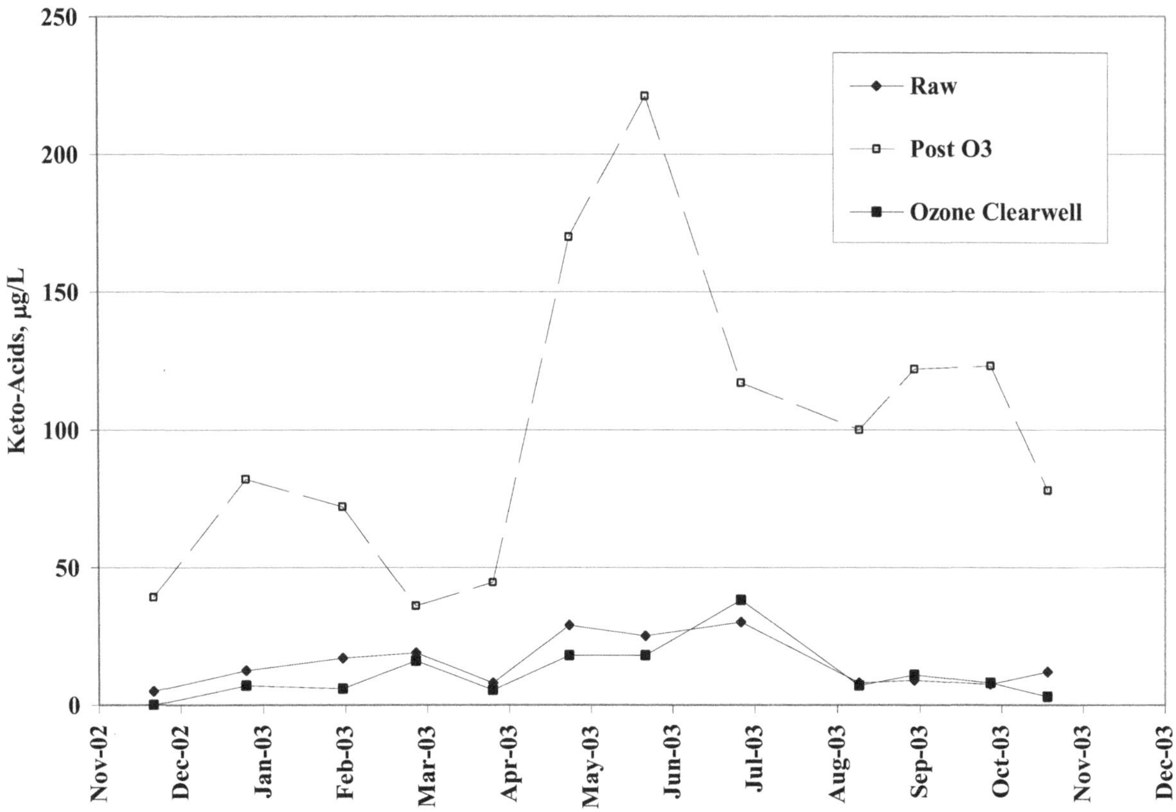

Figure I.7 MWRA keto-acid data

Newport News, VA

Since the conversion to ozone at the Harwood Mills plant, Newport News has noticed a decrease in the levels of halogenated disinfection by-products in the parts of the distribution service served primarily by water from this plant as compared to the water in the distribution system served by the Lee Hall plant. Average trihalomethane levels in the Harwood Mills served areas have decreased 25 to 30% compared to the same location in the period from 1998 to 2001. Figure I.8 presents trihalomethane data from areas served by the Harwood Mills plant. Average haloacetic acid levels decreased from 10 to 30% compared to the same locations in the period from 1998 to 2001. Figure I.9 presents haloacetic acid data from the areas served by the Harwood Mills plant.

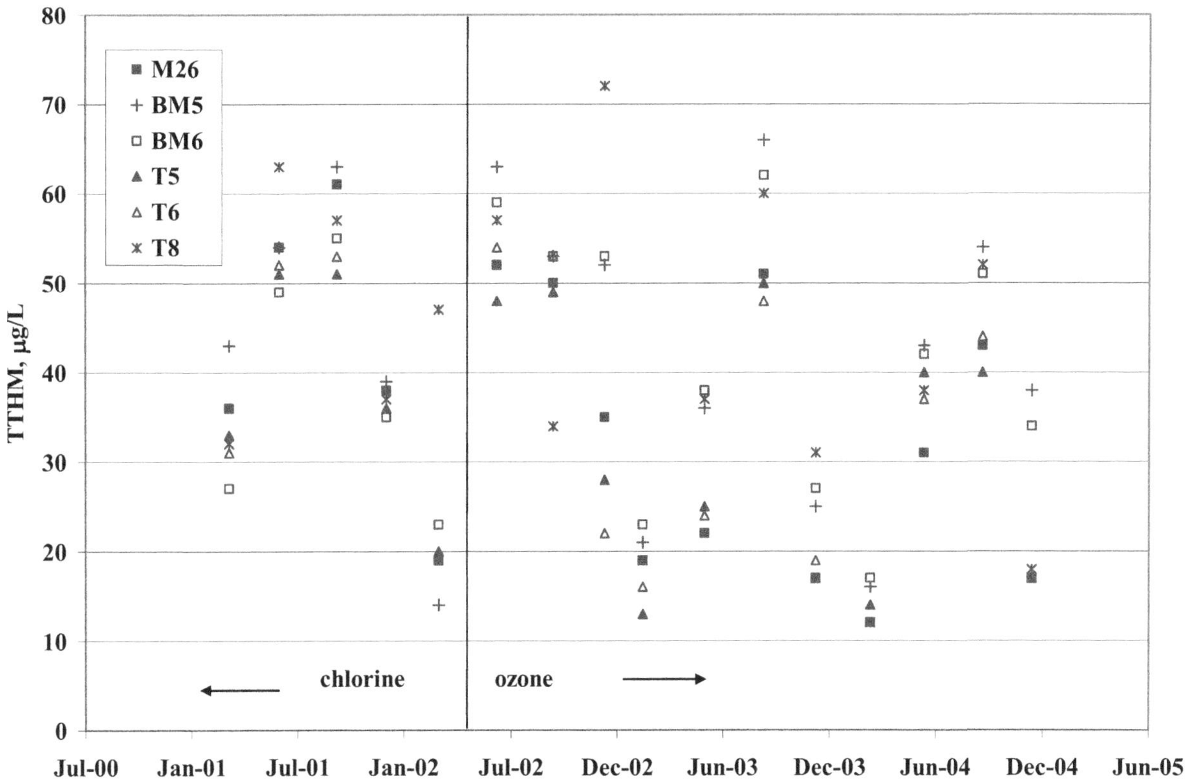

Figure I.8 Newport News,VA TTHM data from selected parts of distribution system

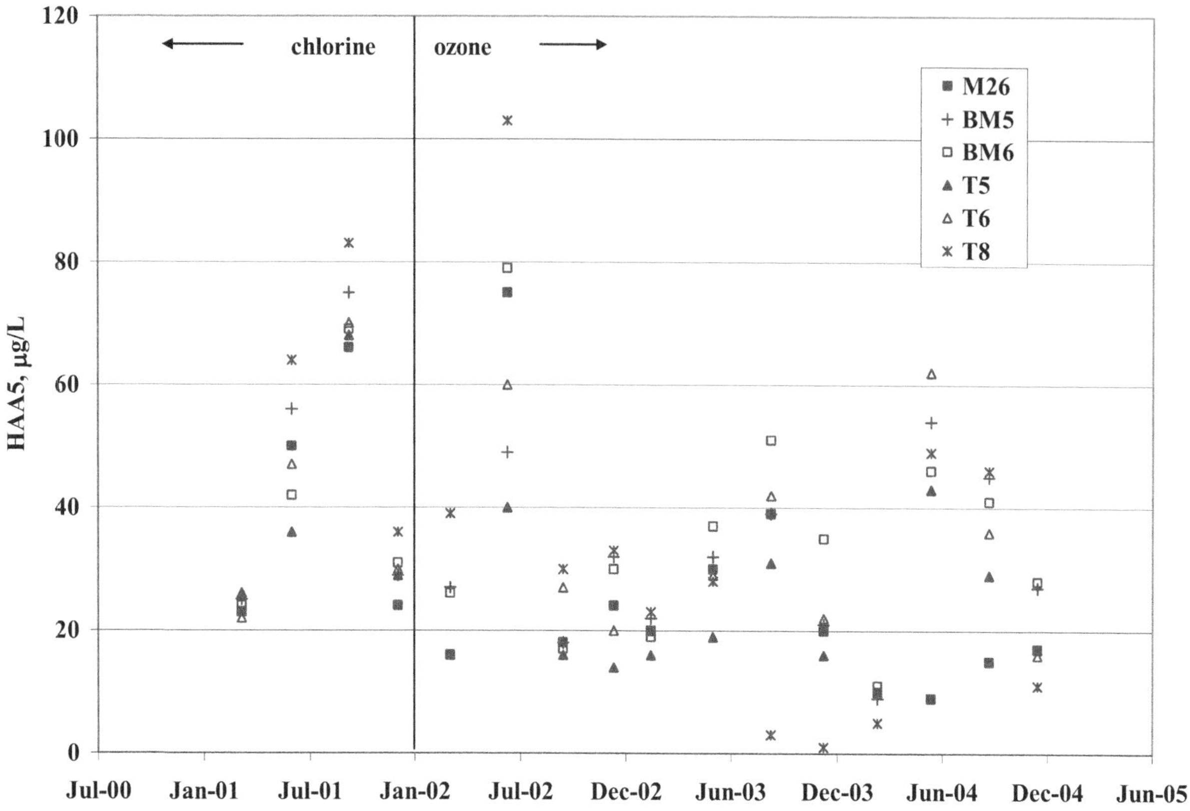

Figure I.9 Newport News, VA HAA5 data from parts of the distribution system

Passaic Valley Water Commission (PVWC), NJ

Over the past few years, the PVWC has made some significant changes to their Little Falls WTP to assure compliance with current and future drinking water regulations. They replaced their conventional sedimentation process with a high-rate clarification process (Actiflo) and added intermediate ozonation. One of the primary reasons for these changes was to reduce DBP levels in the distribution system. The new clarification process was placed in service in January 2003, while the ozonation process was started up in August/September 2004. Figures I.10 and I.11 show TTHM and HAA5 levels (RAA) for the PVWC system. TTHM levels exhibited an initial drop (from generally around 80 µg/L to almost 40 µg/L) after the new clarification process and new GAC/sand filters came on-line, and it appears that an additional drop may have occurred after the ozone was placed in-service. More data are needed to confirm this. HAA5 levels dropped (from about 40 µg/L to about 25 µg/L) after the implementation of the new clarification process and GAC and ozone.

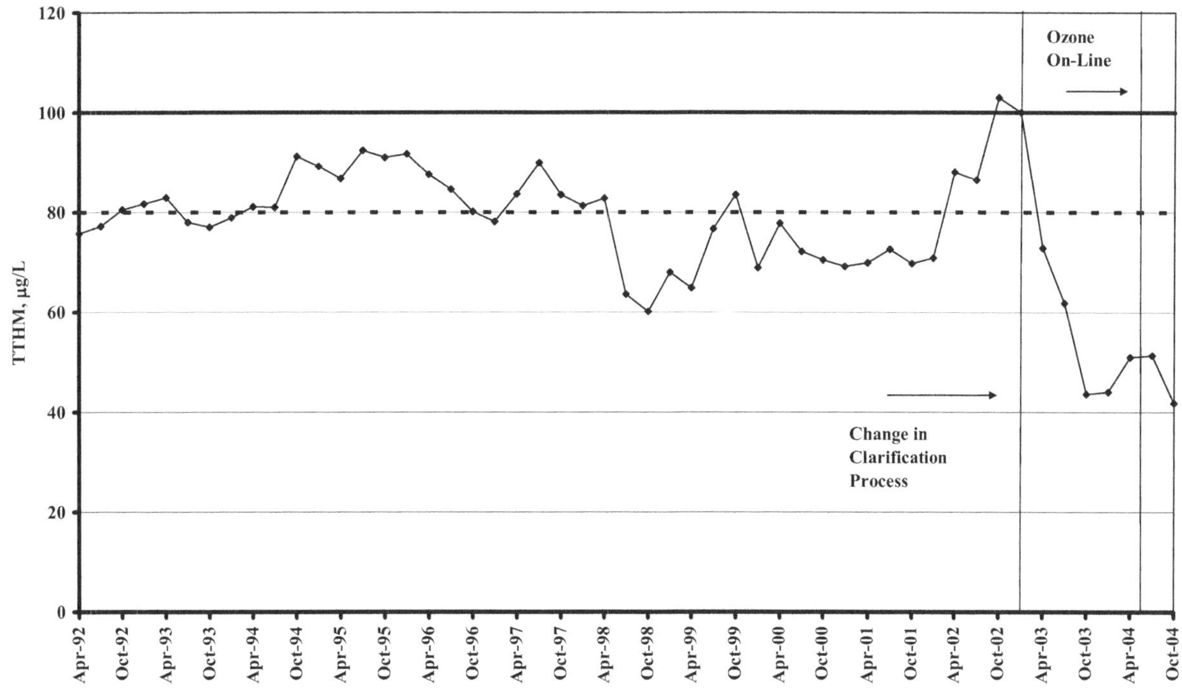

Figure I.10 PVWC running annual average TTHM data

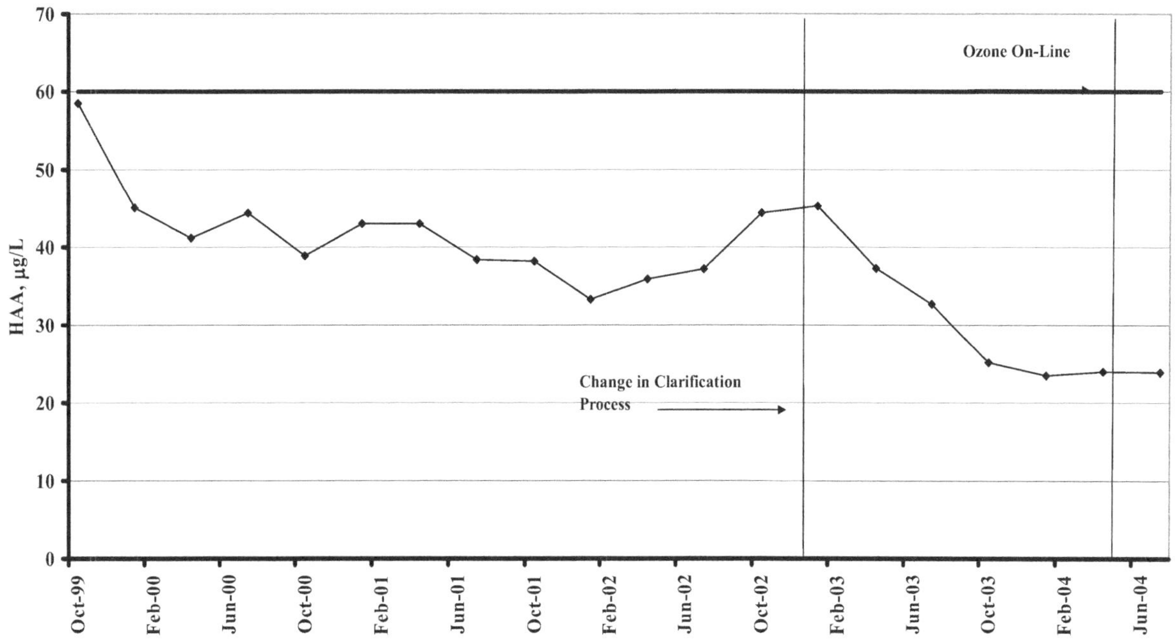

Figure I.11 PVWC running annual average HAA5 data

CHEMICAL QUALITY – CORROSION/METAL RELEASE

MWRA, MA

The ozone trials also encompassed potential effects on corrosion control. Comparison of pre-ozonation versus pre-chlorination treatments was used with pH and alkalinity adjustment to target values of 9.5 and 37 mg/l as $CaCO_3$. The two waters were then introduced into heavily tuburculated pipe sections to model worse case distribution system conditions. Pre-ozonation appeared to reduce the effectiveness of the corrosion control and result in higher iron release (red water) compared to the pre-chlorinated pipe rack. This observation has not been confirmed full scale to see if there is some antagonistic effect between the high pH and ozonation on iron release. Figure I.12 presents the iron release data over the ten month trial.

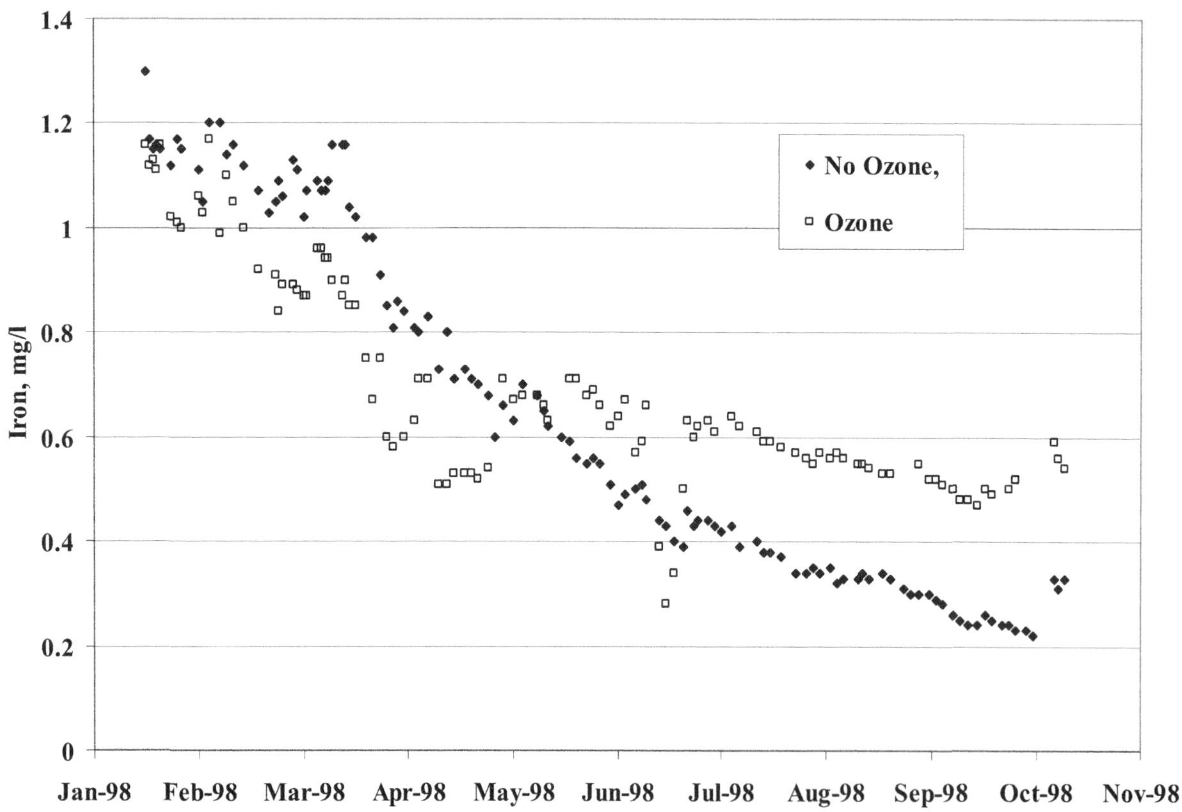

Figure I.12 MWRA iron release data

Newport News, VA

Newport News continued to refine finished water pH and alkalinity target values for the system. The addition of ozone to the treatment process did not appear to affect the ability of the system to achieve the pH and alkalinity targets. Figures I.13 and I.14 present distribution system pH and alkalinity data respectively.

Newport News also noted that they observed manganese release from their filters when they moved the point of chlorination to just prior to the clearwell when the primary ozone disinfection was installed. The manganese release resulted in some customer complaints of discolored water. Moving the point of chlorination to just ahead of the filters resolved the problem.

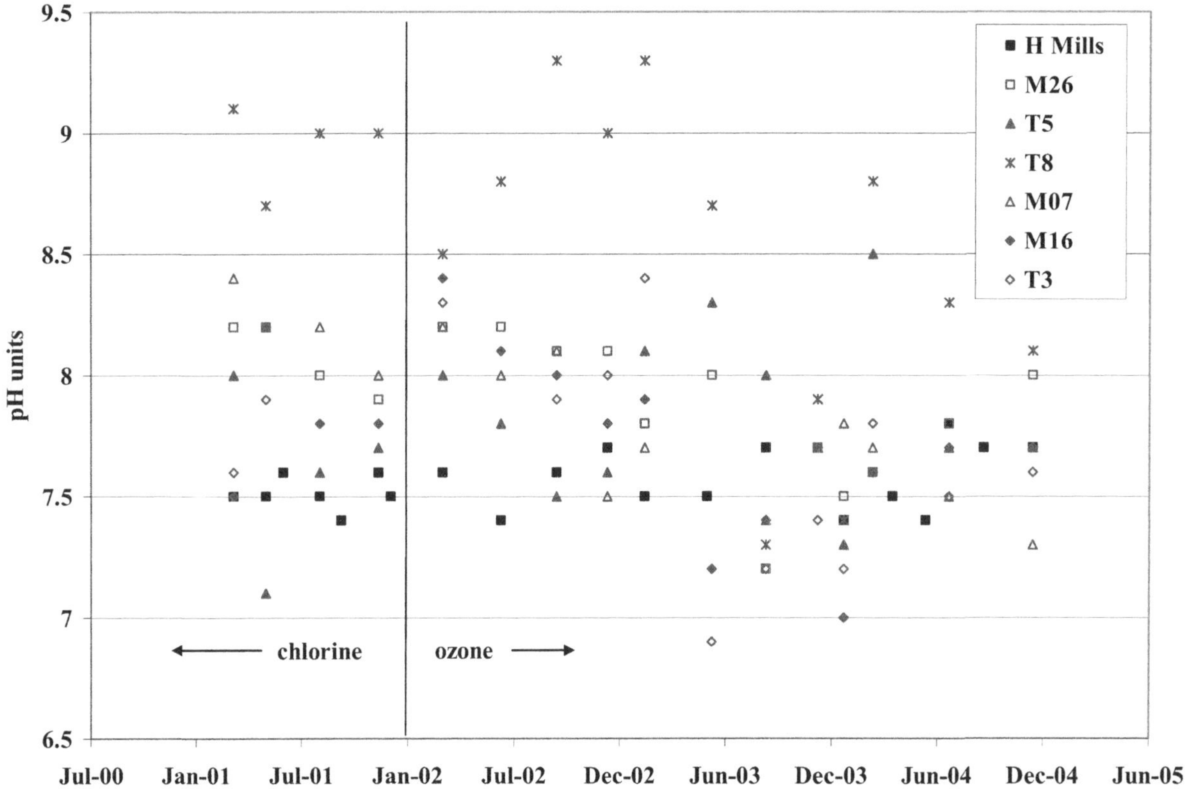

Figure I.13 Newport News pH data

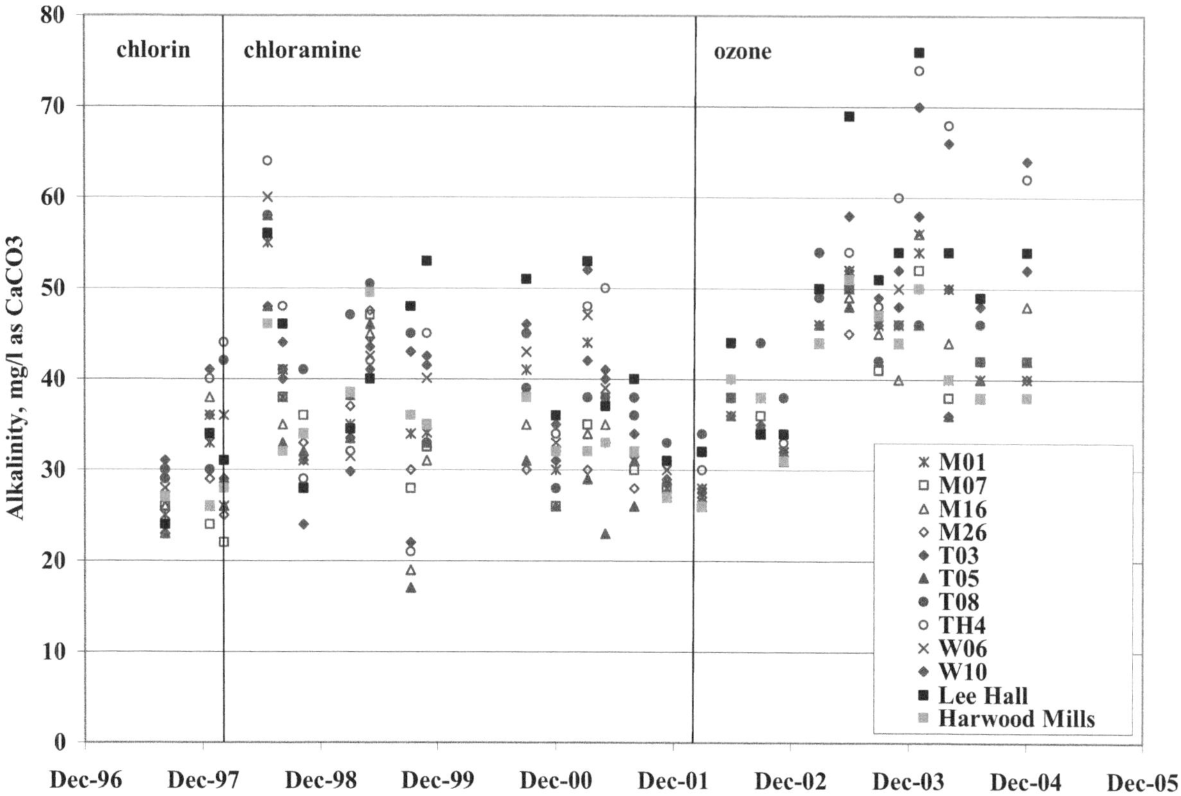

Figure I.14 Newport News alkalinity data

AESTHETIC QUALITY – T&O

Newport News, VA

The system reported an initial decrease in taste and odor complaints in 1999 after conversion to chloramines and a continuing decline after ozone was installed for primary oxidation/disinfection at the Harwood Mills plant. There were some drought related water quality complaints in 2002 (earthy-musty taste and odor and red water complaints) but overall taste and odor complaints were low in 2003, the year after conversion to ozone. Figure I.15 presents these data.

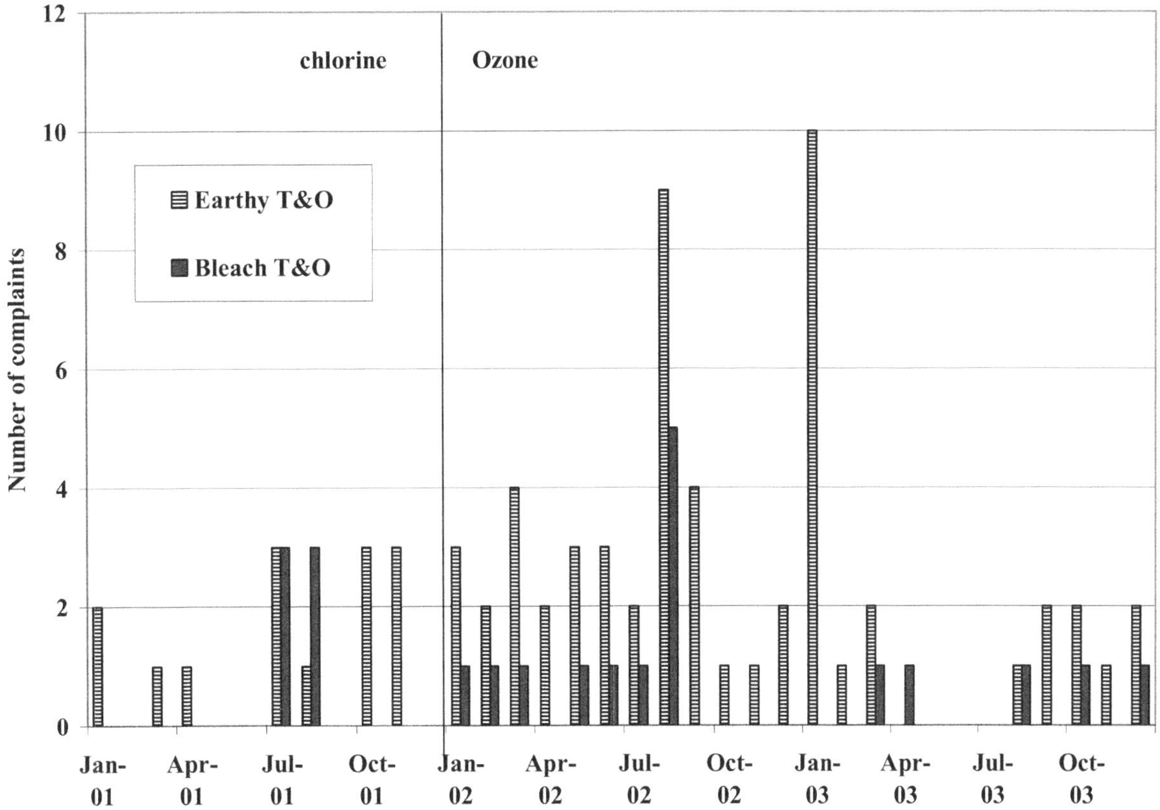

Figure I.15 Newport News taste and odor complaint data

So. Nevada Water Authority

SNWA provided data regarding water quality customer complaints which are summarized in Table I.1. The data show a significant drop in water quality complaints from 2003 to 2004, with ozone being added in July 2003. No information was provided on the type of complaints.

Table I.1
SNWA water quality customer complaint summary

Year	Total Complaints	Total Field Visits	Number of Water Samples	Average Field Calls/Month
1995	511	425	412	35.4
1996	508	388	126	32.3
1997	480	361	102	30.1
1998	463	310	81	25.8
1999	397	248	86	20.7
2000	370	209	61	17.4
	429	245	91	20.4
2001				
2002	370	184	83	15.3
2003	372	183	71	15.3
2004	75	39	10	13

REFERENCES

Abernathy, C.G. and A.K.Camper. 1997. Interactions between Pipe Materials, Disinfectants, Corrosion Inhibitors, Organics and Distribution Biofilms. In *Proc. AWWA Proc. Water Quality Technology Conference*, Denver, CO Nov. 19-12: AWWA

Abernathy, C. and A.K. Camper. 1998. *Interactions between Pipe Materials, Organics, Corrosion Inhibitors, and Disinfectants on Distribution Biofilms.* Volume 1. Bench and Pilot-scale Studies. National Water Research Institute, Fountain Valley, CA.

Anderson, W.B., C.I. Mayfield, D.G. Dixon and PM. Huck. 2003. Endotoxin inactivation by selected drinking water treatment oxidants. *Water Res.* 37:4553-4560.

Andrews, R.A., B. Karney, and G. Ranger. 2001. Chlorine Dioxide as a Post Disinfectant in Wiarton, Ontario. *Proc. Fourth International Symposium on Chlorine Dioxide*, Las Vegas, NV, February 15-16, 2001.

Andrews, R.C., Z. Alam, R. Hoffman, L. Lachuta, R Cantwell, S Andrews, E Moffett, G Gagnon, J. Rand, C. Chauret, 2005. *Impact of Chlorine Dioxide on Transmission, Treatment, and Distribution System Performance.* Denver, Colo: AwwaRF, AWWA, and EBMUD.

Anselme, C., I.H. Suffet, and J. Mallevialle. 1988. Effects of Ozonation on Tastes and Odors. *Jour. AWWA*. 80(10) 45-51.

American Public Health Association (APHA), American Water Works Association, and Water Environment Federation. 1998. *Standard Methods for the Examination of Water and Wastewater,* 20 edn. American Public Health Association, Washington, DC.

Arias, M., J. Caulfield, and R. Summers. 2002. Optimizing Preoxidation for DBP Control and Its Impact on Filter Performance. In *Proc. AWWA Annual Conference*, New Orleans, June 10-15: AWWA

Aston, R.N. 1947. Chlorine Dioxide Use in Plants on the Niagara Border. *Jour. AWWA, 39*(7):687-690

American Water Works Research Foundation (AwwaRF) and Lyonnaise des Eaux. 1987. *Identification and Treatment of Tastes and Odors in Drinking Water.* Report 90518, Editors: J. Mallevialle and I.H. Suffet, Denver, CO, AWWA and AwwaRF

Awwa Research Foundation (AwwaRF) and Lyonnaise des Eaux. 1995. *Advances in Taste-and-Odor Treatment and Control.* editors: I.H. Suffet, J. Mallevialle, and E. Kawczynski, Report 90610, Denver, CO, AwwaRF and AWWA and Lyonnaise des Eaux, Paris , France

AwwaRF and DVGW, 1996. *Internal Corrosion of Water Distribution Systems, 2nd Ed.*, AwwaRF, Denver, CO.

Baribeau, H., M. Prevost, R. Desjardins, and P. Lafrance, 2001a. Changes in Chlorine and DOX Concentrations in Distribution Systems. *Jour. AWWA*, 93:12:102-114.

Baribeau, H., Mackey, E.D., Booth, S.D.J., Crozes, G.F., Suffet, M., and Piriou, P. 2001b "Public Perception of Tap Water Chlorinous Flavor." Paper presented at the American Water Works Association 2001 Water Quality Technology Conference, Nashville, TN, November 11-15, 2001.

Barrett, S.E., M.K. Davis, and M.J. McGuire 1984. Blending Chloraminated Water with Chlorinated Water: Considerations for a Large Water Wholesaler. In *Proc. AWWA Annual Conference,* Dallas, TX June 10-14, Denver, CO;AWWA

Barrett, S.E., M.K. Davis, and M.J. McGuire. 1985. Blending Chloraminated and Chlorinated Waters *Jour. AWWA*. 77(1):50-60

Barrett, S., C. Hwang, and Y. Guo 2003. Occurrence of NDMA in Drinking Water : A North American Survey, 2001-2002. In *Proc. AWWA Annual Conference,* Anaheim, CA June 15-19, 2003, Denver, CO; AWWA.

Bartels, J.H.M., B.M. Brady, and I.H. Suffet, 1989. *Taste and Odor in Drinking Water Supplies –Phase 1 & II*, Philadelphia Water Department, Philadelphia Suburban Water Company, Report 90542, Denver, CO, AwwaRF and AWWA and and Lyonnaise Des Eaux-Dumez, Le Pecq, France.

Batte, M., B. Koudjounou, P. Laurent, J. Coallier and M. Prevost. 2000. Does phosphate-based anticorrosion treatment enhance biofilm resistance to disinfection? In *Proc. AWWA Water Quality Technology Conference.*, Denver, CO Nov 5-8: AWWA

Berg, J.D., J.C. Hoff, P.V. Roberts and A. Matin. 1988. Resistance of bacterial subpopulations to disinfection by chlorine dioxide. *Jour. AWWA* 80(9):115.

Berman, D., E.W. Rice and J.C. Hoff. 1988. Inactivation of particle-associated coliforms by chlorine and monochloramine. *Appl. Environ. Microbiol.* 54:507-512.

Black and Veatch 1996 Groundwater Treatment Study to the City of Portland, Oregon

Black & Veatch 2000 Technical Report to the City of Irving Texas on Relationship of Red Water and Nitrification

Boccelli, D., M. Tryby, M. Koechling, R. Summers, J. Uber, L. Rossman, 1998 Booster Chlorination: Residual Maintenance and Trihalomethane Formation, In *Proc. AWWA Water Quality Technology Conference*, San Diego, CA Nov. 1-5:Denver, CO; AWWA

Botzenhart, G.M., G.M. Tarcson and M. Ostruschka. 1993. Inactivation of bacteria and coliphages by ozone and chlorine dioxide in a continuous flow reactor. *Water Sci. Technol.* Vol. 27, No. 3/4, pp. 363-370.

Boulos, L., H. Baribeau, P.C. Singer, C. Nichols, S. Schlesinger, S. Williams, R. Gullick, 2003 Formation and Decay of THMs and HAAs in Distribution Systems: Results of Five Case Studies, In *Proc. AWWA Water Quality Technology Conference* Philadelphia, PA, Nov 2-6: AWWA

Boozarpour, M., A. Wilczak, and E. So. 2005. Water Quality in San Francisco's Distribution System One Year after Changes from Chlorine to Chloramine. In *Proc. of the AWWA Annual Conference.* San Francisco, CA June 12-16: AWWA.

Bruchet, A., E. Costentin, M.F. Legrand, and J. Mallevialle. 1992. Influence of Natural Nitrogenous Organic Compounds on Taste and Odors in Finished Waters. *Water. Sci. Techn.* Vol. 25. No. 2, pp. 323-333.

Buffle, M.O., G.Galli, U. von Gunten, 2003 Enhanced Bromate Control During Ozonation: The Chlorine-Ammonia Process, In *Proc. AWWA Water Quality Technology Conference,* Philadelphia, PA, Nov 2-6: AWWA

Bull, R.J. and R.C. Kopfler. 1991. *Health Effects of Disinfectants and Disinfection By-Products.* Denver, CO: AwwaRF and AWWA

Buls, Judel. 2000. Making Humic Substances More Biodegradable. Master's Thesis, Civil/Environmental Engineering, Montana State University, Bozeman, MT.

Burlingame, G.A., J.J. Muldowney, and R.E. Maddrey. 1992. Cucumber Flavor in Philadelphia's Drinking Water. *Jour. AWWA,* 84(8): 92-97.

Burlingame, G.A., E.D. Mackey, and J. Choi. 2003. Choosing an Optimal Chloramine Residual Goal for Philadelphia's Tap Water. In *Proc. AWWA Water Quality Technology Conference,* Philadelphia, PA., Nov. 2-6: AWWA

Burlingame, G.A., E. D. Mackey, and J. Choi 2003. In *Proc. AWWA Annual Conference,* Anaheim, CA June 15-19, 2003, Denver, CO; AWWA.

Burr, M., L. Goodrum, J. Buls and A.K. Camper. [N.d.] Effect of UV light on humics and biofilm formation in a model drinking water system. *Water Res*. Submitted.

Camper, A.K. 1996. *Factors Limiting Microbial Growth in the Distribution System: Pilot and Laboratory Studies.* Denver, CO AwwaRF and AWWA

Camper, A.K. and W.L. Jones 2000. *Factors Affecting Microbial Growth in Model Distribution Systems*. Denver, CO AwwaRF and AWWA

Camper, A. K. 2003. Factors Influencing Microbial Water Quality in Distribution Systems. In *Proc. AWWA Annual Conference,* Salt Lake City, UT, June 15-19

Camper, A.K., K. Brastrup, A. Sandvig, J. Clement, C. Spencer and A.J. Capuzzi. 2003. Impact of Distribution System Materials on Bacterial Regrowth. *Jour. AWWA* 95(7):107-121.

Chaiket, T., P.C. Singer, A. Miles, M. Moran, and C. Pallotta, 2002, The Effectiveness of Coagulation, Ozonation, and Biofiltration in Controlling Disinfection By-Products in Drinking Water, *Jour. AWWA* 94(12): 81-95

Chang, S.D. and P.C. Singer, 1991, The Impact of Ozonation on Particle Stability and the Removal of TOC and THM Precursors, *Jour. AWWA*, 83(3): 71-79

Chauret, C.P., C.Z. Radziminski, M. Lepuil, R. Creason and R.C. Andrews. 2001. Chlorine dioxide inactivation of *Cryptosporidium parvum* oocysts and bacterial spore indicators. *Appl. Environ. Microbiol.* 67(7) :2993-3002.

Chen, C.-I., T. Griebe, R. Srinivasan and P.S. Stewart. 1993. Effects of various metal substrata on accumulation of *Pseudomonas aeruginosa* biofilms and the efficacy of monochloramine as a biocide. *Biofouling*. 7 :241-251.

Chen, X. and P.S. Stewart. 1996. Chlorine penetration into artificial biofilm is limited by a reaction-diffusion interaction. *Environ. Sci. & Technol*, 30(6):2078-2083.

Chen, X. and P.S. Stewart. 2000. Biofilm removal caused by chemical treatments. *Water Res*. 34(17):4229-4233.

Choi, J. and R.L. Valentine, 2002, Formation of N-Nitrosodimethylamine from Reaction of Monochloramine: a New Disinfection By-Product. *Water Research*, 36(4): 817-824

Clark, R.M. and M. Sivaganesan. 1999. Characterizing the effect of Chorine and Chloramines on the Formation of Biofilm in a Simulated Drinking Water Distribution System. Cincinnati, OH, United States Environmental Protection Agency EPA/600/R-01/024.

Clement J., M. Hayes, P. Sarin, W.M. Kriven, J. Bebee, K. Jim, M. Beckett, V.L. Snoeyink, G.J. Kirmeyer, and G Pierson, 2002, *Development of Red Water Control Strategies*, Denver, CO: AwwaRF and AWWA

Clement J., A.K. Camper, and A. Sandvig, 2003, *Influence of Distribution System Infrastructure on Bacterial Regrowth*. Denver, CO, AwwaRF and AWWA

Cowman, G.A. and P.C. Singer, 1996, Effect of Bromide Ion on Haloacetic Acid Speciation Resulting From Chlorination and Chloramination of Aquatic Humic Substances, *Environ. Sci. Technol*, 30(1): 16-24

Crozes, G. , M. Marshall, J. Hagstrom, J. Meyerhofer, and I.H. (Mel) Suffet. 1997. In *Proc. AWWA Water Quality Technology Conference*, Denver, Colo., Nov. 9-12

Davis J and Hayes K., 1986, *Geochemical Processes at Mineral Surfaces*. Washington DC, American Chemical Society 323

Dietrich, A.M., R.C. Hoehn, L.C. Dufresne, L.W. Buffin, D.M.C. Rashash, and B.C. Parker, 1995, Oxidation of Odorous Algal Metabolites by Permanganate, Chlorine, and Chlorine Dioxide, Presented at the Fourth International Symposium on Off-Flavors in the Aquatic Environment, Adelaide, South Australia, October 2-7, 1994. *Water Science and Technology*, Vol. 13 (11): 223-228.

Donlan, R.M., and W.O. Pipes. 1988. Selected Drinking Water Characteristics and Attached Microbial Population Density. *Jour. AWWA*. 80 (11): 70-76

Dowbiggen W and Singer P., 1989, Effects of Natural Organic Matter and Calcium on Ozone Induced Particle Destabilization. *Jour. AWWA* 81 (6): 77-85

Driedger A., E. Staub, U. Pinkernell, B. Marinas, W. Koster, U. von Gunten, 2001, Inactivation of Bacillus Subtilis Spores and Formation of Bromate During Ozonation, *Water Research*, 35(12);2950-2960

Dykstra, T.S., K.C. O'Leary, C. Charet, R.C. Andrews and G.A. Gagnon. 2002. Impact of UV Disinfection on Biological Stability in Distribution Systems. In *Proc. AWWA Water Quality Technology Conference*, Seattle, WA, Nov 10-14

Eisnor, J.D., G.A. Gagnon, C. Volk, C.P. Chauret and R.C. Andrews. 2001. Impact of Chlorine Dioxide on Corrosion in a Model Distribution System. In *Proc. AWWA Water Quality Technology Conference*, Nashville, TN, Nov 11-15

Falkinham, J.O. III, C.D. Norton and M.W. LeChevallier. 2001. Factors influencing numbers of *Mycobacterium avium, Mycobacterium intracellulare*, and other mycobacteria in Drinking Water Distribution Systems. *Jour. Appl. Environ. Microbiol.* 67:1225-1231.

Gagnon, G. A. 2003, Disinfectant Type and Levels in Distribution System. Sunday Workshop, Presented at the *AWWA Water Quality Technology Conference*, Philadelphia, PA, Nov. 2-6

Gagnon, G.A., R.C. Andrews, C Chauret, C Volk, T.S. Dykstra,and K.C. O'Leary, 2004, *Impact of UV Disinfection on Biological Stability.* Denver, Colo: AwwaRF and AWWA

Gagnon, G.A., K.C. O'Leary, C.J. Volk, C. Chauret, L. Stover and R.C. Andrews. 2004. Comparative analysis of chlorine dioxide, free chlorine and chloramines on bacterial water quality in model distribution systems. *Jour. Environ. Eng.* American Society of Civil Engineers 130 (11): 1269-1279

Glaze, W.H., 1986 Reaction Products of Ozone: A Review *Environmental Health Perspectives* 69:11:151-7

Glaze, W.H., R. Schep, W. Chauncey, E.C. Ruth, J.J. Zarnoch, E. M. Aieta, C.H. Tate, and M.J. McGuire, 1990, Evaluating Oxidants for the Removal of Model Taste and Odor Compounds from a Municipal Water Supply, *Jour. AWWA* 82(5):79-84

Glaze, W.H. and H.S. Weinberg 1993, *Identification and Occurrence of Ozonation By-Products*, Denver, CO; AwwaRF and AWWA

Goel S., R. Hozalski and E.J. Bouwer. 1995. Biodegradation of NOM: Effect of NOM Source and Ozone Dose. *Jour. AWWA* 87(1):90-105

Gordon, G. & A. Rosenblatt, 1995. Gaseous, Chlorine-Free Chlorine Dioxide for Drinking Water Treatment. In *Proc. AWWA Water Quality Technology Conference,* New Orleans, LA, Nov 12-16

Gu, B, and J. Schmitt., 1994 Adsorption and Desorption of Natural Organic Matter on Iron Oxide: Mechanisms and Models, *Envir. Sci & Technology*, 28(1):38-46

Harrington, G.W. and R. Noguera 2003, *Ammonia from Chlorine Decay: Effects on Distribution System Nitrification*, Denver, CO: AwwaRF and AWWA

Harris, G.D., V.D. Adams, D.L. Sorensen and M.S. Curtis. 1987. Ultraviolet Inactivation of Selected Bacteria and Viruses with Photoreactivation of the Bacteria. *Water Res.* 21 (6):687-692

Hoehn, R.C., A.M. Dietrich, W.S. Farmer, M.P. Orr, R. G. Ramon, E.M. Aieta, D.W. Wood, and G. Gordon.. 1990. Household Odors Associated With the Use of Chlorine Dioxide During Drinking Water Treatment, *Jour. AWWA,* 82(4):166-172.

Hoehn, R.C., C.S. Ellenberger, D.L. Gallagher, E.V. Wiseman, Jr., R.W. Benninger, and A. Rosenblatt. 2002. ClO_2 and By-Product Persistence in a Drinking Water System. *Jour. AWWA.* 95(4): 141-150.

Hoigné, J. and H. Bader, 1976, The Role of Hydroxyl Radical Reactions in Ozonation Processes in Aqueous Solutions. *Water Res.*, 10(5): 377-386

Hoigné J., H. Bader, 1988, The Formation of Trichloronitromethane (Chloropicrin) and Chloroform in a Combined Ozonation/Chlorination Treatment of Drinking Water. *Water Res.*, 22(3): 313-319

Hoigné, J. and H. Bader, 1994, Kinetics of Reactions of Chlorine Dioxide (OClO) in Water. I. Rate Constants for Inorganic and Organic Compounds, *Water Res.*, 28(1): 45

Hrudey, S.E., A. Gac, and S.A. Daignault. 1988. Potent Odour-Causing Chemicals Arising from Drinking Water Disinfection, *Water Science and Technology*, Vol. 20, No. 8/9, pg 58-61

Huang, X., I. Wei, W. Sung, 2005, Pilot Study of Greater Boston Drinking Water Quality Changes – Monochloramine and pH Change. In *Proc. of the AWWA Annual Conference.* San Francisco, CA, June 12-16, Denver, CO;AWWA.

Hudson, L.D., J.W. Hankins and M. Battaglia, 1983, Coliforms in a Water Distribution System: A remedial approach. *Jour. AWWA* 75(11): 564-568

Hulsey, R.A., J.J. Neemann, S.N. Foelimi, and R.C. Hoehn, 2000 Chlorine Dioxide for DBP Control and Filter Enhancement, In *Proc. AWWA Water Quality Technology Conference* Session M5, Paper 3. Salt Lake City, UT, Nov. 5-9, AWWA

Hunsinger R, 2000, Personal Communication, East Bay Municipal Utility District, Oakland, CA

Iatrou, A. and W.R. Knocke, 1992 Removing chlorite by addition of ferrous iron, *Jour. AWWA*, 84(11): 63-68

Jacangelo J.G., N.L. Patania, K.M. Reagan, E.M. Aieta, S.W. Krasner, M.J. McGuire, 1989, Ozonation: Assessing its Role in the Formation and Control of Disinfection By-products. *Journal AWWA*, 81(8):74-84

Jekel M., 1983, The Benefits of Ozone Prior to Flocculation Processes, *Ozone Sci. and Eng.* 5(1), 21-35

Johnson, J. D. and J.N. Jensen, 1968, THM and TOX Formation: Routes, Rates, and Precursors, *Jour AWWA* 78(4):156-162

Kelley, J., G. Kinsey, R. Paterson, D. Brayford, and H. Rossmoor, 2003, *Growth of Fungi That Produce Specific Odorous Compounds*, Denver, CO, AwwaRF and AWWA.

Kelly, D. 2004. Replacement of Chlorine with Chlorine Dioxide for Minimization of Disinfection By-Products in the Pre-Treatment of Surface Water. In *Proc. of the AWWA Annual Conference.* Orlando, FL, June 13-17; Denver,CO AWWA.

Kerneis, A., A. Deguin and M. Feinberg, 1994, Modeling Applications of the Number of Microorganisms according to the Residence Time of Drinking Water in a Distribution System. In *Proc. International Seminar on Biodegradable Organic Matter*, Montreal, Quebec, June 15-17

Khiari, D., S.E. Barrett, and I.H.(Mel) Suffet, 1997. Sensory Analysis of Decaying Vegetation and Septic Odors, *Jour. AWWA.* 89(4):150-161

Khiari, D., S Barett, R Chinn, A Bruchet, P Piriou, L Matia, F Ventura, I.H. Suffet, T Gittelman, P Leutweiler, 2002. *Distribution Generated Taste-and-Odor Phenomena.* Denver, CO, AwwaRF and AWWA

Kieber, R. J., X. Zhou and K. Mopper, 1990, Formation of Carbonyl Compounds from UV-induced Photodegradation of Humic Substances in Natural Waters: Fate of Riverine Carbon in the Sea. *Limnol. Oceanog* 35:1503-1515

Kirmeyer, G.J., G.W. Foust, G.L. Pierson, J.J. Simmler, and M.W. LeChevallier, 1993, *Optimizing Chloramine Formation*. Report 90620, Denver, CO; AwwaRF and AWWA

Kirmeyer, G.J., M. LeChevallier, H Baribeau, K Martel, G. Thompson, L. Radder, W Klement, and A Flores, 2004, *Update of Optimizing Chloramine Treatment,* Report 90993, Second Edition, Denver, CO: AwwaRF and AWWA

Klimkina, N.V., G.V. Tsyplakova, G.M. Trukhina, G.P. Ampleeva, and I.S. Tiuleneva. 1987. Hygienic evaluation of the decoloration of natural water by the oxidation-sorption method. *Gigiena i Sanitaria*, 1:16-19

Knocke, W.R., J.E. Van Benschoten, M. Kearney, A. Soborski, D.A. Reckhow. 1990. *Alternative Oxidation for the Removal of Soluble Iron and Manganese*. Denver, CO: AwwaRf and AWWA

Kool, J.L., J.C. Carpenter and B.S. Fields, 1999, Effect of Monochloramine Disinfection of Municipal Drinking Water on Risk of Nosocomial Legionnaire's Disease. *Lancet*, Vol. 353(9149):272-277.

Kouame, Y. and C.N. Haas, 1991, Inactivation of *E. coli* by Combined Action of Free Chlorine and Monochloramine, *Water Res.* 25(9):1027-1032.

Krasner, S.W. and S.E. Barrett. 1984. Aroma and Flavor Characteristics of Free Chlorine and Chloramines, In *Proc. of the 12th Annual AWWA Water Quality Technology Conference*, AWWA, Denver, CO, Dec. 2-5

Krasner, S.W., S.E. Barrett, M.S. Dale, C.J. Hwang, 1989a, Free Chlorine Versus Monochloramine in Controlling Off-Tastes and Off-Odors in Drinking Water. *Jour. AWWA*, 81(2): 86-93.

Krasner S.W., M.J. McGuire, J.G. Jacangelo, N.L. Patania, K.M. Reagan, M. Aieta, 1989b, The Occurrence of Disinfection By-products in US Drinking Water, *Jour. AWWA*, 81(8) :41-53

Krasner, S.K., C.J. Hwang, T.K. Lieu, M.J. West, 1991, Development of a Bench-Scale Method to Investigate the Factors that Impact Cyanogen Chloride Production in Chloraminated Waters, In *Proc. AWWA Water Quality Technology Conference*, Orlando, FL, Nov 10-14

Krasner, S.K., J.P. Croue, J. Buffle, and E.M. Perude, 1996, Three Approaches for Characterizing NOM, *Jour. AWWA*, 88(6):66-79

Krasner, S.W. et. al, 2002. Relationship Between the Different Classes of DBPs; Formation, Speciation, and Control. In *Proc. AWWA Water Quality Technology Conference,* Seattle WA, Nov 10-13, AWWA

Krasner, S.W., R. Yates, M.J. Sclimenti, S. Pastor, and S. Liang, 2004, Oxidation and Adsorption of Fishy/Swampy/Grassy Odors in Drinking Water. In *Proc. AWWA Annual Conference*, Orlando, FL, June 13-17

Kruithof, J., 2000, Pesticide Control and Disinfection by UV/H2O2 Treatment. In *Proceedings of the International Conference on Applications of Ozone*. Berlin, Germany, International Ozone Association

Kuch A,. 1984 Untersuchungen zum Mechnaismus der Aufeisenung in Trinkwasservertveilungsystemen. Ph.D.diss., University of Karlsruhe, Germany.

Lalezary, S., M. Pirbazari, and J.J. McGuire. 1986. Oxidation of Five Earthy-Musty Taste and Odor Compounds. *Jour. AWWA*, 78(3):62-69

Langlais, O.B., D.A. Reckhow, and D.R. Brink. 1991. *Ozone in Water Treatment: Application and Engineering.* AWWARF and Lewis Publishers, Inc. Chelsea, Mich. .

Leal, J.L., 1909. The Sterilization Plant of the Jersey City Water Supply Company at Boonton, N.J. *Jour. AWWA,* Issue 100.

LeChevallier, M.W., C.D. Cawthon and R.G. Lee. 1988. Inactivation of biofilm bacteria. *Appl. Environ. Microbiol.* 54(10):2492-2499.

LeChevallier, M.W., C.D. Lowry and R.G. Lee. 1990. Disinfecting biofilms in a model distribution system. Jour AWWA 82(7):87-99.

LeChevallier, M.W., C.D. Lowry, R.G. Lee and D.L. Gibbon. 1993. Examining the relationship between iron corrosion and the disinfection of biofilm bacteria. Jour. AWWA 85(7):111-123

LeChevallier, M.W., N.J. Shaw and D.B. Smith. 1996a. Factors Limiting Microbial Growth in Distribution Systems: Full-Scale Experiments. AWWA, Denver, CO

LeChevallier M.W., N.J. Welch and D.B. Smith. 1996b. Full-scale studies of factors related to coliform regrowth in drinking water. *Appl. Environ. Microbiol.* 62(7):2201-2211

Lehtola, M.J., I.T. Miettinen, T.Vartiainen, P. Rantakokko, A. Hirvonen and P.J. Martikainen. 2003. Impact of UV disinfection on microbially available phosphorus, organic carbon, and microbial growth in drinking water. *Water Res.* 37(5):1064-1070.

Liang, L. and P.C. Singer, 2003, Factors Influencing the Formation and Relative Distribution of Haloacetic Acids and Trihalomethanes in Drinking Water, *Environ. Sci. Technol.*, 37(13): 2920-2928

Limoni, B and B. Telsch. 1985. Chlorine dioxide disinfection of drinking water - an evaluation of a treatment plant. *Water Res.* 19(12):1489-1495.

Linden, K., E. Rosenfeldt, S. Johnson, and B. Melcher. 2002. Direct UV and UV Oxidation Processes for Treatment of Taste and Odor Causing Compounds in Water. In *Proc. 2002 AWWA Water Quality Technology Conference*, Seattle, Wash., Nov. 10-14, AWWA, Denver, Colo.

Liu, W., S.A. Andrews, C. Sharpless, M. Stefan, K.G. Linder, and J.R. Bolton. 2002. Bench-Scale Investigations into Comparative Evaluation of DBP Formation from Different UV/H_2O_2 Technologies. In *Proc. 2002 AWWA Water Quality Technology Conference*, Seattle, Wash., Nov. 10-14, 2002. AWWA, Denver, Colo.

Logsdon, G., M. Ferguson, D. Neden, S. LaBonde, S. Andrews, P. Huck and R. Dawson, 1992. Filtration Studies for Greater Vancouver's Water Supply. *Proceedings, Annual Conference of the American Water Works Association*, Vancouver, British Columbia, June 18-22, Denver, CO. AWWA

Lund , V. and D. Hongve. 1994. Ultraviolet irradiated water containing humic substrates inhibits bacterial metabolism. *Water Res.* 28(5):1111-1116.

Lund, V. and K. Omerod. 1995. The influence of disinfection processes on biofilm formation in water distribution systems. *Water Res.* 29(4):1013-1021.

Mackey, E.D., P. Piriou, and A. Bruchet. 2003. A Comparison of French and American Consumer Sensitivities to Chlorinous Flavors. In *Proc. 2003AWWA Water Quality Technology Conference*, Philadelphia, PA Nov. 2-6, 2003. AWWA, Denver, Colo.

Malleret, L. and A. Bruchet, 2003. A Taste-and-Odor Episode Caused by 2,4,6-Tribromoanisole, Jour. AWWA, 94(7): 84-95

Malley, J.P., 1999, Control of Disinfection By-Product Formation Using Ultraviolet Light, In: Formation and Control of Disinfection By-Products in Drinking Water. Editor, Phillip C. Singer, Denver, CO: American Water Works Association; pp. 223-235

Masschelein, W.J. 1991. Chlorine Dioxide In *Chemical Oxidation*: *Technologies for the Nineties*. Proc. First Int. Symposium on Chlorine Dioxide. Nashville, TN, Feb. 20-22, Technomic Pub. Co., 1992

McKnight, Andrew and Reckhow, David A., 1992, Reactions of Ozonation Byproducts With Chlorine and Chloramines. In *Proc. American Water Works Association Annual Conference* ; Vancouver, British Columbia, Canada. June 18-22, pp. 399-409

McGuire, M.J., N.I. Lieu and M.S. Pearthree. 1999. Using chlorite ion to control nitrification. *Jour. AWWA* 91(10):52-61

McNeill, L.S. and M. Edwards 2001. Iron Pipe Corrosion in Distribution Systems, JAWWA Vol. 93, No 7, pgs 88-100

Means, E.G. III, T.S. Tanaka, D.J. Otsuka and M.J. McGuire. 1986. Effects of Chlorine and Ammonia Application Points on Bactericidal Efficiency. *Jour. AWWA* 78(1): 62-69

Mitch, W.A. and D.L. Sedlak, 2002, Formation of N-Nitrosodimethylamine (NDMA) from Dimethylamine during Chlorination. *Environmental Science and Technology*, 36(4): 588-595

Momba, M.N.B. and M.A. Binda. 2002. Combining chlorination and chloramination processes for the inhibition of biofilm formation in drinking surface water system models. *J. Appl. Microbiol.* 92(4):641-648

Morin, P. and A.K. Camper. 1997. Attachment and fate of carbon fines in simulated water distribution system biofilms. Wat. Res. 31(3):399-410

Mounsey, R.J. and M.C. Hagar. 1946. Taste and Odor Control With Chlorine Dioxide. *Jour. AWWA.* 38(9):1051-1056

Muylwyk, Q., J. MacDonald and M. Klawunn. 2001. Success! Switching from chloramines to chlorine in the distribution system: Results from a one year full-scale trial. In Proc. AWWA Water Quality Technology Conference, Nashville, TN, Nov 11-15, AWWA, Denver, CO

Najm, I.N., L. Boulos, M.W. LeChevallier, C. Norton, C. Volk, A. Randall, I. Escobar, L. Keine and C. Campos. 2000. *Case Studies of the Impacts of Treatment Changes on Biostability in Full Scale Distribution Systems.* Denver, CO AwwaRF and AWWA

Najm, I. and R.R. Trussel, 2001, NDMA Formation in Water and Wastewater. Jour. AWWA, 93(2):92-99

Neden, D.G., R.J. Jones, J.R. Smith, G.J. Kirmeyer, and G.W. Foust. 1992. Comparing Chlorination and Chloramination for Controlling Bacterial Regrowth. Jour. AWWA, 84(7): 80-88.

Neemann, J., P. Zhou, R. Swartfager, and R.C. Hoehn. 2002. Full-Scale Application of Chlorine Dioxide Pre-oxidation for Disinfection with Ozone. *Proc. AWWA Water Quality Technology Conference.* Seattle, WA, Nov. 10-14, Denver, CO; AWWA

Nickols, D. and A.J. Varas. 1992. Ozonation. Chapter 7 in: *Disinfection Alternatives for Drinking Water*, New York, Van Nostrand Reinhold.

Obolensky, A., B. Davis, K. Narangajavana, L. Ceci, G.Oalickal, and A. Eyring. 1999, Disinfection Byproducts in Three Water Treatment Plant Distribution Systems, In *Proceedings of the AWWA Water Quality and Technology Conference.* Tampa, FL, Oct 31-Nov 3, Denver, CO; AWWA

Obolensky, A., and M. Frey, 2002, Changes in DBPs from Finished Water to ICR Distribution Systems and Evaluation of SDS Results, in *Information Collection Rule Data Analysis Report*, edited by M. J. McGuire, J. McLain, and A. Obolensky, , Denver, CO AWWARF and AWWA

Ollos, P.J., R.M. Slawson and P.M. Huck. 1998. Bench Scale Investigations of Bacterial Regrowth in Drinking Water Distribution Systems. Water Sci. Technol. 38:275-282

Owen, C.A. and S. Johnson. 2002, Achieving Multiple Water Quality Goals Using Ozone, BAC, and Peroxide. , *Proc. AWWA Water Quality Technology Conference*, Salt Lake City, UT, Nov. 5-9, Denver, CO; AWWA.

Parrotta, M.J. and F. Bekdash. 1998. UV-disinfection of Small Groundwater Supplies. Jour. AWWA 90(2):71-81.

Price M.L. 1994. Ozone and Biological Treatment for DBP Control and Biological Stability. Amer. Water Works Assoc.; AWWARF, Denver, CO.

Rand, J., Z. Alam, G. Gagnon, C. Chauret, R. Hofmann, and R. Andrews. 2005. Effects of UV Treatment on Biological Fouling in a California Aqueduct. In *Proc. of the AWWA Annual Conference*. San Francisco, June 12-16, Denver, CO: AWWA.

Reckhow, D. A. and P. C. Singer, 1985, Mechanisms of Organic Halide Formation During Fulvic Acid Chlorination and Implications with Respect to Pre-Ozonation, pp. 1229-1257 in *Water Chlorination: Chemistry, Environmental Impact and Health Effects, Vol. 5*, edited by R.L. Jolley . Chelsea, MI: Lewis Publishers, Inc.

Reckhow, D. A., B. Legube and P. C. Singer, 1986, The Ozonation of Organic Precursors: Effect of Bicarbonate, *Water Research*, 20(8);987-998

Reckhow, D.A., P.C. Singer and R.L. Malcolm, 1990a, Chlorination of Humic Materials: By-Product Formation and Chemical Interpretations, *Env. Sci. Tech.*, 24(11) 1655-1664

Reckhow, D.A. and P.C. Singer, 1990b, Chlorination By-products in Drinking Waters: From Formation Potentials to Finished Water Concentrations, Journ. AWWA 82(4)173-180

Reckhow, D.A. 1996. Disinfection Alternatives In *Proc. Annual AWWA Conference*, Toronto, Ontario, Canada,June 23-27, Denver, CO; AWWA

Reckhow, David A., 1999, Control of Disinfection By-Product Formation Using Ozone. In *Formation and Control of Disinfection By-Products in Drinking Water*. Edited by Philip C Singer, Denver, CO: AWWA

Reiber S. Blue Water Production. *Proceedings of the New England Water National Conference on Integrating Corrosion Control*, May 20-21, Boston MA, NEWWA, Milford MA 1996.

Reiber S. 1993, Chloramine Effects on Distribution System Materials. Awwa Research Foundation; American Water Works Association, Denver CO.

Richardson, S.D., A.D. Thruston, T.W. Collette, 1994 Multispectral Identification of Chlorine Dioxide Disinfection Byproducts in Drinking Water, Env. Sci. Tech., 28:4:592-599

Romain, D., B. Kuslikis, and M. Carlson, 2003, Advanced Oxidation of Geosmin Using UV, Hydrogen Peroxide, and Aluminum-Based Coagulant. In *Proc. AWWA Water Quality Technology Conference,* Philadelphia, PA, Nov. 2-6, Denver, CO; AWWA

Routt, J.C. 1989. Chlorine Dioxide Trihalomethane Control and Distribution System Odor Problem: A Utility's Five-Year Experience. In *Proc. AWWA Seminar on Impact of Changing Disinfectants*, Los Angeles, June 18, Denver, CO;AWWA

Salerno, J.J. 2005. *Legionella*: Impact of Drinking Disinfectant Conversion from Chlorine to Monochloramine. A Local State and Federal Agency Collaboration between SFPUC, CDC, CEIP, SFDPH and CDHS. In *Proc. of the AWWA Annual Conference.* San Francisco,CA, June 12-16, Denver, CO: AWWA.

Schechter, D.S.and P.C. Singer, 1995, Formation of Aldehydes During Ozonation,_*Ozone Sci. Eng.*, 17(1):53-59

Schreppel, C.K., D.W. Fredericksen and A.A. Geiss. 1997. The Positive Effects of Corrosion Control on Lead Levels and Biofilms. In *Proc. AWWA Water Quality Technology Conference*, Denver, CO, Nov 9-12, Denver, CO AWWA

Schick R., 2002, *Untersuchungen zum Einfluss der Oxidation auf die Anderrung der Partikelstabilitat und deren Auswirkung bei der Aufbereitung von Oberflachenwasser*, Ph.D. diss., University of Berlin, Berlin, Germany

Schock M, D. Lytle and J. Clement, 1995, *Effect of pH, DIC, Orthophosphate and Sulfate on Drinking Water Cuprosolvency*. Washington, D.C. USEPA EPA/600/R-95/085

Schock, M. R., J. Swertfeger, S. M. Harmon*, R. Lohmann, AND J. DeMarco. Tetravalent Lead: A Hitherto Unrecognized Control Of Tap Water Lead Contamination. Presented at AWWA Water Quality Technology Conference, Nashville, TN, 11/11-14/2001

Schock, M.R. and R.Giani. 2004. Oxidant/Disinfectant Chemistry and Impacts on Lead Corrosion. In *Proc. AWWA Water Quality Technology Conference*, San Antonio, TX, Nov. 14-18, 2004, Denver, CO, AWWA

Servais P., G. Billen and M.C. Hascoet. 1987. Determination of the Biodegradable Fraction of Dissolved Organic Matter in Water. *Water Res.* 21:445-450.

Sharp, R.R., A.K. Camper, J.J. Crippen, O.D. Schneider and S. Leggiero. 2001. Evaluation of Drinking Water Biostability using Biofilm Methods. *Jour. Environ. Eng.* 127(5):403-410

Sharpless, C.M. and K.G. Linden, 2001, UV Photolysis of Nitrate: Effects of Natural Organic Matter and Dissolved Inorganic Carbon, and Implications of Water Disinfection, *Env. Sci. Tech.*, 35(14), 2949-2955

Shaw, J. P., J.P. Malley Jr. and S.A. Willoughby. 2000. Effects of UV-irradiation on Organic Matter. *Jour. AWWA* 92(4):157-167

Shen, Y., M. Bergen, L. Yoo and S. Fitzsimmons, 1997. Effect of Residual Chlorine on the Threshold Odor Concentrations of MTBE in Drinking Water, In *Proc. AWWA Water Quality Technology Conf.*, Denver, CO, Nov. 9-12, Denver, CO; AWWA

Sheiham, I, and Jackson P,. 1981, The Scientific Basis for Control of Lead in Drinking Water by Water Treatment. *Jour Inst Water Engr. And Scientists.* Volume 35(6): 491-515

Singer, P. C. and W. K. O'Neil, 1987,Technical Note: The Formation of Chlorate from the Reaction of Chlorine and Chlorite in Dilute Aqueous Solution, *Jour. AWWA*, 79(11) 75-76

Singer, P.C., 1988, *Alternative Oxidant and Disinfectant Treatment Strategies for Controlling Trihalomethane Formation*, Report No. EPA/600/S2-88/044, US Environmental Protection Agency, Washington, DC (Oct. 1988).

Singer, P.C., J. Shi, and H.S. Weinberg, 1998, Impact of pH Adjustment and Chloramination for the Control of Haloacetic Acid Formation in North Carolina Drinking Water, In Proc. AWWA Water Quality Technology Conf, San Diego, CA, Nov 1-5,Denver, CO, AWWA

Singer, P.C. and D.A. Reckhow, 1999, Chemical Oxidation, Ch. 12 in *Water Quality and Treatment: A Handbook of Community Water Supplies,* 5th edition, edited by R.D. Letterman, New York, NY;McGraw Hill and Denver, CO; AWWA

Singer, P.C., G.W. Harrington, G.A. Cowman, M.E. Smith, D.S. Schechter, and L.J. Harrington, 1999, *Impacts of Ozonation on the Formation of Chlorination and Chloramination By-Products*, Report No. 90766, Denver, CO, AwwaRF and AWWA

Singer, P.C. and C.J. Johnson, 2003, Reduction in Ozone Demand and Bromate Formation Using a Magnetic Ion Exchange Resin, In *Proc. Water Quality Technology Conference*, Philadelphia, PA, Nov 2-6, Denver, CO, AWWA

Smith, C.D., J.F. Smith, and B. Milosky. 1996. Excessive Loss of Chloramine Residual: A Case Study, In *Proc. AWWA Annual Conference,* Toronto, Ontario, Canada, June 23-27, Denver, CO; AWWA

Snicer, G.A., J.P. Malley, A.B. Magolin and S.P. Hogan. 1996. Evaluation of ultraviolet technology in drinking water treatment. In *Proc. AWWA Water Quality Technology Conference* , Boston, MA, Nov 17-21, Denver, CO, AWWA

Snoeyink V. 2003, Mechanism for the Release of Iron from Iron Pipe Scales. Proceedings of the *IWA Conference on Corrosion and Scaling*. March, 2003 Milton Keyes, England

Song, R. P. Westerhoff, R. Minear, and G.L. Amy, 1997, Bromate Minimization During Ozonation. *Jour AWWA* 89(6):69-78

Speight, V.L. and P.C. Singer, 2003, Association Between Residual Chlorine and Reduction in Haloacetic Acid Concentrations in Distribution Systems, *Jour. AWWA* 97(2): 82-91

Speitel G.E. Jr., J.M. Symons, A.C. Diehl, H.W. Sorensen and L.A. Cipparone. 1993. Effect of Ozone Dosage and Subsequent Biodegradation of DBP precursors. *Jour. AWWA* 85(5):86-95

Speitel, G.E., 1999, Control of Disinfection By-Product Formation Using Chloramines, In *Formation and Control of Disinfection By-Products in Drinking Water* edited by Philip C Singer, Denver, CO: AWWA

Staehelin, J. and J. Hoigne, 1982, Decomposition of Ozone in Water: Rate of Initiation by Hydroxide Ions and Hydrogen Peroxide, *Env. Sci. Tech.*, 16:10:676-681

Stewart, P.S., T. Griebe, R. Srinivasan, C.-I. Chen, F.P. Yu, D. deBeer and G.A. McFeters. 1994. Comparison of Respiratory Activity and Culturability during Monochloramine Disinfection of Binary Population Biofilms. *Appl. Environ. Microbiol.* 60:1690-1692.

Stewart, P.S. and J.B. Raquepas. 1995. Implications of Reaction-Diffusion Theory for the Disinfection of Microbial Biofilms by Reactive Antimicrobial Agents. *Chem. Eng. Sci.*, 50(19):3099-3104

Stolarik, G. and P. Liu. 2000. Chlorine Dioxide for Algae and DBP Control: The Los Angeles Experience, In *Proc. AWWA Water Quality Technology Conference*, Salt Lake City, UT, Nov. 5-9, Denver, CO:AWWA

Stumm, W and J.J. Morgan.1985, *Aquatic Chemistry*, 2nd Edition, New York, NY: John Wiley and Sons

Suffet, I.H., and A.K. Meng, 1996. *Taste and Odor in Drinking Water Supplies – Phase III,* Philadelphia Water Department, Philadelphia Suburban Water Company,Report 90585, Denver, CO, AwwaRF and AWWA and and Lyonnaise Des Eaux-Dumez, Le Pecq, France.

Switzer, J.A., V. V. Rajasekharan, S. Boonsalee, E. A. Kulp, and E. W. Bohannan, 2006. "Evidence that monochloramine disinfectant could lead to elevated Pb levels in drinking water," *Environ. Sci & Technol.* 40, 3384-3387.

Taylor, R.H., J.O. Falkinham, C.D. Norton and M.W. LeChevallier. 2000. Chlorine, Chloramine, Chlorine Dioxide and Ozone Susceptibility of *Mycobacterium avium. Appl. Environ. Microbiol.* 66(4):1702-1705.

Tryby, M., D. Boccelli, M. Koechling, J. Uber, S. Summers, L. Rossman, 1999, Booster Chlorination for Managing Disinfectant Residuals, *Jour. AWWA*, 91(1):95-108

Tryby, M., D. Boccelli, M. Koechling, J. Uber, S. Summers, L. Rossman, 2002, Facility Location Model for Booster Disinfection of Water Supply Networks, *Journ. Water Resources Planning and Management*, 128(5)322-333

Valentine, R.L.,Vikesland,P.J.,Angerman,B.D.,Hackett,S.A.,Shoup,M., and Slattenow, S., 2000, *The Role of the Pipe-Water Interface in DBP Formation and Disinfectant Loss,* Denver, CO;AwwaRF and AWWA

Valentine, R., K. Ozekin and P.J. Vikesland. 1998. *Chlorine Decomposition in Distribution Systems and Model Waters.* Denver, CO; AwwaRF and AWWA

van der Kooij D., A. Visser and W.A.M. Hijnen. 1982. Determining the Concentration of Easily Assimilable Organic Carbon in Drinking Water. Jour. AWWA 74(10):540-545

van der Kooij D., W.A.M. Hijnen and J.H.C. Kruithof. 1989. The Effects of Ozonation, Biological Filtration, and Distribution on the Concentration of Easily Assimilable Organic Carbon (AOC) in Drinking Water. Ozone Sci. & Engrg. 11(3):297-311

Videla H, 1996, *Manual of Biocorrosion* Chelsea, MI, Lewis Publishers

Vikesland, P.J., R. Ozekin and R.L. Valentine. 1998. Effect of Natural Organic Matter on Monochloramine Decomposition: Pathway Elucidation through the use of Mass and Redox Balances. *Environ. Sci. Technol.* 32(10):1409-1416.

Von Gunten, U., Y. Oliveras. Advanced Oxidation of Bromide-Containing Waters: Bromate Formation Mechanisms. Environmental Science and Technology, 32(1): 63-70. (1998)

Walker R, and Oliphant, R., 1982 Corrosion of Lead in Drinking Water. *Anti Corrosion Methods and Materials*. 29(4): 8-11

Walker, G. and F.P. Lee. 1983. Oxidative Control of Taste and Odour in Drinking Water: The Edmonton Experience, In *Proc. 35th Annual Convention, Western Canada Water and Sewage Conference,* Edmonton, Alberta, Sept 21-23, pp. 111-123. The Conference, Calgary, Alberta.

Walker, J.T., C.W. Mackerness, D. Mallon, T. Makin, T. Williets and C.W. Keevil. 1995. Control of *Legionella pneumophila* in a Hospital Water System by Chlorine Dioxide. *Jour. Indus. Microbiol.* 15(4):384-390

Ward, N.R., R.L. Wolfe and B.H. Olson. 1984. Effect of pH, Application Technique, and Chlorine-to-Nitrogen Ratio on Disinfectant Activity of Inorganic Chloramines with Pure Culture Bacteria. *Appl. Environ. Microbiol.* 48:508-514

Weinberg, H.S., S.W. Krasner, S.D. Richardson, and A.D. Thruston, 2002, *The Occurrence of Disinfection By-Products (DBPs) of Health Concern in Drinking Water: Results of a Nationwide DBP Occurrence Study*, EPA/R600-02/068, US Environmental Protection Agency, Washington DC

Werdehoff, K. S. and P. C. Singer, 1987, Chlorine Dioxide Effects on THMFP, TOXFP, and the Formation of Inorganic By-Products, *Jour. AWWA*, 79(9)107-113

Werner P. and B. Hambsch. 1986. Investigation of the Growth of Bacteria in Drinking Water, Water Supply. 4(3):227-234

Westerhoff, P., G. Aiken, G. Amy, J. Debroux, "Relationships Between the Structure of Natural Organic Matter and its Reactivity Towards Molecular Ozone and Hydroxyl Radicals," Water Research, 33:10:2265-2276 (1999)

White, Geo. Clifford. 1999. *Handbook of Chlorination and Alternative Disinfectants*, 4th ed. New York, NY, John Wiley & Sons, Inc.

Wilczak, A., W.R. Knocke, R.E. Hubel, E.M. Aieta, 1993, Manganese Control Issues Associated With the Ozonation of Surface Waters, *Jour. AWWA*, 85(10);98-104

Wilczak, A., A. Assadi-Rad, C. Wong, L. Hoover, R. Berger, R. Hunsinger, J. Smith, F. Rodigari, L. Lazzelle, C. Heaney, H. Lai, A. Sheikholeslami, 2002, Screening of Treatment Processes for NDMA Control. In *Proceedings of the AWWA Annual Conference*, 2002, New Orleans, LA, June 16-20, Denver, CO: AWWA.

Wilczak, A., C. D. Smith, Y.K. Cohen, and P.B. Martin. 2003. Strategies for Combining Free Ammonia and Boosting Free Ammonia and Boosting Chloramines in Distribution Systems – Survey of Utility Practice, In *Proc. AWWA Water Quality Technology Conference,*Philadelphia, PA, Nov. 2-6, Denver, CO; AWWA

Williams S.L., R.L. Williams, and A.S. Gordon, 1995, Degradation of Haloacetic Acids (HAA) at Maximum Residence Time Locations (MRTLs), In *Proceedings of the AWWA Water Quality Technology Conference*, New Orleans, LA, Nov. 12-16, Denver, CO; AWWA

Williams, M.M. and E.B. Braun-Howland. 2003. Growth of *Escherichia coli* in Model Distribution System Biofilms Exposed to Hypochlorous Acid or Monochloramine. *Appl. Environ. Microbiol.* 69:5463-5471

Winnzenbacher R.,R. Schick, H Stabel, and M. Jekel, 2000, The Role of Preozonation in an Optimized Solid-Liquid Separation Process Used for Drinking Water Treatment, In *Proceedings of the International Conference on Applications of Ozone and also on UV and Related Ozone Technologies*. Berlin, Germany, Oct 23-26, publ. Paris, France, International Ozone Association

Wolfe, R.L. 1990. Ultraviolet Disinfection of Potable Water, Current Technology and Research Needs. *Environ. Sci. Technol.* 24(6):768-773.

Xu, X., P.S. Stewart and X. Chen. 1996. Transport Limitation of Chlorine Disinfection of *Pseudomonas aeruginosa* entrapped in Alginate Beads. *Biotech. Bioeng*, 49(1):93-100

Zaitlin, B., S. Watson, J Ridal, T. Satchwill, and D. Parkinson, 2003. Actinomycetes in Lake Ontario: Habitats and Geosmin and MIB Production, *Jour. AWWA*, 95 (2): 113-118.

ABBREVIATIONS

AOC	assimilable organic carbon
AOP	advanced oxidation processes
AWWA	American Water Works Association
AwwaRF	Awwa Research Foundation
BDOC	biodegradable organic carbon
BrOX	brominated organic halides
cfu/ml	colony forming units per milliliter
Cl_2	chlorine
ClO_2	chlorine dioxide
ClO_2^-	chlorite
CNCl	cyanogen chloride
CT	disinfectant residual over time in minutes
Cu	copper
D/DBP Rules	Disinfectant/Disinfection By-Product Rules
DBPs	disinfection by-products
DIC	dissolved inorganic carbon
DO	dissolved oxygen
DOC	dissolved organic carbon
EBMUD	East Bay Municipal Utility District
E. coli	*Escherichia coli*
EPA	Environmental Protection Agency
Fe	iron
FPA	flavor profile analysis
GCWA	Gulf Coast Water Authority
GW	ground water
HAA	haloacetic acids
HAA5	the five regulated haloacetic acids
HPC	heterotrophic plate count
H_2O_2	hydrogen peroxide
I	intensity
ICR	Information Collection Rule
LCR	Lead and Copper Rule
LP	low pressure ultraviolet lamps

MCL	Maximum Contaminant Limit
mg/l	milligrams per liter
MIB	2-methylisoborneol
mJ/cm^2	milliJoules per square centimeter
Mn	manganese
MP	medium pressure ultraviolet lamps
MUA	Municipal Utilities Authority
mWs/cm^2	milliWatts per square centimeter
μ/l	micrograms per liter
NA	not available
NDMA	nitrosodimethyl amine
nm	nanometers
ng/l	nanograms per liter
NH_2Cl	monochloramine
NOM	natural organic matter
OTCs	odor threshold concentrations
O_2	ozone
PAC	powdered activated carbon
Pb	lead
PbO_2	lead oxide
pH	negative log of the hydrogen ion concentration
Preox	preoxidation
PVC	polyvinyl chloride
redox	reduction-oxidation
SOC	synthetic organic chemicals
sp.	specie
TCR	Total Coliform Rule
THM	trihalomethane
TOX	total organic halides
TTHM	total trihalomethanes
T&O	taste and odor
UV	ultraviolet light
WTP	water treatment plant
#	number
%	percent

IWA Publishing's authorised EU representative for General Product Safety Regulations is Diane D'Arras, 15 rue Duret, 75116 Paris, France, e-mail: safety@iwap.co.uk.

Printed and bound by CPI Group (UK) Ltd, Croydon, CR0 4YY
17/09/2026
02218793-0009